DICTIONARY OF ENTOMOLOGY, PLANT PATHOLOGY AND NEMATOLOGY

About the Authors

Dr. O.P. Singh is Associate Professor of Entomology in Jawaharlal Nehru Agricultural University, Sehore campus. He has about 24 years of teaching and research experience in various capacities and has extensively worked on insect pests of important crops. . His work on soybean pests has earned wide recognition world over. Dr. Singh is already author of 4 books entitled "Soybean Ke Keet" (Asian Publishers), "Insect Pests of Soybean in India" (International Book Distributors), "Soybean" (Agro-Botanical Publishers) and "Vistas in Insect Pests and Diseases of Soybean in India" (Int. Book Dist.); and has published over 200 papers in reputed journals of India and abroad. He is receipient of **Dr. Rajendra Prasad Award** of I.C.A.R. Dr. Singh is member and Fellow of many learned societies of the country and is in the Editorial Board of many journals.

Dr. S.K. Srivastava is Associate Professor of Plant Pathology at Sehore campus of Jawaharlal Nehru Agricultural University. Dr. Srivastava is deeply associated with the research and teaching of Mycology and Plant Pathology from the last 23 years. He has worked on diseases of different crops and has special interest on mustard, sugarcane and soybean diseases. He has published about 75 research papers in Indian and foreign journals of repute and is also author of two books entitled "Soybean" and "Vistas in Insect Pests and Diseases of Soybean in India" with Dr. Singh. He is co-receipient of **Dr. Rajendra Prasad Award** of I.C.A.R. He is fellow of many professional societies of the country, and is on the panel of referees of many journals.

DICTIONARY OF ENTOMOLOGY PLANT PATHOLOGY AND NEMATOLOGY

O.P. SINGH
Associate Professor
Department of Entomology
Jawaharlal Nehru Agricultural University
College of Agriculture
Sehore 466 001 (India)

S.K. SRIVASTAVA
Associate Professor
Department of Plant Pathology
Jawaharlal Nehru Agricultural University
College of Agriculture
Sehore 466 001 (India)

CONCEPT PUBLISHING COMPANY NEW DELHI - 110059

ISBN 81-7022-673-2 (H.B.)
ISBN 81-7022-674-0 (P.B.)

First Published 1998

Published and Printed by
Ashok Kumar Mittal
Concept Publishing Company
A/15-16, Commercial Block, Mohan Garden,
NEW DELHI - 110059 (India)

Ph.: 5504042, 5554042
Fax : 011-559-8898

PREFACE

While remaining intimately associated with the profession of teaching, research and extension and frequent interactions with the students, teachers and all others dealing with entomology, plant pathology and nematology gave us a firm impression that a dictionary of technical terms in these fields is essential. The urgency is felt more deeply in view of the fact, that many a times persons of these professions come across such technical terms which appear to be of common use, but are difficult to define precisely. Furthermore, sources of literature to the definitions and explanations of these terms are widely and vastly scattered and instantaneous access to them is usually very poor for most of the people, except those few who are working at institutions with a rich library. Our experience of this practical problem, inspired us to take up this project of writing a "Dictionary of Entomology, Plant Pathology and Nematology".

Although, innumerable publications have come into existence on these subjects, yet an endeavour has been made to incorporate the missing links and prepare a compact source of readily available and up-to-date information for entomologists, plant pathologists, nematologists and others in a simple and lucid language. This dictionary will be of immense use to students, teachers, research workers and others having a special interest towards learning of entomology, plant pathology and nematology and will be quite helpful in understanding the fundamentals of these highly technical and intricate branches of agricultural sciences.

The dictionary includes about 6500 technical terms; 3700 being on entomology, 2400 on plant pathology and 400 on nematology, drawn from various text books, reference books and research literature. With a view to make this work more comprehensive, the terms used in veterinary and medical sciences have also been included. The list of insect pests, diseases and nematodes is enormous and demands an independent and voluminous publication; however, in the present book, we have attempted to define some of the important and common insect pests, diseases and nematodes, particularly those, occuring in India and Asia. The terms on various aspects of these branches such as morphology, anatomy, physiology and taxonomy, etc. have generally been defined in a concise manner except those which required more elaborae explanation. The plurals for most of the words have invariably been mentioned within parentheses and cross references of the words or their interrelationships, with each other have also been given at the end of the definition, either by using the word "cf" or "also see". Usually, both common names and chemical names of the insecticides, fungicides, nematicides, etc. have been given at the same place, along with the trade names, wherever applicable. In case of terms having more than one meanings or explanations, different meanings have been given at the same place either using sequential numbers or semicolons.

A dictionary is never an original work, but a collection from the others work, and we too have consulted a number of publications, and are immensely grateful to all the scientists and authors whose work has been freely explored from the literature. Obviously, a publication of this nature has a number of limitations. The creeping in of mistakes as well as lapses in the completeness are inevitable. The authors would be grateful to the readers for the suggestions and healthy criticism towards the further improvement of this publication.

It is our proud privilege to express the sinceremost gratitude towards Dr. K.S. Johar, Honb'le Vice Chancellor, Jawaharlal Nehru Agricultural University, Jabalpur for inspiration and constant encouragement. We are also grateful to Dr. R. K. Gupta, Director Research Services; and Dr. S.M. Vaishampayan, Prof. & Head (Entomology), J.N. Agricultural University, Jabalpur and so also to the Dr. V. S. Tomar, Dean, JNKVV, College of Agriculture, Sehore for their abounding interest, encouragement and help in various ways.

The constant encouragement and help of Dr. S.R. Sharma, Cane Commissioner, Govt. of Madhya Pradesh, Bhopal; Dr. S.C. Agrawal, Dr. R.K. Joshi, Dr. S.N. Verma and Dr. K.K. Nema, JNKVV, College of Agriculture, Sehore is also gratefully acknowledged. We are thankful to M/S Concept Publishing Company, New Delhi for publishing this book. We are especially thankful to Mrs. M. Singh and Mrs. Samriddhi Srivastava, who willingly accepted the sacrifices demanded by this work.

Sehore O.P. SINGH
S.K. SRIVASTAVA

SECTION - I

ENTOMOLOGY

A

Abdomen: This is the third and terminal part of the insect body, at the posterior end of which are located anus and aperture of the reproductive organs. There are only a few appendages on the abdominal segments, typically 11, which are movable, joined together by the intersegmental membranes.

Abdomen of embryo: According to the degree of development of the segmentation and appendages, the abdomen is distinguished in three embryonic stages. In the earliest or *Protopod* phase, segmentation is absent or indistinct and there are rudimentary appendages only on the head and thorax. The *Polypod* phase has a clearly segmented abdomen and each segment of the body bears a pair of appendages, while the *Oligopod* embryo is clearly segmented but lacks abdominal appendages.

Abdominal gills: These are tracheal gills, usually borne on the abdomen. They are filiform or more or less lamellate structures that are well supplied with tracheae and tracheoles. They are present in the majority of aquatic larvae and in some aquatic pupae. In many cases, they are the only organs of respiration but in Culicid larvae, they are accessory in function and coexist with open spiracles.

Abdorsal line: Among caterpillars, a longitudinal line a little to one side of the dorsal.

Above-ground or non-root feeder nematodes: These nematodes feed on plant parts above the soil, e.g., wheat gall nematodes, *Anguina* spp.

Absolute alchohal: Pure C_2H_5OH. It is much more costly, and is brought in very small quantities as a pharmaceutical preparation. It is normally used as a dehydrating agent, for making slide mounts, or in the dry preparation of larvae.

Absorption: The entrance of a chemical into a plant or microorganism.

Acaricide: A chemical that kills mites and ticks; miticide.

Acariform: Shaped-like a mite (Acarina).

Acarina: An order of the class Arachnida. It includes ticks and mites, in which the cephalothorax and abdomen are broadly joined and the segmentation is not distinct.

Accepted daily intake: A concentration expressed in ppm of residue in or on a food when first offered for comsumption. It is calculated from (i) the accepted daily intake (ADI), (ii) the food factor, and (iii) the average weight of consumer. The ADI is daily dose which during an entire life will not cause any injury. It is set at a safe level as determined by the data obtained.

Accessory cell: A close cell in the front wing of Lepidoptera formed by the fusion of two branches of the radius, usually the R_3 cells.

Accessory glands: In the female insect, a pair of glands opening primarily on the ventor of the ninth abdominal segment, secreting an adhesive substance or material forming a covering or case (Ootheca) for the eggs; in the male, mucous glands opening ino the ejaculatory duct.

Accessory hearts: Pulsatory organs consisting of ampullae with valves, sometimes present near the bases of antennae, wings or legs.

Accessory sexual characters: The structures and organs (except the gonads) of which the genital tract is composed , including accessory glands and external genitalia.

Accessory vein: An extra branch of a longitudinal vein; the most posterior vein in the anal area of the front wing of certain Hymenoptera.

Accessory ventral condyle: An additional condyle or articulating process on the ventral side of the mandible and located mesad of the regular ventral condyle.

Accumulate: The build-up; add; store; pile up.

Accumulative pesticide: The chemical which tends to build up.

Acephalous: Larvae apparently without head or having head that is much reduced and not separated from the thorax, e.g. housefly larvae.

Acerous: Without antennae. Only Protura are entirely acerous.

Acetabulum: The articulating cavity in which an appendage or condyle of an appendage operates; a cuplike cavity within the sucking mouth of a maggot.

Acetometaarsenite: It is the first arsenical insecticide to be used in large quantity. It is a compound theoretically containing 55-58% of arsenic trioxide.

Acetophos: It is a contact organophosphorus insecticide and is similar to trichlorofon. Its chemical formula is o, o-diethyl-S-carboxymethyl thio-phosphate.

Acethion: It is an analog of malathion having low mammalian toxicity. Its chemical formula is o, o-diethyl-S-carbothioxy-methyl-dithiophosphate.

Acetylcholine: A substance that helps in transmission of impulses in the nervous system.

Acetylcholine esterase: An enzyme that hydrolyses the acetylcholine.

Acid gland: The principal poison gland of the bee, secreting a complex mixture of histamine-liberating enzymes and proteins of low molecular weight, collectively known as *apitoxin*.

Acidic: Acid in reaction resulting from excess of hydrogen ions over hydroxyl (OH) ions in solution; pH less than 7.

Acidophilic: Acidophilus; thriving in an acid medium; readily stained with acid dyes.

Acinus: (Pl. Acini). One of the small sac-like dilations of a compound gland.

Acone eye: A compound eye in which crystalline cones are absent or poorly developed, e.g. in the Crane fly.

Acraein: A protective or distasteful secretion of certain butterflies.

Acridoidea: One of the four super-families of Orthoptera. It includes short-horned grasshoppers and locusts. The insects are hemimetabolous with straight wings and with biting mouth parts, having the head bent at right angles to the body. The fore wings are horny and thickened and thus called *tegmina.* The femora of the hind legs are lengthened for jumping.

Acrosternite: The narrow marginal flange anterior to the antecosta of a definitive sternal plate that includes the preceding primary intersegmental sclerotization; characteristics of abdominal sterna of insects, but absent on thoracic sterna.

Acrostichal bristles: One or more longitudinal rows of small bristles along the centre of the mesonotum (In Diptera).

Acrotergite: The anterior precostal part of the tergal plate of a secondary segments; usually a narrow flange, but some times greatly enlarged, and frequently reduced or obliterated.

Acrotrophic egg tube (ovariole): A type of egg tube in which the trophic cells (food cells) remain in the apical chamber or tip of each ovariole and may be connected to the developing egg by strands of protoplasm acting as nutritive cord, e.g. in certain bees and plant bugs.

Actinic light: Usually the violet and ultra violet light proportions of the electromagnetic spectrum.

Actispray: It is the market product of an antibiotic cycloheximide.

Activator: A chemical added to a pesticide to increase, either directly or indirectly, its toxicity.

Actual dosage: The amount of active ingredient (not formulated product) which is applied to an area or other target.

Active ingredient: The actual amount of pesticides present in the formulations. The chemical or chemicals in a product responsible for the desired effects, which are capable, in themselves, of preventing, destroying, repelling, or mitigating insects, fungi, rodents, weeds, or other pests. The toxic component of a formulated pesticide.

Aculea (Pl. Aculeae): A fixed hair of the cuticula; Minute spines on the wing membrane of Lepidoptera.

Aculeata: A division of Hymenoptera which includes the stinging forms, bees, ants and wasps in contrast to *Parasitica* which are parasites of other insects.

Aculeate: With a sting (Hymenoptera); With minute bristles under the scales of the wing (Microlepidoptera).

Acuminate: Tapering to a long point.

Acute: In the Medical sense, it means short duration; As applied to material inhaled or absorbed through the skin, it refers to a single exposure; When applied to materials ingested, it refers to a single dose; In structural sense, it means pointed, forming an angle of less than 90°.

Acute dermal toxicity: Poisoning from a single dose of a chemical absorbed through the skin.

Acute inhalation toxicity: How poisonous a single dose (or exposure) of a pesticide is when breathed into the lungs.

Acute oral toxicity: Poisoning from a single dose of a chemical taken by mouth.

Acute poisoning: Poisoning which occurs after a single dose (or exposure) of a pesticide.

Adaptability: Capacity for evolutionary change.

Adaptation: The condition of showing fitness for a particular environment, as applied to characteristics of structure, function or entire organism; also the process by which such fitness is aquired.

Adaptive radiation: Evolutionary divergence of members of a single phyletic line into a series of rather different niches or adaptive zones.

Additive: Any material that is added to a pesticide (not necessarily a wetting agent or surfactant).

Adecticous pupa: Pupa in which mandibles are not functional, often reduced and not used for escaping from the cocoon. It is further divided into two types: the *Exarate*, with free appendages and the *obtect*, with appendages glued to the rest of the pupal body.

Adenine: A pure nitrogenous base occuring in DNA and RNA. It pairs normally with thymine in DNA.

Adenotrophic viviparity: A phenomenon in certain insects where the larvae are retained in the parents body after hatching and develop by feeding on 'uterine' secretion, moult twice and are finally deposited in a mature form ready for pupation, e.g. found only in Tsetse fly (*Glossina* spp.)

Adephaga: A suborder of order Coleoptera, comprising families in which both adults and larvae are usually active and predatory, e.g. Ground beatle, Tiger beetle.

Adherence: The property of a substance to adhere or stick to a given surface.

Adhesive: A substance that will cause a spray material to stick to the sprayed surface. It is often referred as sticking agent.

Aedes aegypti: A tropical mosquito (Diptera). It spreads yellow and dengue viruses. Formerly known as *Stegomyia faciata.*

Adfrontal areas: A pair of narrow oblique sclerites on the head of lepidopterous larvae or structures adjoining the frons or forehead.

Adfrontal setae: Setae borne on the adfrontal areas and usually numbered according to their proximity to the vertex, the closest being termed the first adfrontal seta.

Adipokinetic hormone: The

harmone which induces changes in the lipids of the blood and fat body.

Adjacent: Next to neighbouring.

Adjuvants: Accessory substances; ingredients that improve the properties of pesticide formulations. They include wetting agents, spreaders, emulsifiers, dispersing agents, etc.

Adrenergic blocking agents: Agents that block the nerve fibres which librerate sympathin-an adrenaline substance.

Adsorption: The binding of a chemical to the outside surface of small soil particles or to plant parts.

Adventitia of heart: The heart is lined externally by connective tissue sheaths. This sheath is known as adventitia.

Adventitious vein: A secondary vein, neither accessory nor intercalary, usually a result of number of cross veins lined upto form a continuous vein.

Aedeagal apodeme: An apodeme of the aedeagus.

Aedeagus: The distal part of the phallus, usually the principal part of the intromittent organ; typically a sclerotic tube.

Aerate: To bring in contact with more air; to loosen or stire up the soil mechanically or use additives such as peat; to let in or pull in fresh air after fumigation.

Aerial application: Treatment applied with the use of an airplane or helicopter.

Aerobic: Living or functioning in air or free oxygen. The opposite of anaerobic.

Aeropyle: Aeropyles are special channels present in the chorion of egg through which inner chorionic layer communicates respiratory exchange with the atmosphere. These are sometimes filled with a porous, spongy, air filled protein and may be restricted to certain areas of the shell. Over 180 such channels occur in the rim of the egg-cap of *Rhodnius*. Specialized pores on the egg surface for oxygen uptake.

Aeroscopic plate: A strip of modified chorion acting as an air-channel. In most endoparasitic larvae a connection is established between the atmosphere and the open tracheal system of the parasite. In 1st instar larva of the Encyrtidae this is brought about by the larva retaining a close association with the egg-shell which is connected with the surface of the host by a stalk-like prolongation and bears the so-called aeroscopic plate.

Aerosols: An aerosol is a system of particles of ultra-microscopic size (0.5 to 5 μ diameter) dispersed in a gas. Insecticidal aerosols are produced by either incomplete combustion of tobacco or other organic material containing an insecticide or by spraying a solution of an insecticide such as DDT in fuel oil on a heated surface or by spraying a solution of an insecticide from a container of liquid gas.

Aerosol spray: A fine spray produced by pressurized gas that leaves very small droplots of pesticides suspended in the air.

Adult: A full grown, sexually mature insect, mite, nematode, or other animal. This stage is often the one to migrate and begin new colonies or infestations.

Adulterated: When strength or purity of a pesticide falls below standard claimed by or if any substance has been added or any part substracted.

Adulticide: An insecticide which is toxic to the adult stage of insects.

Aestivation: A dormancy (period of inactivity) during a warm or dry season in insects.

Afferent nerve: A nerve that conducts from the periphery towards a nerve centre; the axon of a sensory neuron.

Affinity: Relationship; sometimes misleadingly employed as synonym for phenetic similarity.

Aflatoxins: mycotoxins Toxins produced by moulds are mycotoxins. Of the various such toxins, the aflatoxins, produced by a fungus *Aspergillus flavus* are highly toxic, carcinogenic to mammals, and most widespread contaminants of the stored food materials, particularly groundnut, maize and cotton seed.

AGA mixture: 95% alcohal sparts, distilled water 5 parts, glycerine 1 part and glacial acetic acid 1 part.

Agamic: A species reproducing parthenogenetically, i.e. without mating.

Agamospecies: A species without sexual reproduction; an asexual species.

Aggregation: Grouping by a species resulting from some attractive stimulus.

Agile: Able to act quickly and easily; active; nimble.

Agitator: A mechanical device in the sprayer to ensure uniform distribution of chemical and to prevent sedimentation.

Aglutinin: An antibody which produces clumping (agglutination) of the antigenic structures.

Agroecology: The study of ecology in relation to agricultural systems.

Agroecosystem: It is a complex of organisms and the environmental factors found in an agricultural field. It differs from a natural ecosystem in that it is relatively artificial (also refer ecosystem).

Agromyzidae: A family of Diptera order which includes leaf and stem minors. These are minute flies whose small white larvae (maggot) burrow into the leaves, young shoots of wide range of plants, e.g. Soybean stem fly (*Ophiomyia phaseoli*), Tur pod fly (*Melanagromyza obtusa*) are the serious pests of soybean and *tur* in India.

Agrotis flammatra (Lepideptra: Noctuidae): Commonly known as Gram cutworm is also important pest of chickpea in India. It has also been reported from Europe, Syria, Persia, Western Siberia and Pakistan. Being a polyphagus pest, it also feeds on potato, cucurbits, pea, okra, tobacco, wheat, etc. The larvae cut the plants at night and drag them in the soil and eat foliage there. The adult moth is greyish brown. Its fore wings have characteristic markings and smoky patches, two-third of the proximal areas being pale. Each wing have a semicircular spot below the pale area and a yellowish brown kidney shaped spot towards the apical area.

Agrotis ypsilon **Rott.** (Lepidoptera: Noctuidae): It is distributed throughout the world and cause damage to potato, gram, linseed, soybean, tobacco, peas, etc. The young larvae feed on the leaf epidermis. The grownup larvae live in cracks and holes in the ground and come out at night and fell the plants by cutting their stems, either below the surfaces or above the ground. The cut branches are seen to have been dragged into the holes, where the leaves are eaten at leisure. The larvae are black or brown with grey body. The adult is dark or blakish with some greyish patches on the back and dark streaks on the four wings.

Air blast sprayer: A machine which can deliver high and low volumes of spray. It is used for orchards, shade trees, vegetables and fly control.

Air sac: A thin walled dilation of the trachea. It lacks taenidia in its walls. It is capable of inflation and supposed to lessen specific gravity, and to aid in flight.

Akinesis: A motionless state of an insect resembling sleep, particularly after loss of or, damage to, sensory organs such as antennae.

Alae cordis: Ring like bands of fibrous tissues suspending the heart from the pericardium in insects and other Orthopods.

Alarm pheromone: These are simple compounds, found in many formicidae, *Apis, Trigona* and a few Isoptera. They are produced by the mandibular glands and various abdominal glands according to the species concerned. Alarm pheromones cause the insects perceiving them to run or fly more actively and to show aggressive behaviour, but their effects vary considerably and may depend on the concentration of the pheromons.

Alary (alar): Wing-like; pertaining to wing.

Alary (Aliform) muscles: A series of triangular wing—like muscles inserted on to the membrane that separates the perivisceral from the pericardial cavity of an insect. Contraction of these causes blood to flow into pericardial cavity and thence through small openings or ostia into the long dorsal tubular heart.

Albinism: In zoology, the absence of pigmentation, and particularly of melamins, in an animal.

Albino: Generally applied to animals where skin, hair, and eyes are unpigmented and whose skin and eyes are pinkish because of blood vessels showing through. Also used to describe plants in which a particular pigment (usually the chlorophyll) is absent.

Aldicarb: See Temic.

Aldrin: (HHDN; 1, 2, 3, 4, 10, 10, -hexachloro-1, 4, 4a, 5, 8, 8a-hexahydro-exo 1, 4, -endo, 5, 8-dimethano-naphthalene; C_{12} H_8 Cl_6). It is a chlorinated hydrocarbon insecticide produced by the Diels-Alder reaction. It has minimum of 95% HHDN (active material).

Aleukia: The absence or extremely decreased number of leucocytes (white blood cells) in circulating blood.

Alienicole: Viviparous wingless female aphids that produce parthenogenetically in enormous numbers on herbaceous plants throughout the spring and summer. Lateron they give rise to winged forms that fly to another plant in the autumn and reproduce by normal sexual process.

Alifom: Wing-like.

Aliform (alary) **muscles:** These are composed of striated fibres arise from the terga and spread out fanwise over the surface of the dorsal diaphram. The fibres of one alary muscle attach to the heart or meet, beneath the heart, those of the corresponding muscle of the opposite side of the body.

Alimentary canal: The food tube transversing the body; in insects consisting of an endodermal mesenteron, an ectodermal stomodaeum and an ectodermal proctodaeum.

Alimentary castration: Sterlity of the worker and soldier castes of some social insects, brought by a deficiency in their diet.

Alinotum: The wing bearing (notal) plate of the dorsum of the mesothorax or metathroax of pterygote insects.

Alitrunk: The two wing bearing segments of insects' thorax (usually second and third segments). In some Hymenoptera, the first segment (prothorax) is also involved in wing attachment.

Alkaline: Having the reaction of an alkali resulting from an excess of hydroxyl ions (OH) over hydrogen ions in solution; pH more than 7.

Alkaline gland: Dufour's gland; one of the glands connected with the sting of a bee, producing an oily alkaline secretion whose function is obscure, but may be lubricatory.

Alkaloids: These are naturally occurring heterocyclic, optically active substances with nitrogenous base of relatively high molecular weight and having marked physiological activity, e.g. nicotine. They combine with acids to form salts.

Allatectomy: A technique for removing the corpora allata.

Allele (Allelomorph)**:** One member of a group of alternative forms of a gene that can occur at a particular locus on homologous chromosomes.

Allergy: It is a stage in which cells of the body are hypersensitive to certain substances, introduced into it. The reaction of the body is exhibited as oedema, inflamation and destruction of tissue.

Allethrin: (Divinyl pyrethrin I, DVPY-I, Allylrethrum allyl analog of cinerin-S). It is a natural organic insecticide and is an ester made from alcohol having the allyl side chain (allethriolone). Allethrum is a mixture of four racemic forms.

Allochromic species: Species which do not occur at the same time level.

Allochthonous: Exotic; introduced; acquired from elsewhere (opp. autochthonous).

Allometric growth: Disproportionate growth; growth in which the growth rate of one part of an organism is different from that of another part, or of the body as a whole.

Allopatric: Of populations or species, occupying mutually exclusive (but usually adjacent) geographical area.

Allopatric hybridization: Hybridization between two allopatric populations (species or subspecies) along a well-defined contact zone.

Allopatric speciation: Species formation during geographical isolation.

Allopolyploid: A polyploid possessing whole chromosome sets from different species.

Allosematic resemblance: Batesian mimicry; the resemblance of a harmless animal to a poisonous or distasteful one, giving protection to the former, since predator will tend to avoid both, e.g. the close resemblance of a clearwing moth to a wasp.

Allotype: A paratype of the opposite sex to the holotype.

Alluminium oxide: It is used as corrective in the formation of arsenical insecticides to prevent arsenical injury.

Almond moth: *Cadra (Ephestia) cautella* (Walker) (Lepidoptera: Phycitidae). It is also known as dried currant moth and fig moth. It is distributed worldwide and is serious pest of dried fruits, such as apples, dates, berries, figs, almonds, walnuts, tamarind seeds, etc. and has also been reported on lac, malted milk, dried mango pulp, garlic bulbs, various cereal grains and grain products. The adult moth has greyish wings with transparent stripes on outer region. The caterpillars make tunnels in the food materials.

Alpha taxonomy: The level of taxonomy concerned with the characterization and naming of species.

Alternation of generation: The alternation of a bisexual with unisexual (parthenogenetic) generation.

Alternative: One or two or more choices; another way to do something.

Alula (Pl. Alulae): A pair of membe-ranous scales above the halters; a pair of lobes at posterior angle of the wing base, particularly developed in some Diptera; also called calypters.

Alveolate: Having cells or alveoli; deeply pitted.

Alveolus: A hair socket.

Amatidae (Syntomidae)**:** A family of tropical moths (Lepidoptera). These are small, dayflying and brightly coloured. Some species resemble wasp. Larvae of some moths are agricultural pests, e.g. *Syntomis fortunei*.

Ambulatory leg (Running)**:** The coxae are will developed; with these type of legs insects move about swiftly, e.g. cockroaches.

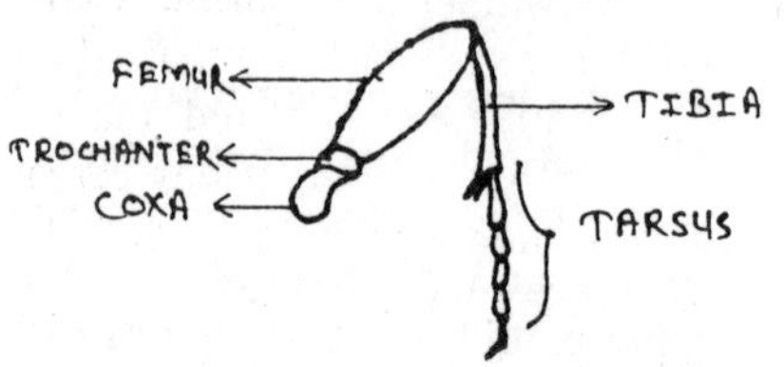

Ambulatory leg

Ametabola: Primitive wingless insects that donot undergo metamorphosis, e.g. springtails and bristletails. The group comprises Diplura, Thysanura, Collembola and Protura.

Ametabolous: The insects which develop without metamorphosis.

Amine: Any organic compound containing nitrogen derived from ammonia.

Amino acid: A class of organic compounds containing an amino ($-NH_2$) group and a corboxyl ($-COOH$) group, plus a side chain; the basic constructional unit of proteins.

Aminoparathion: (O, O, Diethyl - O, 4-amino-phenylthiophosphate). It is the compound formed by reacting parathion with reducing agents like metals in acid medium. It is not toxic to animals as well as to insects.

Ammochaeta: A stiff type of bristle occurring in bunches on the heads of ants, used for cleaning the legs, etc.

Ammophos: (O, O-dimethyl-S-2-acethyl-amide ethyldithio-phosphate). It is a new systemic organophosphorus insecticide. It has systemic effect and is used in aquous emulsion.

Amnion: The cellular, membranous coverning of the embryo formed from folds of the extraembryonic blastoderm or specifically, the inner layer of each amniotic fold, or of the complete covering.

Amnion: (Pl. Amnions or Amnia). The cellular membranous covering of the embryo from folds of the germ band.

Amniotic cavity: The cavity between the amnion and the embryo.

Amphidiploid: A tetraploid individual having two sets of chromosomes from each of two known ancestral species.

Amphiploid: A polyploid produced by the chromosome doubling of a species hybrid, that is, of an individual with two rather different chromosome sets.

Amphipneustic larvae: Having two

pairs functional spiracles only on the first abdominal segment, and the last one, or two, e.g. maggot of house fly.

Amphitoky: The parthenogenetically developing eggs may have either a haploid or a diploid set of chromosomes and may give rise to both sexes, is known as amphitoky.

Amplexiform (wing coupling): Brought about by extensive overlapping of two wings.

Amylases: They are polysaccharases, and are widely distributed in insects. These catalyse the hydrolysis of the more complex carbohydrates, acting upon starch.

Anabasine: See neonicotine.

Anaemia: Any condition in which number of red blood cells and amount of haemoglobin are less than normal. It is frequently manifested by pallor of the skin and mucous membranes.

Anaerobic: Living or functioning in absence of air or free oxygen.

Anaerobic respiration: Releasing energy within the cell without oxygen.

Anal: Pertaining to the last abdominal segment (which bears the anus); the posterior basal part of the wing.

Anal angle: Tornus. The angle between the outer and anal morgins of a wing.

Anal area of the wing: The posterior portion of the wing, usually including the anal veins.

Anal cell: A cell in the anal area of the wing; cell 1 A in Diptera.

Anal crossing: Where vein A branches posteriorily from Cu+A (In Odonata).

Anal fold: See plica vannalis.

Anal glands: Ectodermal glands opening close to the anus.

Anal lobe: A lobe in the posterior basal part of the wing.

Anal loop: A group of cells in the hind wing of dragonflies, between Cu_2, 1 A and 2 A, which may be rounded, elongate or foot shaped.

Analogous: A similar feature in two or more taxa which can not be traced back to the feature in the common ancestor of these taxa.

Anal papillae: Thin walled outgrowths in the anal region of some aquatic insects, e.g. mosquito larva. They may have respiratory function but they appear to act chiefly as organs for absorption of chloride and other ions from the water.

Anal veins: All the veins between the cubitus and the jugal region; according to the Comstock-Needham system, the veins here called post cubitus and vannals; the most posterior veins of insects' wing.

Anamorphic development: Development by adding segments to the body.

Anamorphosis: The formation of extra abdominal segments on an insect after it has hatched from the egg, e.g. in primitive group of Protura.

Anaphase: The stage of nuclear division characterized by movement of chromosomes from the spindle equator to spindle poles. It begins with division of the centromeres and closes with the end of poleward movement of chromosomes.

Anapleurite: The dorsal supracoxal sclerotization of generalized thoracic pleuron.

Anatomy: The science of internal morphology, as revealed by dissection.

Anatrepsis: During embryonic development, the germ band of many insects undergoes a sequence of movements. This movement is completed in two stages-*anatrepsis* and *Katatrepsis*. In the first stage or *anatrepsis*, the germ band immersed in the yolk.

Androconia: Modified wing scales and connected scent glands of some male butterflies, producing a sexually attractive secretions.

Androconia

Anecdysis: A long passive period between two moults of an insect during which there appears to be no preparation for the next moults.

Anellus: a sclerotization of the inner wall of the phallocrypt or phallotheca, often forming a ring or tube about the base of the aedeagus; a chitinous ring supporting parts of male genitalia in some insects.

Anemophilous: Plants which are pollinated by winds.

Anemotaxis: A tendency of some insects to fly into a current of air.

Anepimeron: The upper part of the epimeron or posterior sclerite of the subcoxa at the base of an insect's leg.

Anepisternum: The upper part of the episternum or anterior sclerite of the subcoxa at the base of an insect's leg.

Angoumois grain moth: *Sitotrega cerealella* (Olivier) (Lepidoptera: Gelechiidae): It is a serious pest of stored wheat, maize, sorghum, barley, oats, etc. throughout the world. They cause damage by feeding on the grain kernels. The larva is white with yellow brown head. The adult is buff, grey-yellow, brown or straw coloured moth having narrow pointed fringed wings with long hairs.

Anisometric growth (disharmonic or disproportionate): The type of growth where there is a different rate of increase in different parts of the body, with a resulting change of shape.

Annular: Shaped like or forming a ring.

Annulate: Bearing rings or arranged in the form of an annulus or ring.

Annulus: A ring-like marking or a ring of hard cuticle.

Anolog: A compound that is very similar in both structure and formula to other compound.

Anopheles maculipennis: Most widely distributed malaria mosquito, characterised by the appearance of spots on the wings due to concentration of scales in the region of cross veins.

Anoplura (Siphunculata)**:** An order of wingless blood sucking lice, mostly parasites of mammals.

Ant: An insect of the superfamily Formicoidea with a single family Formicidae. The insects with elbowed antennae and with a very narrow abdominal construction or 'waist'; sometimes known as 'petiolc'. They live in highly organised colonies containing many castes. Only queen lay eggs after nuptial flight in which several colonies take part. After mating the females cast off their wings and start colonies in the ground.

Antagonism: The phenomenon which results in a depression of compound activity when two or more are in close proximity or mixed together.

Anteapical: Just proximad of the apex.

Anteapical cell: A cell in the distal part of the wing in leaf hoppers.

Anteclypeus: In Psocopteran insects the clypeus is partially or completely divided by a transverse suture into two sclerites—the postclypeus and the anteclypeus.

Antecosta: The anterior submarginal or marginal ridge on the inner surface of a tergal or sternal plate corresponding to the primary intersegmental fold, on which typically the longitudinal muscles are attached.

Antecostal suture: The external groove of the antecosta.

Antecoxal sclerite: A sclerite of the metasternum, just enterior to the hind coxa.

Antegen: A substance capable of inducing the formation of antibodies when introduced into the blood-stream of an animal.

Antenna (Pl. Antennae)**:** The appendicular organs of the procephalic region of the head innervaed from the deutocerebral lobes of the brain, called first antennae or antennules in Crustacea; absent in

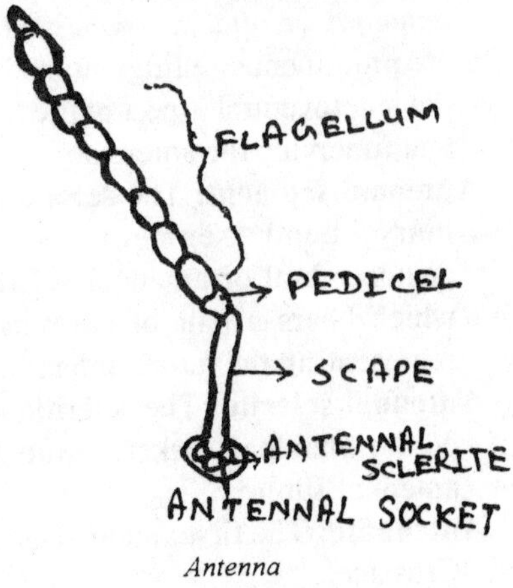

Antenna

Chelicerata, usually sensory in function. This is very varied in structure and segmentation.

Antenna-cleaner: In the Apocritan Hymenoptera, the enlarged apical spur of the anterior tibia fits against a pectinated semicircular pit in the first tarsal segment, and the antennae are cleaned by being passed between these two organs.

Antennal club: The enlarged distal segments of a clubbed antenna.

Antennal comb: A row of bristles forming a comb—like structure at the proximal end of the anterior tarsus in bees, used for removing pollens, etc. from the antennae.

Antennal fossa: A cavity or depression in which the antenna is located.

Antennal groove: A groove in the head capsule into which the basal segment of the antenna fits.

Antennal (olfactory) **lobe:** The deuto-cerebrum represents the fused ganglia of the antennal segment. It is chiefly composed of the paired *antennal* or *olfactory lobes* which are prominent swellings situated on the anteroventral aspect of the brain and innervate the antennae.

Antennal segment: The second primary head segment is the deutocerebral or antennal segment which bears a pair of outgrowths representing the future antennae.

Antennal sclerite: The scleritic rim of the antennal socket within the antennal suture.

Antennule: The first antennae of the Crustacea.

Antenodal cross veins: The cross veins along with costal border of the wing, between the base of the wings and the nodus, extending from the costa to the radius (in Odonata).

Antennal suture: The line of inflection in the cranial wall sorrounding the antennal socket.

Antepenultimate: The second from the last.

Antepygidial bristle: One or more larger bristles on the apical margin of the seventh (next to the last) tergum in Siphonaptera.

Anterior: Front; infront of or nearer to the head of the insect.

Anterior cross vein: The r-m cross vein in Diptera.

Anterior intestine: The part of the proctodaeum between the ventriculus and the rectum, or between the pylorus and the rectum when a proctodaeal pyloric region is distinct.

Anterior mesenteron rudiment: The anterior group of cells of the ventral endoderm remnant that regenerates the mesenteron.

Anterior notal wing process: The anterior lobe of the lateral margin of the alinotum supporting the neck of the first axillary.

Anterior pharynx: the precerebral part of the pharynx in insects having also a pharyngeal section of the stomodaeum behind the brain.

Anterodorsal: In the front and at the top or upper side.

Anteromesal: In the front and along the midline of the body.

Anteroventral: In the front and underneath or on the lower side.

Anthio: Formothion. O, O-Diethyl-S-(N-Methyle-formylcarbamoyl methyl)-dithiophosphate). It is an organic phosphorus compound similar to dimethoate but is more stable than dimethoate. It is marketed as 25% emulsion concentrate.

Anthogenesis: The reproduction of males and females parthenogenetically, e.g. plant lice.

Anthophilous: Attracted by flowers; feeding on flowers.

Antibiosis: It is an adverse effect of a plant upon a pest owing to harmful effect of a specific chemical, or lack of specific nutrient in the plant infested by the pest.

Antibiotic: It is a substance produced by micro-organism which inhibits the growth of a poisonous or toxic chemical or radioactive materials.

Antibody: A serum globulin which is produced in the blood of an immunized animal in response to the introduction of a foreign antigen.

Anticlypeus: An anterior division of the clypeus.

Anticoagulant: A substance which prevents the normal clotting of blood.

Antidote: A substance given to a patient to counteract the effect produced by the ingestion of a poisonous or toxic material.

Antidrift agent: A chemical used to reduce spray drift during the actual spraying operation by various physical factors.

Antifeedant: A chemical possessing the property of inhibiting the feeding of certain insects.

Anti-feeding compound: A compound which will present the feeding pests on a treated material without necessarily killing or repelling them. It is not a repellent.

Antigen: A foreign substance (usually protein or protein polysaccharide complex in nature) which elicits the formation of specific antibodies with an organism.

Antigenic: With the properties of an antigen.

Antiserum: Blood serum containing specific antibodies.

Antlia: The tubular proboscis of butterflies; A dilated part of the post-pharynx in some Diptera. The Oesophageal pump.

Ant-lion: It is a predatory larva of a winged insect of family Myrmeleontidae in the order Neuroptera resembling a dragonfly. It digs a pit in sandy soil and lies with its large jaws partly exposed as it waits for passing insects particularly ants. It devours a large number of these, when fully grown pupates and eventually become a four winged insect.

Anus: The posterior opening of the alimentary tract (canal).

Aorta: The anterior non-chambered part of the dorsal blood vessel. It starts from the heart and opens in the head.

Apanteles: A genus of small ichneumon flies of the family Braconidae whose larvae parasitize and destroy the caterpillars and chrysalids. The yellow-white silken cocoon of these insects are often to be seen immediately after pupation which takes place on emerging from the dead caterpillar of the various lepidopterous pests.

Aphelinidae: Minute chalcid flies parasitic in the bodies of aphids. They have been introduced successfully into Newzealand and elsewhere in order to parasitize and control the wooly aphis of apple trees.

Aphidovorous: Feeding on aphids.

Aphis (Aphid): Plant lice. A minute insect whose nymphs and adults suck the cell sap of various cultivated and ornamental plants and exude a sticky or waxy secretion from the abdomen. They belong to the order Homoptera and have very complex life. Wingless parthenogenetic females breed viviparously in very large numbers. The following terms are used in connection with the life history. *Fundatrix*—the first, usually wingless female of a season, *Fundatrigeniae*-Parthenogenetic female off spring of the fundatrix, *Alicicolae*-Parthenogenetic, viviparous females, generally many generations mostly on the secondary host, both winged and wingless forms. *Migrantes*—Migraory winged females produced parthenogenetically, *Sexuparae*—winged females which produce several males and females, *Sexuales*-normal sexual males and females, e.g. mustard aphid *(Lipaphis erysimi)*.

Apiary: A place where colonies of bees are intentionally kept for the benefit of honey producing and pollinating activities.

Apical: At the end; tip or outermost part.

Apical cell: (i) A special, usually large, trophic cell in the upper end of the testicular tube in some insects. (ii) A cell near the wing tip.

Apical cross vein: A cross vein near the apex of the wing in Plecoptera, Homoptera.

Apices: The ends of various structures and appendages.

Apiculture: Pertaining to the care and culture of bees.

Apidae: A family of Hymenoptera order; Aculeate Hymenoptera. It includes bumble bees, carpenter bees. A normal hive contains one queen, a few hundred drones and many thousand workers. Bees differ wasps in the following ways: (i) They build brood cells and honey comb of wax instead of wood pulp (ii) They feed entirely on pollen and nectar whereas wasps are carnivorous.

Apitoxin: The poison of bees' stings consisting chiefly of histamine producing enzymes and proteins of low molecular weight.

Apneustic: Without specific external brething organs, either spiracles or gills; the tracheal system either absent or rudimentary.

Apocrita: It is suborder of order

Hymenoptera. It includes bees, wasps, ants and ichneumon flies and the members are recognized from their constricted or petiolated abdomen. The honey bee, *Apis indica* (Apidae), the red wasp, *Vespa orientalis* (Vespidae), are the common examples. The larvae are grub-like and without legs.

Apodeme: Any cuticular invagination of the body wall, usually formed in a multicellular matrix, but sometimes in a single cell, forming rigid processes for the attachment of muscles or support organs.

Apodous (Apodae) **larva:** This type of insect larva has no legs or abdominal processes and the head is also not distinct. The worm-like larvae are common among the dipterous flies.

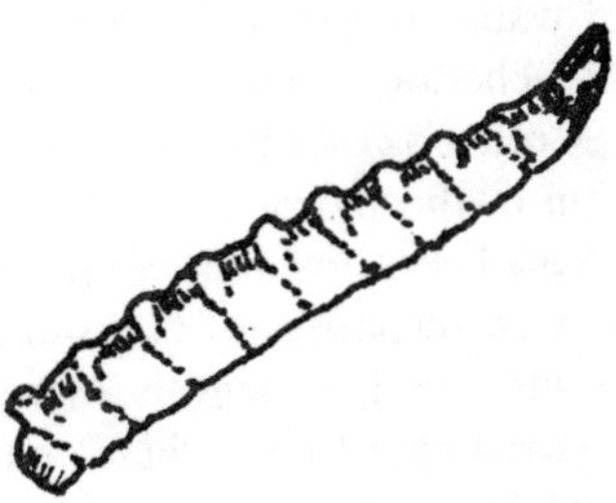

Apodous larva

Apolysis: During moulting, changes take place in the epidermal cells which divide mitotically and so become close-packed and columner in form. As a result of the changes in cell shape, a tension is generated at the surface of the epidermal cells which results in their separation from the cuticle. This separation of the cuticle from the epidermis is called Apolysis.

Apomorph: A derived character.

Apophysis (Pl. Apophyses)**:** A tubercular or elongate process of the bodywall, either external or internal.

Aposematic: Having a warning colouration, either indicating that an animal is dangerous, e.g. wasp or in the case of harmless insects giving some degree of protection by mimicking the colours of a more dangerous perdators.

A posteriori weighting: The weighting of taxonomic characters on the basis of their proved contribution to the establishment of sound classification, i.e. of monophyletic taxa.

Appendage: Any attachment to the body that is hinged to the head, thorax or abdomen by a joint, e.g. antennae, legs, wings and so on.

Appendiculate: Bearing appendages.

Appendiculate marginal cell: With the vein forming the posterior side of the marginal cell extending a short distance beyond the apex of the cell in Hymenoptera.

Appendix: A supplementary or additional piece or part of the homopteran wing; any attachment; also used for a short, stump-like vein.

Apple wooly aphis: *Eriosoma lanigerum* (Hansmann) (Hemipera: Aphididae). It is a serious pest of apple, pear and crab-apple in India and Pakistan. The aphids suck cell-sap from the twigs and from the roots underground. Swelling or knots appear on the roots which hinder the normal plant functions. The aphids reproduce parthenogenetically.

Application: The placing of a pesticide on a plant, animal, building, or soil or its release into the air or water to prevent damage or destroy pests.

Applicator: A person or piece of equipment which applies pesticides to destroy pests or prevent damage by them.

Applied control: The measures adopted by human beings to control pests; also known as artificial control.

Applied entomology: The application of pure entomology to the control of insects.

Apposed image: An image build in eyes adapted solely for seeing by day, by apposed points of light falling side by side and not overlapping; mosaic vision (also called apposition image).

Apposition eyes: In diurnal insects, a compound eye which absorbs oblique rays light in the pigmented walls of the ommatidium and form mosaic image, i.e. an errect image composed of large number of dots or points or light. In this case the crystalline cones of the eye are sorrounded by a sheath of pigments. This prevents the light from passing from one cone to another and allows only a small point of light to impringe on the retina from each cone.

A priori weighting: The weighting of taxonomic characters on the basis of preconceived criteria, i.e. their physiological importance.

Aptera: A name formerly used for all apterous insects, but now best avoided, since wingless occur both primitively and secondarily in a wide range of different orders.

Apterous: Wingless.

Apterygote: Insects belonging to the subclass Apterygota or insects whose ancestors have always been wingless. These comprise Diplura, Protura, Thysanura and Collembola.

Aquatic: Living in water, either fresh or salt; in practice salt water forms are usually described as marine.

Aqueous: Indicating the presence of water in a solution. A solution of a chemical in water.

Aquired characters: The alternation in the morphology (form) or physiology (function) of an organism in response to its ecological factors (environment) is known as aquired characters. These characters are not heritable.

Archidictyon: Archedictyon; An irregular network of a small veins, like those of a leaf in the wings of the Palaeodictyoptera primitive fossil insects of the Carboniferous period. This name was given by Tillyara in 1918.

Arachnida: A class of the phylum Arthopoda. It includes scorpions, spiders, ticks and mites. Their body has two distinct regions, viz. Cephalothorax and Abdomen. They donot have antennae and compound eyes but have simple eyes. They have well developed palps on the head and possess four pair of legs on the thorax.

Aramite: 2-choloroethyl-1-methyl-2(p-tort-butyl phenoxy ethyl sulphate). It is a chlorinated hydrocarbon insecticide and commonly used as Acaricide.

Araneida: An order of the class Arachnida. It includes the spiders in which cephalothorax and abdomen are joined by a short stalk.

Arborizations: The fine branching terminal fibres of axons or collaterals.

Archenteron: The gatrocoel, or cavity of endoderm.

Archetype: A hypothetical ancestral type arrived at by the elimination of specialized characters.

Archicephalon: The primitive Annelid—Arthopod head, or prostomium.

Archicerebrum: The primitive suprastomodaeal nerve mass of the prostomium; the ganglion nerve mass of the prostomium in annelids.

Arctiidae: A family of order Lepidoptera which includes several species of hairy caterpillars. The moths are brightly coloured. The larvae are hairy and pests of cultivated crops, e.g. Bihar hairy caterpiller, *Spilosoma obliqua*.

Arcuate: Bent like a bow or arched.

Arcuate vein: See vena arcuate.

Arculus: An arch-shaped vein forming cross connection between the other veins; a basal cross vein between the radius and the cubitus in Odonata.

Areole: An accessory cell; a cell of the wing in some Lepidotera, between R_3 and R_4.

Areolet: A small cell in the wing; in the ichneumonidae, the small submarginal cell opposite the second recurrent vein.

Arista: A specialised bristle or process on the antennae of certain Diptera, formed from the numerous segments of the flagellum.

Aristate: Bristle-like; with an arista; Aristate antenna.

Aristate antenna

Army worm: *Mythmina separata* (Haworth) (Lepidoptera: Noctuidae). It is a serious sporadic pest of graminaceous crops all over the world. In India, it is a sporadic pest of wheat, sugarcane, maize, sorghum and other graminaceous crops. The larvae feed voraciously and migrate from one field to another. Army worms are also called 'cutworms'.

Arolium (Pl. Arolia)**:** The terminal pad between claws in Orthoptera; A pad-like structure at the base of each tarsal claws in Hemiptera. It is probably a remnant of the dactylopodite.

Aromatic Odour: Flavour like that of a compound belonging to the closed chain class of organic compounds, or benzene derivatives.

Aromatics: Oil or oil-like materials which kill plant and animal tissue similar to burning. Chemically, a compound having a closed ring structure.

Arrhenotoky: Production of males from unfertilized eggs, e.g. Drone bees.

Arsenic Pentaoxide: This is a white amorphus solid formed by the oxidation of arsenious oxide with nitric acid. Due to its high toxicity to plants, arsenic acids are not used as insecticides. However, they are used for preparation of arsenates such as lead arsenates and calcium arsenates.

Artenkreis: Super species.

Arthropodin: The main structural proteins of the insect cuticle are sometimes known collectively as arthropodin.

Articulation: A movable point of contact between too sclerotic parts of the body wall.

Arthropoda: A phylum of the invertebrate animals. The arthropods possess jointed legs, each terminating into a claw, with chitinous exoskeleton, and their body is divided into segments which are discernible externally, but may be fused internally. Besides insects it includes crustaceans, arachnids, millipedes and centipedes.

Arthropods: Animals belonging to the phylum Arthropoda. (also see Arthropod).

Artificial classification: Classification based on convenient and conspicuous diagnostic characters without attention to characters indicating relationship; ofen a classificaion based on a single arbitrarily chosen characters instead of an evaluation of the totality of characters.

Asexual reproduction: Reproduction not involving the fusion of nuclei of different gametes.

Aspidiotus: See san jose scale insect.

Aspirator: A device with which insects may be picked up by suction.

Assembly and aggregation pheromones: These induce the formation of large, temporary or persistent aggregations of a species and include the phenomone that causes hibernation aggregation of the coccinellid beetle, *semadalia undecimnotata*.

Asymmetrical: Not alike on the two sides.

Asymptote: The point in the growth of a population at which numerical stability is reached.

Atlas: In taxonomy, a method of

presenting taxonomic materials primarily by means of comparative illustrations rather than by comparative descriptions.

Atomiser: A device for breaking up liquid into very fine droplets.

Atomization: Process of breaking liquid into fine spray.

Atrial orifice: The external opening of the spiracular atrium.

Atrium: The spiracular chamber formed by a secondary invagination of the body wall external to the primary tracheal orifice; a chamber; a chamber just inside a body opening.

Atrophied: Reduced in size; rudimentary.

Atropine: A specific antidote against poisoning by organophosphate pesticides.

Attapuglite: A magnesium silicate clay that is used as a carrier for the insecticides.

Attenuated: Very slender and gradually tapering distally.

Attractant: Material with an odour that attracts certain insects. Lure

Auditory organs: Organs by which an insect can hear sound or perceive vibrations.

Augur arrangement: It is a spiral mechanism formed about a turning shaft for moving the material.

Auricle: A small lobe or ear like structure in Hymenoptera.

Auricular valve: The wall of the insect heart is reflected inwards and forwards at each ostium to form an *auricular valve*, which precludes the return flow of the blood into the dorsal sinus.

Australasian region: Australia and Newzealand, and the Island groups of the South Pacific.

Autecology: The study of a single species.

Authority citation: The custom of citing the name of the author after a scientific name.

Autochthonous: Aboriginal; Native; indigenous; formed where found (opp. allochthonous).

Autocidal control: The method is used to reduce pest population by sterilizing insects and genetic manipulation; reproduction control.

Autocides: The control of a pest by the sterile male technique.

Automictic parthenogenesis: In parthenogenesis, the eggs are formed without meiosis (apomictic parthenogenesis) or, when this occurs, there is a later doubling of the chromosome number, either by fusion of cleavage nuclei or oocyte and polar body. This is termed as automictic parthenogenesis.

Autopolyploid: A polyploid all of whose sets of chromosomes are those of the same species.

Autosome: A chromosome not associated with the sex of the individual and therefore, possessed in matching pairs by diploid members of both sexes.

Auxiliae: Small plates beneath the bases of the pretarsal claws, bearing the pulvilli.

Auxiliary vein: The subcosta in Diptera.

Avenin: O, O-diethyl-N-(isopropoxy

carbamoyl phosphate). It is a systemic organophosphorus insecticide with low activity and narrow selective action.

Avicide: A substance to control bird pests. Generally not designed to kill but repel or to so affect a few individuals that others are frightened away.

Avoidance: The use or manipulation of environmental factors in crop production to take advantage of the stage in the life cycle of pests, when it can be most easily interrupted or when it is ineffectual, scarce or absent.

Axilla (Pl. Axillae): A triangular or rounded sclerite laterad of the scutellum, and usually just cauded of the base of the front wing in Hymenoptera.

Axillary cell: A cell in the anal area of the wing.

Axillary cord: The thickened, corrugated posterior edge of the articular membrane of the wing base, continuous with the posterior marginal fold of the alinotum.

Axillary plate: The posterior sclarite of the wing base in Odonata, supporting the subcostal, radial, medial, cubital and vannal vein.

Axillary region: The region of the wing base containing the axillary sclerites.

Axillary sclerites: The sclerites of the axillary region in the wing flexing insects, partly differentiated in Ephemerida, represented by the axillary plate in Odonata.

Axillary vein: A vein in the anal area of the hind wing (Hymenoptera, Symphyta).

Axon: Neurite; The principal process, or nerve fibre, of a neurone.

Azinophos methyl: See Guthion.

B

Bacillus thuringiensis: Berliner. Biobit. Dipel. This biological compound in a suspension of spores is an inert power. This is specific in action to lepidopterous larvae, on which it acts as a stomach poison only. This bacterium, in addition to forming resistant spores, also produces a crystal of thermolabile endotoxin called a parasporal body and it is this parasporal body which is the insecticidal agent in any preparation of *B. thuringiensis.*

Back Cross: The cross of a progeny individual with one of its parents.

Bacteria: These are microscopic, achlorophyllous, single celled living forms which occur in almost each and every abiotic (e.g. air, water and earth) and biotic (e.g. plants and animals) habitat of earth.

Bacheriophase: Virus which infects and may kill bacteria.

Bacterium: Small, typically one celled micro-organism, characterized by the absence of a formed nucleus.

Bagworms: Moths come in a family Psychidae, mostly tropical and subtropical. The caterpillars construct bag of silk, leaves, twigs, etc. around themselves. These bags hang from trees and may differ in size. In most cases the female moths are wingless and stay in the bags throughout their life, eventually laying eggs in them.

Bait: Food stuffs used to attract pests. It is usually mixed with a poisonous chemical to make a poison bait.

Bait-shyness: The discarding of a poison bait by the animals after the initial mortality of their fellows.

Bait stations: Containers that are used for keeping bait for the feeding of rats.

Bal: It is one of the specific antidotes for arsenical poisoning since it readily complexes and solubilises trivalent arsenic.

Balancers: Halteres. These are the reduced hindwings in Diptera, shaped like short drumsticks. During flight these balancers vibrate in a constant place so that any change of attitude by the fly cause on activation of tension receptors at their bases and reflex correction by the muscles controlling the wings. If the halteres are removed the insects soon loses its balance and falls to the ground.

Banana weevil, *Cosmopolites sordidus* **(Germer)** *(Coleoptera Curculionidae):* It is distributed in Australia, the Hawaii Islands, tropical and south Africa and tropical America. This insect attacks only member of the genus *Musca.* The damage is done by weevil.

Band application: Application as a band, usually made over or beside a crop row.

Band: A transverse marking broader than a line.

Bar eye: A name given to one of the mutations of *Drosophila* in which the number of optical units in the compound eye is greatly reduced.

Barrier application: The use of pesticide or another agent to stop pest from entering container, area, field or building.

Basad: Towards the base.

Basal: At the base; near the point of attachment of an appendage.

Basalare: The episternal epipleurite (sometimes double) giving insertion to the anterior pleural muscles of the wing; often represented by an undetached or partially detached lobe of the episternum before the pleural wing process.

Basal anal cell: An anal cell near the wing base; a cell at the base of the wing between 1 A and 2 A (Plecoptera).

Basal areole: A small cell at the base of the wing; the cell at the base of the wing between Sc and R (Lepidoptera).

Basal cell: A cell near the base of the wing, bordered at least in part by the unbranched portions of the longitudinal veins; In the Diptera, one of the two cells proximal to the anterior cross vein and the discal cell.

Basal fold: See *plica basalis*.

Basal vein: A vein in about the middle of the front wing, extending from the median vein to the subcostal or cubital vein (Hymenoptera).

Basal treatment: A treatment applied to the stems or trunks of plants at and just above the ground line.

Basement membrane: A non-cellular membrane underlying the epidermal cells of the body wall.

Basiconic receptor: An olfactory and other sense organ of an insect, consisting of a small projecting cone whose base is connected to a nerve fibre.

Basicostal suture: The external groove of a leg segment forming the basicosta.

Basicoxite: The usually narrow basal rim of the coxa proximal to the basicostal suture and its internal ridge.

Basipodite: The basal segment of the telopedite, or second segment of a generalized appendage in Crustacea bearing the exopodite.

Basisternum: The principal area of the sternum anterior to the roots of the sternal apophyses or the sternacostal suture.

Basitarsus: The first segment (proximal) of the tarsus; sometimes erroneously, called the metatarsus.

Basivalvulae: Small sclerites sometimes occuring at the bases of the first valvulae, often confused with first valvifers.

Basket-forming leg: These are slender and hairy and, when joined together, they form a sort of basket in which mosquitoes, flies, etc. are caught as their prey, e.g. Dragonflies.

Basket worm: See Bagworm.

Batesian mimicry: The resemblance of a harmless animal to a poisonous or a dangerous one giving protection to the former, since predators will tend to avoid both, e.g. the clearwing moth resembling a bee.

Baytex: See Fenthion.

Beak: The protruding mouth partstructures of a sucking insect; proboscis.

Bean aphid: *Aphis craccivora*. The brown black aphid that infests beans and many other plants. Like other aphids, the nymphs and adults suck the cell sap and reproduce parthenogenetically.

Bean weevil: A common name for a number of small weevils of the family Lariidae (Brachidae); common pest of leguminous crops.

Bedbug: *Cimex lectularium Linn.*, *C. hemipterus* Fabr. (Hemiptera: Cimicidae). This pest is worldwide. The nymphs and the adults suck the blood of human beings and other mammals and poulty. The bedbug is a small bristly mahogany-brown, wingless, pear shapped insect, having a distinct odour.

Bee giant: *Apis dorsata*; the largest among the honey bee, about 2 cm long, occurring throughout south-east Asia; builds a single honey comb upto 2 m long attached to a tree, building or a rock.

Bee milk: An old name for royal jelly, a predigested food substance on which the larvae of bees must be fed in order to develop into queen bees. It is produced from the pharyngeal glands of the worker bees and contains a high proportion of proteins.

Bee moth: *Aphomia* sp.; a small moth of the family Pyralidae whose larva lives in the nests of bees and wasps.

Bee-purple: A mixture of yellow and ultraviolet colours, which can be distinguished by the honey bee, *Apis mellifera*.

Bee sting: It is modified ovipositor on the hind end of the abdomen. It rests in a pocket in the seventh abdomen segment and can be protruded as needed. The stylet consists of a number of valves form from long drawn-out horny outgrowth of the 8th and 9th abdominal segments. These interlock to form a channel down which venom flows from the acid gland. The venom consists of histamine producing enzymes and proteins of low molecular weight. In addition to venom gland there is an alkaline gland secreting an oily, non-toxic liquid which may help to clean and lubricate the sting.

Bee stingless: These are the bees which do not contain sting, found in tropical countries, and resemble the bees in many ways but differ from it by the fact that they are much smaller and does not have ability to sting. These also live in colonies, collect nectar and pollen and build honey comb, nesting usually in hollow trees, rocks, etc.

Beetles: Coleopterous insects with thickened forewings (elytra) and biting mouth parts. The adults as well as larvae are of various size and shapes; some are carnivorous, others hervivorous; some terrestrial, others aquatic. They undergo complex metamorphosis (Holometabolous). Some larvae are *Campodeiform* and others are *scarabaeoid* type.

Bee wax: The wax secreted by the worker bees from the glands situated under the abdomen and used for honey comb. Chemically it is a mixture of esters and paraffin hydrocarbons all of high molecular weight.

Belostomatidae: A group of tropical water bugs of Hemiptera order.

Bellows: A structure made of rubber, leather, or plastic for producing a current of air.

Beneficial: Useful or helpful to man. A lady beetle is beneficial because it feeds on other insects that damage plants.

Benzene: It is not good insect killing agent, but is a useful fluid to possess for some other purposes (e.g. for cleaning). It can be used occassionally in the killing-bottle if nothing else is at hand. Benzene is *highly inflammable*.

Benzene Hexachloride: BHC; Hexachlorocyclohexane; Gammexane; HCH; 1, 2, 3, 4, 5, 6-haxachlo-rocyclohaxane, $C_4H_6\ Cl_6$. It is a chlorinated hydrocarbon insecticide, first synthesized by Micheal Faraday in 1825. Its use on agricultural crops in India has been banned since 1997.

Ber beetles: *Adoretus pallens* Arrow and A. *nitidus* Arrow (Coleoptera: Scarabaeidae). The two leaf defoliating beetles are widely distributed in northern India and Pakistan and are polyphagous but they prefer the ber tree (*Zizyphus jujuba*) and the grapewine. Only the adult beetles are destructive and can be recognised from their bright yellow colour and yellowish brown shining wings. The beetles are attracted to light.

Ber fruit fly: *Carpomyia vesuviana* Costa (Diptera: Tephritidae). The pest is distributed in India, Pakistan and southern Italy. Damage is caused by the larvae which are creamy white and slightly smaller than those of the other fruit flies. The adults are smaller than the housefly, are brownish yellow in colour, with brown longitudinal strips on the thorax.

Berlese's organ: A glandular organ in the bed bug (*Cimex lectularius*), situated in the base of the sac that receives sperm cells from the male during copulation and probably secretes a substance that help to active them.

Beta taxonomy: The level of taxonomy concerned with the arranging of species into a natural system of lower and higher taxa.

Betonite: It is a colloidal native hydrated aluminium silicate (clay) varying in colour from yellowish white to almost black. It is used as an emulsifier of oil.

Bibliographical reference: For nomenclatural purposes, the citation of the name of the author and date of publication for a scientific name; a full bibliogrpahical reference includes in addition, the citation of the exact place of publication of a scientific name (i.e. title of book or journal, volume, page etc.).

Bidrin: (3-dimethoxy phosphinyloxy) N, N-diethyl crotonamide. It is systemic vinyl phosphate insecticide. Bidrin is the name applied to the pure cis isomer of the compound.

Bifid: Forked or divided into two parts.

Biforous spiracles: Spiracles of coleopterous larvae having two pouches of the atrium originally supposed to open separately to the exterior.

Bifurcate: Forked; divided into two, but usually only partly.

Bihar hairy caterpillar: *Spilosoma (Diacrisia) obliqua* Walker (Lepidoptera: Arctiidae). This is a polyphagous pest and distributed in Orient. In India it is very serious in Bihar, Madhya Pradesh, Uttar Pradesh and the Punjab. Caterpillars eat leaves and soft portions of stems and branches. The larva is covered with long greyish brown hair.

Bilateral symmetry: Symmetry in which the body can be divided by a medium plane in mirror images of one another.

Bilobed: Divided into two lobes.

Bimodal: Having two peaks.

Binary: Refers to designations consisting of two kinds of names.

Binocular microscope: This term is used for microscopes that have two objectives (i.e. the lenses close to the object), as well as two eye pieces, and which are suitable for looking at solid objects, such as dried insect specimens, by reflecting light.

Binomen: The scientific designation of a species, consisting of a generic and specific name.

Binomial nomenclature: The system of nomenclature first standardized by Linnaeus and now generally referred to as binomial nomenclature. It is adopted by the international congress of Zoology, by which the scientific name of an animal is designated by both a generic and specific name.

Binucleate eggs: Eggs containing two nuclei which may have different gene contents and may be fertilized by two different spermatozoa. This phenomenon is known to occur in a number of inscets and accounts for existance of *gynandromorphs* (insects having some male and some female features).

Bioactivity: Pertains to the property of affecting life.

Bioallethrin: It is a powerful contact insecticide, and by nature a systemic pyrethroid with a rapid knock-down affect. It is much more persistent than the natural pyrethrins. It is insoluble in water, but misrible with most organic solvents. It is mostly used against house hold pests.

Bioassay: The qualitative or quantitative determination of a substance by the systemic measurement of the response of living organisms as compared to measurement of the response to a standard or standard series.

Biocide: A chemical which has a wide range of toxic properties, usually to members of both the plant and animal kingdom.

Bioconcentration (Biological magnification): When a small organism feeds on several microorganisms containing residue of pesticide, the pesticide may get deposited in its body. This increase in chemical concentration, as it posses through the food chain is called Biological magnification or bioconcentration.

Biodegradable: Capable of getting broken down in the biological system.

Biological classification: The arranging of organisms into taxa on the basis of inferences concerning their genetic relationship.

Biological control: The action of parasites, predators, or pathogens (usually present in an ecosystem) on pest populations which produces a lower general equilibrium level than would prevail in the absence of these agents. When naturally occurring or new biological control agents are deliberately introduced, manipulated or modified in an agroecosystem, the term *Applied Biological Control* is used.

Biological insecticide: A biological agent such as *Bacillus thuringiensis,* which kills insects like a chemical insecticide, and then rapidly dissipates in the environment.

Biological species concept: A concept of the species category stressing reproductive isolation, and the possession of a genetic programme effecting such isolation.

Biological races: Non-interbreeding sympatric populations which differ in biology but not or scarcely, in morphology; supposedly prevented from interbreeding by preference for different food plants or other hosts.

Biology: In broadest sense it includes zoology, botany, physiology, morphology, genetics, embryology, but in common usage it emplies the study of the principles of widest application to the origin, development, structure, function, and distribution of plants and animals and the more generally occurring phenomena accompanying their life, growth and reproduction. It has also been used in narrow sense as equivalent to ecology or bionomics.

Bioluminescence: The production of cold light by certain insects. Detail process involved in it is not known, however, it is believed that the oxidation of substance *luciferin* with the help of a specific engyme *luciferase,* e.g. the glow worm *(Lampyris sp.),* West Indian Cucujo beetle *(Pyrophorus sp.).* The Pyrophorus sp. is the most brilliant of all known luminescent animals.

Biomass: The amount of living matter.

Biome: Major communities of animals and plants having a similar life form.

Bionomics: The branch of biology which deals with the relations of organisms to their environment, ecology.

Biota: The flora and fauna of a region.

Biotic: Relating to life.

Biotic community: The plants and animals of a given habitat.

Biotic pesticides: Formulations containing microorganisms known as pathogens. They are applied in the same manner as conventional pesticides to control pest species; microbial pesticides.

Biotin: It is one of the B-complex vitamins, usually needed for normal insect growth.

Biotype: Groups of insects primarily distinguishable on the basis of interaction with relatively genetically stable varieties or clones of host plants.

Biramous: Divided into two branches consisting of an endopodite and exopodite (Crustacea).

Bird repellent: A substance which drives away birds or discourages them from roosting.

Birlane: Chlorfenvinphos; 2-chloro-(2, 4-dichlorophenyl) vinyl diethyl phosphate. It is a moderately persistent contact and stomach insecticide. Of the two isomers the trans isomer is insecticidally most active.

Bipectinate: Having branches on two sides, like the teeth of a comb.

Bisexual: Of a population composed of functional males and females: sometimes also applied to a individual possessing functional male and female reproductive organs (hermaphrodite).

Bituberculate: With two tubercles or swellings.

Bivalent: A pair of synapsed homologous chromosomes, whether or not these have yet replicated to form chromatids.

Bivalved: With two valves of a part: clamlike.

Blade: A flat, thin structure; especially used of mouth parts and antennae.

Blastocoel: The cavity of blastula.

Blastoderm: The surface layer (periferal) of cells of the blastula before gastrulation.

Blastokinesis: During embryonic development the germ band of many insects undergoes a sequence of movements, this movement is known as blastokinesis.

Blastomeres: The cleavage cells, or cells produced by the division of the egg or its nucleus that form the blastoderm.

Blastospores: The mouth of the gastrulation cavity.

Blastula: The early stage of the embryo in which the cell layer is the blastoderm.

***Blatta orientalis*:** The common European and Oriental cockroach; a dark brown species about 25 mm long, now a pest all over the world.

Blattodia (Blattaria, Blattidae): A

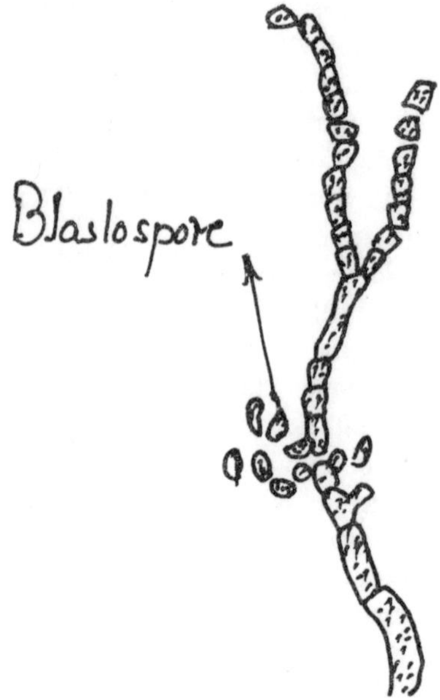

Blastospores

family of cockroaches (insects) in the order Dictyoptera (formerly classed as Orthoptera), having dark brown, oval, flattened bodies with long legs and antennae. The head is hidden under a large shield like pronotum, and the mouth-parts are biting type.

Blister beetles: Insects of the family Meloidae in the order Coleoptera, with short wings and the adults exude an evil smelling slimy oil in self-defence. Their blood contains the drug, *Cantharidin*. The adults are hervivorous but the larvae are parasitic. The first stage larvae is known as *Triungulin*.

Blood: The body cavity liquid and its contents. The blood of insects is more properly called haemolymph, since it combines some of the fuctions of the blood and lymph. Its main functions are nutritional and excretory, not the transport of oxygen which is done by the tracheal system. It is colourless liquid with several white corpuscles. The circulatory system is open type.

Blood cells: The cellular elements of the blood like Haemocytes, Leucocytes.

Blood circulation: The blood circulatory system in insects is open type, i.e. the blood fills all the cavities (haemocoels) bathing every organ. The dorsal tubular heart (true blood vessel) have a number of compartments, its anterior extension, the aorta and sometimes a pair of an antennary arteries. The part of haemocoel sorrounding the heart is known as the pericardial cavity. Blood from here enters the heart through ostia (openings), and then pumped forward along the aorta and antennary arteries and ultimately reenters the main body cavity.

Blood gills: Hollow, nontracheated, usually filamentous respiratory evaginations of the body wall or the hind gut (proctodaeum) of some aquatic insects particularly the larvae of Chironomidae and some endoparasitic insects. They contain the red blood formerly thought to be respiratory organs, but it has been proved recently that respiration in these insects takes place over the whole body surface and now it is believed that the blood gills are chiefly responsible with the absorption of water and salts.

Blood gills

Blood worm: The work-like aquatic larvae of a Chironomid fly (a non-biting midge belonging to the group Nematocera). The larvae are bright red due to the presence of haemoglobin in blood, a rarity in insects.

Blow fly: *Calliphora erythrocephala* (Meigan), *Lucilia* sp., *Chrysomyia bezziana* Villen. (Diptera: Calliphoridae). These flies have bright matallic appearance, being shining blue and green and known as blue beetles or green beetles. Three of the genera of the family Calliphoridae namely *Chrysomyia, Calliphora*, and *Lucilia*, include the largest number of flies that feed on carrion and many of them cause myiasis in man and domestic animals.

Blue bull Nilgai: *Boselaphus tragoeamelus* Pallas. It is commonly found throughout India, and is considered to be a 'cow' and is not molested and shot by hunters. It causes considerable damage to millet, sorghum, sesame and pulse crops.

Blue butterflies: Insects of the family Lycaenidae in the order Lepidoptera. These are small swift-flying butterflies. In some species the male is blue and the female is brown. The larvae are shaped like woodlice, and secrete a liquid much liked by the ants. Some of them are serious crop pests.

Body cavity: The definite cavity of the body and appendages, not strictly equivalent in all animals.

Body louse: *Pediculus humanus*. A wingless blood sucking parasite of the order Siphunculata (Anolura), having flattened body, piercing mouth parts and short legs with large claws. With in the same species there are two distinct races, the head louse (*Pediculus humanus capitis* and the body louse (*Podiculus humanus corporis*). These underbreed under suitable conditions. The nits (eggs) are firmly cemented to the hair of the host. These are the carriers of typhus and other diseases.

Body wall: The integument of the body, formed of the ectoderm, consisting of epidermis, cuticle and basement membrane.

Book gill: The leaf-like gills of a hoseshoe crab.

Book lung: A respiratory cavity containing a series of leaf-like folds (Spiders).

Book louse: *Liposcelis transvaalensis* Enderlein (Psocoptera:

Psocidae). These are tiny, soft bodied, wingless, light coloured insects with chewing mouthparts and have relatively long legs and antennae. It is a minor house hold pest and feeds upon paste and paper of old books and sometimes flour, cereal products in the kitchen.

Boll weevil: *Anthonomus* sp. A small weevil whose larvae feed on the cotton bolls or seeds of the cotton plants and cause widespread damage.

Boll worms: The lepidopterous larvae of different species which cause widespread damage to bolls of cotton, e.g. *Heliothis* spp. *Pectinophora sp.*

Bombycidae: A family of the Lepidoptera order, that includes silk moth, *Bombyx mori*. It is also called Chinese silk worm. It is cultivated in most countries for silk. The adult moth is creamy white and unable to fly due to its heavy body and small wings. the larvae (silkworms) when cultivated commercially are fed on mulbery leaves, but can eat other plants also. The larvae feed and grow for about six weeks and then spin cocoons from which adults emerge out after two weeks. One cocoon may produce 300 meters of silk. The larvae secrete the threads for their cocoon from modified salivary glands through median spinneret near the mouth.

Bombykol: Trans-10, cis-12-hexadecadien-l-ol. This is the sex attractant released by the abdominal glands of females of *Bombyx mori* to attract males. This has been identified chemically, and can now be synthesized.

Bombyx mori: See Bombycidae.

Boom: Horizontal or vertical light frame carrying several spray nozzles.

Boom spray: Horizontal or vertical stand carrying several nozzles.

Bordeaux mixture: It consists of copper sulphate (in lb), lime (in lb) and water (in gallons) in the ratio of 4:4:50. The solution of coppersulphate and finely ground quicklime are prepared separately and both mixed simultaneously in third container with constant agitaion.

Borrow pit: A pit formed by an excavation, where earth has been 'borrowed' for use elsewhere.

Bot and warble flies: *Oestrus ovis* Linn., *Hypoderma lineatum* Devillers (Diptera: Oestridae). The bot flies in larval stage act as parasites on the flesh of various animals. They attack cattle, horse, sheep, goat, elephant, etc. and feed in the alimentary tract, the nasal and pharyngeal cavities and in tumors formed in the subcutaneous tissue. In India, these flies belong to the genus *Cestrus, Hypoderma, Gasterophillus, Cobboldia, Rhinoestrus,* and *Gryostigma*.

Botanical pesticide: A pesticide produced by and extracted from plants. Examples are nicotine, pyrethrum, strychnine and rotenone.

Bouin's fixative: Saturated solution of picric acid in water 30 parts, 40% fomaldihyde 10 parts and glacial acetic acid 2 parts.

Bouton: A small swelling or lobe at the tip of a bee's tongue, also called the flabellum or labellum.

Brace vein: A slanting cross vein. In Odonata, a slanting cross vein just behind the proximal end of the stigma.

Brachycera: A term properly applied to all Diptera other than Nematocera, but more usually refering to Brachycera-Orthorrhapha, two winged flies such as Robber flies and Horse flies. A suborder having the antennae shorter than the thorax and often with a terminal bristles. The larvae are legless maggots with a tiny incomplete head which may be partly retracted into the thorax. The pupae are exarate.

Brachycerous: Having short antennae.

Brachypterous: Insects with short wings that donot cover the abdomen and generally incapable of flight.

Bracon gelechiae: It is an ectoparasite of the full grown larvae of the soybean leaf miner (*Aproaerema modicella*) and sometimes also of the *Hedylepta indicata* (a soybean leaf folder). It belongs to the family Braconidae in the order Hymenoptera.

Braconidae: A family of small parasitic Hymenoptera related to the Ichneumon flies. Eggs are laid on or in the body of some other insects such as a caterpillars and hatch out into parasitic maggots which destroy the body of the victim before pupation (also see *Bracon* sp.).

Brain: The cephalic nerve mass situated above the stomodaeum, including the primitively postoral second antennal ganglia in insects. More correctly a group of suprao-esophageal ganglion, situated in the dorsal side of the collar of nerve fibres that sorrounds the oesophagus. It is formed by the fusion of three pairs of segmented ganglia. The first pair (protocereb-rum) are greatly enlarged laterally to form the optic lobes from which nerves go to eyes. The second pair (deutocerebrum) inner-vates the antennae. The third pair (tritocereberum) are poorly developed. Nerves from the brain go to the principal sense organs of the head and others pass round the nerve collar which connects the brain with the rest of the central nervous system.

Brain hormone: It is a protein or mixture of proteins with molecular weight up to about 40000 and is produced by the neurosecretory cells of the brain, and is, therefore, indirectly responsible for moulting and, in some cases, for the termination of diapause.

Branchia: (Pl. Branchiae). In the immature stages of many aquatic insects special respiratory organs known as gills or branchiae are present. These may or may not co-exist with open spiracles. The gills arise as hollow outgrowths of exoderm.

Brand: The name, number, trade-mark or designation of a pesticide or device made by manufacturer, distributer, importer or vendor. Each pesticide differing in the ingredient statement, analysis, manufacturer or distributor name, number or trade-mark is considered as a distinct and separate brand.

Brinjal fruit borer: *Leucinodes orbonalis* Guenee (Lepidoptera: Pyraustidae). It is distributed in Malaysia, Burma, Sri Lanka, India, Pakistan, Germany, and East Africa. The damage is done by the caterpillars which are creamywhite, when young, but become light pink when full grown. The moth is white with pale brown or black spots on the dorsum of thorax and abdomen. The larvae bore top shoots and fruits also making them unfit for human consumption.

Brinjal lace bug: *Urentius sentis* Distant (Hemiptera: Tingidae). It is found throughout the India. Both the nymphs and adults suck cell sap from leaves and cause yellow spots. The attack is more serious in August-September. The full grown nymphs are pale orchraceous and are stoutly built with very prominent spines. The adult bugs are straw coloured on the dorsal side and black on the vental side.

Brinjal Stemborer: *Euzophora perticella* Rag. (epidoptera: Phycitidae). It is a minor pest but distributed throughout India. It also attacks chillies, potato, and tomato plants. Damage is caused by the caterpillar which is creamy white with few bristly hairs and their body tapers posteriorly, by boring the main stem resulting into drying up of plants. The adult is moth which has head and thorax greyish and pale yellow abdomen.

Bristle: Setae which are particularly stout and rigid and known as *bristle,* and are found in Tachinidae. In certain lepidopterus larvae, they function as the outlet for the secretion of epidermal glands and are termed *glandular bristles.*

Bristle field: The trichoid sensilla are often grouped into *hair plates* or *bristle fields,* where they act as proprioccptors. They perform many different biological functions. They form a cephalic air-flow receptor on the vertex of *ocusta* and *Schistocerca* that helps to regulate flight. In *Gryllus* males they occur on each fore wing and control their stridulatory movements. In *Apis* workers cervical hair plates and sensory hair on the mandibles and antennae are needed for normal comb construction.

Bristle tails: Any member of the order, Diplura and Thysanura. These are small, primitive, wingless insects (Apterygota) with long antennae, long cerci and in the case of Thysanura with a median tail filament. Mouth parts are biting type, e.g. *Lapisma, Petrobius* etc.

Broadcast application: An application over an entire area.

Broad spectrum insecticide: Non-selective, having same toxicity to most insects.

Bromophos: O, O-dimethyl-0-2, 5-dichloro-4-bromophenylthiophosphate. It is a 4 bromoanalog of ronnel and is a persistent general purpose organic phosphate insecticide of low acute toxicity to mammals.

Brood: A group of individuals of a given species of insects that have hatched into young or have become adult at approximately the same time, and live together in a defined and limited area.

Bruchus: A genus of leaf beetles that attack leguminous plants. *Brachus rufimanus* common in bean fields; *Bruchus pisorum,* a pest of dricd peas.

Bubonic plague: An acute infection disease caused by the bacteria *Pasteurella pestis.* It is primarily a disease of rodents and is transmitted to human beings by flies that have bitten the infected animals.

Bucca: (Pl. Buccae). A sclerite on the head below the compound eye and just above the mouth opening (Diptera).

Buccal cavity: The first part of the stomodaeum, lying just within the mouth, its dilator muscles arising on the clypeus, and inserted before the frontal ganglion and its connectives; mouth cavity or oral cavity.

Buccula: (Pl. Bucculae). One of two ridges on the underside of the head, on each side of the beak (Hemip-tera).

Bud moth: *Phycita infusella* (Meyrick) (Lepidoptera: Pyralididae). It is a minor pest of cotton found thoughout India. In the caterpillar stage, it makes characteristic webbings around the terminal young leaves, where it feeds. The full grown caterpillar is green, having black front and with faint brown stripes on the rest of the body. The adult moth is a greyish yellow, speckled with fuscous. The front wings are faint yellow at the base with red wavy lines, whereas the hind wings are white with brown marginal lines.

Bug: Insects of the order Hemiptera. These have very long proboscis adopted for piercing and sucking and is a modified organ formed from mandibles and maxillae in the form of long stylets in a trough like labium; palps are reduced or absent. The mouth parts are directed downward when feeding but may be folded right back under the thorax when at rest. Mostly they belong to the sub-order Heteroptera; most of them are plant feeders but some prey on other insects and a few are predators.

Build up: Accumulation of pesticide in soil, animals or in the food.

Bulk density: It is the weight of

definite volume of dust measured under specific conditions expressed as grams per cubic inch and is dependent upon particle size, shape, and actual density.

Bulla: A blister-like structure; a weak point on a wing vein.

Bunch caterpillar: *Andraca bipunctata* Walker (Lepidoptera: Notodontidae). It is found in India, Indonesia, Formosa and Vietnam on tea. The larvae are gregarious and feed on leaves. The forewings have wavy cross lines and two white spots near the outer margins.

Bursa copulatrix: A copulatory pouch of the female, usually the genital chamber or a part of the latter. This is formed from an infolding of body wall around the genital aperture of female insects etc.; adopted for receiving the intromittent organ of the male.

Bursicon: A peptide or protein hormone produced by neurosecretory cells in the brain and released into the blood from the thoracic or abdominal ganglia for hardening and darkening of the cuticle of *Calliphora, Locusta, Periplaneta* and other species.

Butterflies: Insects of the order Lepidoptera, having clubbed antennae. The Lepidoptera divides into *Homoneura* and *Heteroneura*, the former having almost identical arrangement of veins in the two pairs of wings, the latter having the hind wings markedly different from the fore wings. But more recently the Lepidoptera is divided into *Monotrysia* and *Ditrysia*. Butterflies are day flying and are usually more brightly coloured than moths. The elaborate coloured patterns of the wings are partly due to the pigments in the minute scales that cover the wings and also due to the structural colours.

Butamine stearate (LOVO): It is an additive added to pesticides in aerial spraying for decreasing the evaporation of the drops.

C

Cabbage caterpillar: Cabbage white butterfly; *Pieris brassicae* (Linnaeus) (Lepidoptera: Pieridae). It is a serious pest of cabbage, cauliflower, knol-knol, turnip, radish, mustard and other cruciferous plants and distributed worldwide. Damage is caused by the caterpillars only. The butterflies are pale white and have a smoky shade on the dorsal side of the body. The wings are pale white, with a black patch on the apical angle of each forewing and a black spot on the costal margin of each hind wing.

Cabbage gall weevil: *Ceurhorrhynchus pleurostigma*. A small weevil that lay eggs on the roots of cabbage, turnip and on other cruciferous plants. The young larvae feed on the roots and cause the formation of large round galls about the size of marbles.

Cabbage semilooper: *Plusia orichalcea* (Fabr.), *P. acuta* Fab. and *P. nigrisgna* (Walker) (Lepidoptera: Noctuidae). These above pests cause serious damage to cabbage, cauliflower, mustard, soybean etc. The caterpillars are plump, and palish green, and cause damage by biting round holes. The adults of *Plusia* are light palish brown with a large golden patch on each forewing. The adults of *P. nigrisigna* are darker and have dark-brown and dirty white patches on the forewings.

Cabbage web worm: *Crocidolomia birotallis* zell. A small pyralid moth that lay eggs on leaves and the larvae feed on mustard, cabbage, cauliflower by concealing themselves in a web of silk. It causes serious leaf defoliation leaving only midribs of cabbage and mustard in India.

Cabbage white fly: *Aleurodes brassicae*. A small plant bug of the group Homoptera having the body and wings covered with white powdery wax. Eggs are laid in cabbage leaves to which they are stuck by a thin stalk or thread. The larvae are flattened and at first remain hidden beneath a covering of wax. They suck the sap of the plant and exude a sticky secretion of honey dew.

Caddis worm: Aquatic larva of a caddish fly (Trychoptera).

Caceum: Pl. caeca; A tube-like or sac-like blind structure which arises from the anterior end of the midgut in certain insects.

Calamistrum: One or two rows of curved spines on the metatarsus of the hind legs (in Spiders).

Calculate: To figure out by mathematical process by working with numbers.

Calibrate: To measure or figure out how much pesticide will be applied by the equipment to the target in a given amount of time.

Calibrated: Checked; measured; knowing how much of a pesticide chemical is being applied by each nozzle or opening of a sprayer, duster or granular applicator to a given area, plant or animal.

Calling: A virgin female moth releasing sex pheromones to attract males for the purpose of mating.

Callow: Teneral; An insect newly emerged from the pupa, having a soft body not fully coloured. The term may be used generally but is particularly applied to worker ants.

Callus: Pl. calli; a rounded swelling.

Calopteryx: Damselfly (*Agrion*).

Calotermes: A genus of termites (Isopteral) that live in wood and feed on it. The digestion of wood is performed by large numbers of symbiotic protozoa present in the intestine.

Calyciform: Goblet cells; a type of cells found in the mid gut epithelium. In larvae of Lepidoptera and Trichoptera, they are with less rough endoplasmic reticulum, less regular microvilli and a highly folded plasma membrane that encloses the invaginated or vacuole-like goblet chamber. They may be involved in active transport of potassium ions across the gut wall.

Calypter: Pl. calypters; A pair of membranous scales above the halteres in Diptera; Also called Squama or Alulae.

Calyx: The wider anterior end of the lateral oviduct receiving the ovariole chamber.

Camera-lucida: A device enabling one to make accurate drawing of objects seen through a microscope; when it is attached to the eye piece of the microscope, the observer can see the object under the microscope and his drawing paper at the same time.

Campaniform receptors: Sense organs of insects in the form of bell-like projections from the skin with a nerve cell beneath each. They are believed to monitor stresses in the cuticle end hence act as proprioceptors.

Campaniform sencillum: A sense organ consisting of dome-shaped cuticular area into which the sensory cell process are inserted like the clapper of a bell.

Campodeiform larva (Oligopod larva): These larvae have elongated flattened bodies. They resemble silver fish. Larvae have prominent antennae, cerci, and legs adopted for running, e.g. larvae of some Coleoptera and Neuroptera.

Cannibalistic: Feeding on other individuals of the same species.

Canthariasis: Condition arising when beetle larvae live in the body of another animal; also called "Myiasis".

Cantharidin: A drug extracted from the blister beetles or spanish fly *(Lytta vesicatoria)* and others of the family Meloidae.

***Cantheconidia furcellata* (wolff) (Petatomidae: Hemiptera):** It is a predator of most of the lepidopterous larvae. The nymphs and adults both suck the haemolymph of the various larvae of the lepidopterous pests.

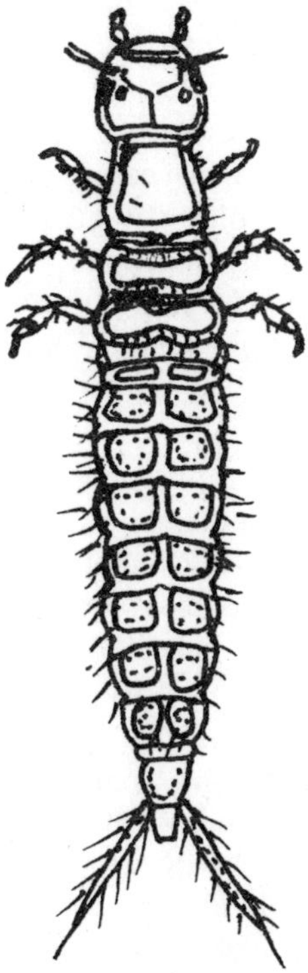

Campodeiform (Oligopod) larva

Cap cell: The distal or Peripheral cell of a sense organ, or one of the component units of the organ, probably corresponding to the tormogen, or socket forming cell of a seta.

Capillarity: The rising or moving liquids in minute tubes by surface tension; capillary action.

Capitate: An antenna with an apical knob-like enlargements.

Capitophorus: A genus of aphids which cause galls or red blister-like patches on the leaves of rose and fruit trees, e.g. *Capitophorus rosarum* on rose trees.

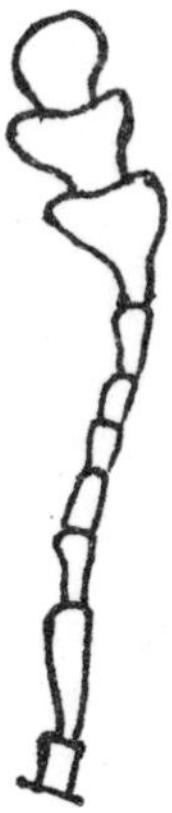

Capitate antenna

Carabaeiform larva: These larvae are elongate, flattened with well developed legs and no filaments on the abdomen or posterior, e.g. larvae of carabid or ground beetles and leaf beetles.

Carapace: A hard dorsal covering consisting of fused dorsal sclerites (Crustacea).

Carbamates: A group of chemicals which are salts or esters of carbonic acid. They may be insecticides, herbicides, and fungicides.

Carbaryl: It is a good contact carbamate insecticide with occasional systemic action.

Carbofenothion: See Trithion.

Carbofuran: See Furadan.

Carbohydrates: Any of a group of chemical compounds, including sugars starches, and cellulose, containing carbon, hydrogen and oxygen only, with the ratio of hydrogen to oxygen atoms usually 2:1.

Carbohydrases: These are polysaccharases, such as amylases, glycosidases, etc. These catalyse the hydrolysis of the more complex carbohydrates.

Carbophos: See Malathion.

Carcinogenic: The term used to describe the cancer producing property of a substance or agent.

Carcinogens: Cancer producing substances.

Cardia: The anterior part of the vertriculus; in many Diptera taking the form of a small spherical sac, often mistaken from a proventriculus.

Cardiac sinus: The dorsal part of the embryonic hoemocoel, a part of which becomes the lumen of the dorsal blood vessel.

Cardiac valve: See Stomodaeal valve.

Cardioblasts: The cells of the upper edges of the embryonic mesoderm that form the dorsal blood vessel.

Cardo: Pl. cardines; the proximal (basal) division of the maxilla by which it articulates with the head; one or two small latero-basal sclerites in the millipede gnathochilarium; Any hinge-like part of an organ.

Carina: (Pl. Carinae); A ridge or keel.

Carinete: Ridged or keeled.

Carnivorous: Feeding on the flesh of other animals, including other insects; 'Cannibalistic' usually only when an insect feeds on others of its own species.

Carpopodite: The fifth segment of a generalized appendage (Tibia).

Carrier: An inert material that serves as a diluent for the active ingredient or toxicant.

Carrion fly: Flesh eating dipterous flies, e.g. blow fly (*Calliphora* sp.), flesh fly (*Sarcophaga* sp.).

Cartridge: The part of a respirator which absorbs fumes and vapours from the air before the applicator breathes them in.

Casein: It is a phosphoprotein occurring in milk, in beans and in nuts. The principal protein is cows' milk contains about 3% casein.

Cashew tree borer: *Plocaederus ferrugineus* Linnaeus (Coleoptera: Cerambycidae). It is most important pest of cashew tree in south India. The adult is a medium sized dark brown beetle. The grub stage feeds inside the tree trunk or the branches, making tunnels.

Castes: The various morphological and functionally distinct types of mature individuals as workers, soldiers and queens with in colony of social insects. Such castes are well developed in Hymenoptera (bees, ants and wasps) and Isoptera (termites). The chief castes are queen (fertile female), workers (sterile females) and drones (fertile males). Many more types are also present.

Castor capsule borer: *Dichocrocis punctiferalis* Guenee (Lepidoptera: Pyralididae). This is a borer pest of castor distributed throughout India. The caterpillar bores into the main stem of a young plant and ultimately into the capsules. The moths are orange yellow, with black markings on both the wings.

Castor hairy caterpillar: *Euproctis lunata* (Walker) (Lepidoptera: Noctuidae). This pest is widely distributed in India with an allied species, *E. fraterna*, is a serious pest of castor, linseed, groundnut, grapevine. The damage is caused by the larvae which feed on leaves. The attacked plants remain stunted and produce little seed. The larvae are covered with long hair all over the body. The moths are pale yellow.

Castor semilooper: *Achaea janata* Linnaeus (Lepidoptera: Noctuidae). It is a serious of castor in all parts of India and Pakistan and has also been reported from Sri Lanka, and Thailand. The larvae feed on foliage whereas adult moths of this species are fruit sucking and cause serious damage to citrus. The adult moth is a pale reddish brown and the wings are decorated with broad zig-zag markings, a large pale area and dark brown patches.

Castration: Castration, in the broad sense, implies any process which inhibits completely or to a considerable extent the production of mature gametes by the organism. whose gonads may be greatly atrophied or well developed but not normally functional.

Catalog: An index to taxonomic literature arranged by taxa so as to provide ready reference to at least the most important taxonomic and nomenclatural references to the taxon involved.

Catalyst: A substance that speeds up the rate of a chemical reaction but is not itself used up in the reaction.

Caterpillar: An eruciform, polypod larva; the larva of butterfly, moth, sawfly or scorpionfly. The soft bodied larvae having in addition to the six true legs on the thorax, a number of prolegs or false legs on the abdomen. The larvae of sawflies are sometimes not regarded as caterpillars, as they differ from butterflies and moths in having a large number of prolegs and in the absence of rings of tiny hooks on these. The number of prolegs on the catterpillar of butterflies or moths may be upto five pairs but in the large group known as looper (Geometridae) only the hind pairs are present. These are hervivorous with powerful side moving jaws.

Cattle: Domesticated animals, particularly cows, oxen, bulls raised for milk, meat or draft purposes.

Cattle louse: *Haematopinus eurysternus* Mitzsch (Siphunculata: Pediculidae). It is a universal parasite of cattle. The adult is wingless and slate grey and has a flat body and pointed legs. The lice cling to the hairs with their claw like legs and thrust their mouth parts into the skin for sucking blood.

Cattle tick: *Hyalomma anatolicum,*

Boophilus microplus Canestrini (Acarina: Ixodidae). Ticks are important external parasites of cattle, sheep, goat, horse, etc. Out of many species, which attack cattle, *B. microplus* and *H. anatolicum* are very common in India. Ticks have their head and thorax fused. Numphs and adults suck the blood near the ears, under the base of the tail, with their head embeded deep into the skin. They also transmit several diseases like tick fever, texas fever.

Caudal: Pertaining to the anal end of the body.

Caudal filament: A thread-like process at the posterior end of the abdomen.

Caudal gills: Respiratory organs in the form of thin walled lamellae or outgrowths from the hind end of the abdomen in some aquatic insects, e.g. In Numphs of damsel flies (Zygoptera).

Cauded: Towards the tail, anus or posterior tip of the body.

Caustic potash: Potassium hydroxide (KOH) is brought in pellets or in sticks and is dissolved in water to give approximately a 10% solution; 5 gram solid in 50 cc is nearly enough. The solid is likely to burn the skin, and rot clothing, and of course is very poisonous if taken. If not kept in a tightly kept jar it absorbs carbon dioxide from the atmosphere and becomes coated with potassium carbonate. It is to soften the specimen and to destroy the soft internal tissues, but not chitinized part of the body, nor the membranous cuticle between them. The specimen may be left overnigh in cold potash.

Caution: A warning to the user of pesticide chemicals. Used on labels of pesticide containers.

Cecidomyiidae: Gall midges, gall gnats. These are minute two winged insects, seldom increasing 3 mm in length, belonging to the suborder Nematocera and having long thin antennae, long narrow wings and very long legs. The larvae are yellow or brown, without legs and cause the galls on many crops and trees.

Cedar wood oil: An alternative to clove oil as a clearing agent. Does not stain, nor harden and will clear from 95% alcohal.

Cell: A unit mass of protoplasm, sorrounded by a cell membrane and containing one or more nuclei or nuclear material; A space in the wing membrane partly (an open cell) or completely (a closed cell) sorrounded by veins.

Cellulose: An inert carbohydrate; the chief component of the cell walls of the plants.

Cement layer: A thin layer of the surface of cuticle of an insect. It is formed by the hardened secretion of the dermal glands.

Centigrade (C.): A thermometer scale in which water freezes at 0 degrees C and boils at 100 degrees C. To change to degrees Fahrenheit, multiply degrees, centigrate nine-fifths and add 32 (9/5+32).

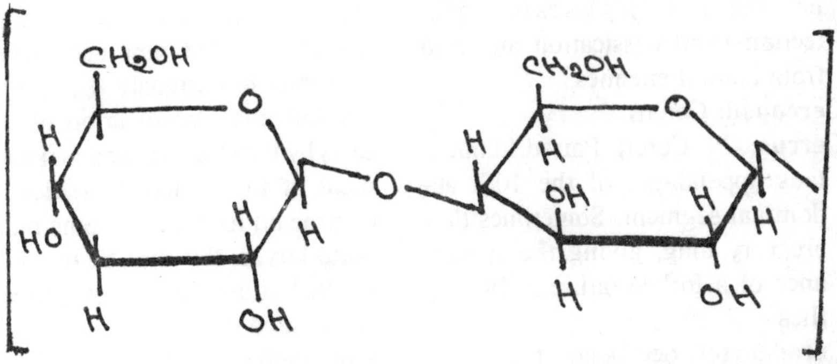

Cellulose

Also, celsius (C.), the preferred form approved by the Ninth General Conference on Weights and Measures in 1948.

Central nervous system: The part of the nervous system containing the motor neurocytes and the synaptic junctions between communicating neurones.

Centrolecithal egg: An egg having the yolk concentrated in the centre, characteristic of insects and other orthopods.

Cephalad: Towards the head or anterior side.

Cephalic: On or attached to the head; anterior.

Caphalic lobes: The head lobes of the embryo, comprising the region of the prostomium and usually that of the tritocerebral somite.

Cephalothorax: The united head and thorax of Crustacea and spiders.

Cerambycidae: Coleopterous longicorn or longhorned beetles. The adults have very long antennae. The larvae which generally grubs have hard brown head and rudimentary legs, do immense damage to several crops by boring the stem and pupating inside it. Some larvae may live four or five years in larval stage. Girdle beetle, *Oberea brevis* Swed. is a serious cerambycid pest of soybean in Madhya Pradesh, India.

Ceratina sp.: It is a gregarious ectoparasite of the girdle beetle *(Obrea brevis* Swed.) and belong to the family Apidae in order Hymenoptera. The adult parasite is bee which usually select hollow tunnels in the stems of soybean plants both for living and breeding. The female parasite also cut down the host larvae in as many pieces as the eggs laid.

Cercopidae: It includes frog hoppers of the group Homoptera whose body is frog-like with broad head and have jumping power.

The nymphs produce and live in a frothy secretion produced by glands near the anus. This gives them protection from dessication and also from natural enemies.

Cercopod: Cercus.

Cercus: (Pl. Cerci); Paired filamentous appendages of the 10th abdominal segment. Sometimes they are very long, giving the appearance of a forked tail, e.g. In May flies.

Cerodecytes: See Oenocytes.

Cervical: Pertaining to the neck or cervix.

Cervical muscles: They originate from the cervix and prothorax and are inserted into the tentorium and epicranium. These control the movements of the head and are classified into levators, depressors, retractors and rotators according to their function.

Cervical sclerites: Cervicalia; The sclerites of the neck, particularly one or two pairs of lateral neck plates joining the head to the prothoracic episterna.

Cervix: The neck, including probably the posterior nonsclerotized part of the labial somite and the anterior part of the prothroax; a membranous region between head and prothorax containing sclerites.

Chaeta: A stiff hair, or thin bristle.

Chaetosema: Jordom's organ. A cluster of sensory bristles on the heads of some adult Lepodoptera, believed to replace the ocelli.

Chaetotaxy: The arrangement and nomenclature of the bristles on the exoskeleton. A term originally used for the pattern produced by certain strong bristles on the thorax of flies, but now used more generally to designate the identification and classification of insects by the structure and arrangement of the principal bristles or chaetae on its body. In some mosquito larvae, this may be the only method of distinguishing between two related species.

Chalcidoidea: A very large superfamily of minute Hymenoptera, nearly all of which are parasitic. Some are hyperparasites, and others form galls and some develop within the seeds. They have *polyembryony* by which an egg may give rise to fifty or more separate embryos by a complex process of budding.

Character displacement: A divergence of equivalent characters in sympatric species resulting from the selective effects of competition.

Character index: A numerical value, compounded of the ratings of several characters, indicating a degree difference of related taxa; also a rating of an individual, particularly a hybrid, in comparison with its most nearly related species.

Checklist: Usually a skeleton classification of a group listed by taxa for quick reference and as an aid in the arrangement of collections.

Cheek: The lateral part of the head between the compound eye and the mouth.

Chelate: Pincer-like, having two opposable claws.

Chelicera: (Pl. Chelicerae); The first pair of appendages of adult Chelicerata (Arachnida) innervated from the tritocerebral ganglion of the brain; equivalent to the second antennae of Crustacea.

Chelicerata: The orthropods (king crabs, sea-spiders, spiders, scorpions, mites, ticks, etc.) that bears chelicere, in place of sensory antennae. Their body is usually divided into cephalothorax and abdomen. Cephalothorax has four pairs of legs and there are no antennae.

Cheliped: A leg terminating in an enlarged pincer-like structure (In Crustacea).

Chemical control: The control of pests by the use of chemical pesticides.

Chemical name: One that indicates the chemical composition and/or chemical structure of the compound.

Chemical pesticides: The chemicals which are used to control pests (insect pests, weeds. diseases etc.).

Chemical reaction: When two or more substances are combined and as a result, undergo a complete change to make new substances or materials.

Chemoreceptor: A sense organs supposedly sensitive to chemical properties of matter (a "taste" receptor or an "odor" receptor).

Chemosterlant: Chemicals used to make an insect sterile.

Chewing-lapping mouth-parts: Found in honeybees. The labrum and the mandibles are of the chewing type, but the maxillae and the labium are united and are elongated to form the tongue, with which they collect nector from flowers.

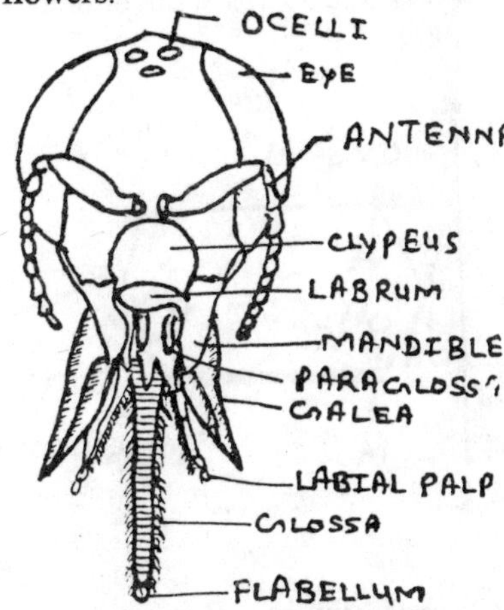

Chewing-lapping mouth parts

Chiasma: The visible connection or crossover between two chromatids seen during prophase-I of meosis.

Chilopoda: A class of the phylum Arthropoda. It includes the centipedes. They are flattened worm like creatures, with a distinct head bearing two antennae and one pair of legs on every segment of the body.

Chironomidae: This includes small two winged insects whose equitic larvae may be red, blue, yellow or colourless. The small red larvae of *Chironomus pulmosus* are called blood worm being red in colour and blood contains haemoglobin a rarity in insects. This is thought to be an adoptation to life in poorly aerated water.

Chitin: A colourless, horny nitrogenous polysaccharide with formula ($C_{32}H_{54}N_4O_{21}$), occuring in cuticle of arthropods. It is insoluble in water, alcohol, and other solvents and is resistant to acid and alkalies.

drogen, carbon and chlorine. They are persistant insecticides that kill insects mainly by contact action. They are insoluble in water and are decomposed by alkaline materials and high temperature.

Cholorinated terpenes: See Toxaphene.

Cholorobenzilate: 2-hydroxy-2, 2-bis (4-chlorophenyl) ethyl acetate.

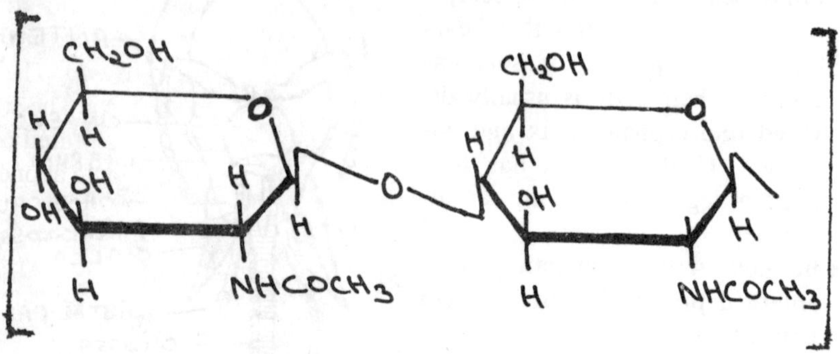

Chitin

Chloral hydrate: Used in 5% solution. The insect is slit open and warmed in this solution, but not boiled, for one minute. Drain on blotting paper, put in fresh solution and change this after one week.

Chlordane: (Chlorondin; 1, 2, 3, 4, 5, 6, 7, 8, -Octachloro-4, 7-methane-3, 4, 7, 7a-tetrahydroin-dene octachloro-4, 7-methane-tetrahydrolindene: $C_{10}H_6Cl_8$); It is a chlorinated hydrocarbon insecticide, first prepared by Haymen by the Diels-Alder reaction.

Chlorfenvinphos: See Birlane.

Chlorinated hydrocarbons: Synthetic compounds that contain hy-

It is a chlorinated hydrocarbon and is primarily an acaricide. The technical material contains 90% of active ingredient.

Choloroform: It is not recommended as a general killing agent for insects, but can be used as an aesthetic. A few drops may be put on to the pinned insects in the collection, if these show being attacked by pests. It should be kept in dark bottle and thrown away if it has not been used for some time.

Chlorondin: See Chlordane.

Chlorophyll: Any of a group of related green pigment found in photosynthetic organisms.

Chlorosis: In plants, the yellowness

of normally green tissues due to partial failure of chlorophyll to develop or removal of chlorophyll.

Cholesterol: A glistening white soapy unystalline substance, $C_{27}H_{45}$ OH, the most common animal sterol, a precursor of a form of vitamin and a universal tissue constituent, occuring notably in bile, gallstones, the brain, blood cells, plasma, egg yolk, and seeds. It seems essential for some carrion-feeding dermestids.

Choline: A natural amine, $C_5H_{15}No_2$, often classed in the vitamin B-complex and a precursor of acetylcholine.

Choline acetylase: An enzyme of the insect body that synthesizes the acetylcholine.

Cholinesterage: An enzyme of the body necessary for proper nerve function. It is inhibited or damaged by the organophosphorus or carbamate insecticides.

Cholinesterase inhibitor: Any organophosphate, carbamate or other pesticide chemical that can instruct the action enzymes which inactivate the acetylcholine associated with nervous system.

Chorda: The stem of vein R_{4+5} in some Lepidoptera.

Chordates: The animals possessing a back bone. These animals comes under "Chordata".

Chordotonal organ: A sense organ of the scolopophorous type, probably detecting changes of tension in muscles and possibly enabling the insect to balance itself while flying. Each consists of a thin fibre attached to the cuticle, passing through a rod called a *Scolopale* to a sense cell beneath the surface, e.g. Johnston's organ in the antennae.

Chorion: The non-chitinous shell of an insect egg, formed in the egg follicle. This may be composed of a hard impenetreble layer of scleroprotein called the *endochorion* covered a thin surface layer *(exochor-ion)*. The layers may be more also. The whole structure may be sculptured or ornamented. A Micropyle at one end is present.

Chorionin: The scleroprotein forming the chief part of the shell of an insect's egg.

Chorology: The study of the geographical distribution of organisms.

Chromatid: One of the two identical, longitudinal halves of a chromosome which shares a common contromere with a sister chromatid; results from the replication of chromosomes during a nuclear division.

Chromatin: The readily stainable nuclear material (DNA, RNA, histones and non-histone proteins) composing the chromosomes.

Chromatophore: A pigment bearing intracellular body.

Chromosomal inversion: Reversal of the linear order of the genes in a segment of a chromosome.

Chromosome: Nucleoprotein structure, generally filamentous or rod like during nuclear division. It is the physical site of nuclear genes which are arranged in linear order; best seen during cell division.

Chromic: In the medical sense, this term is used in contrast to 'acute' and means "of long duration". As applied to material inhaled or absorved through the skin, it means exposure times of long duration. As regard materials injested, it refers to repeated doses.

Chronic toxicity: How poisonous a pesticide is to man or animals after apparantly harmless repeated doses. It is measured as chronic oral toxicity, chronic dermal toxicity, and chronic inhalation toxicity.

Chrysalis: (Pl. Chrysalids or Chrysalides); The pupa of butterfly; the naked pupal stage; a pupa of *obtect type*.

Chrysomelidae: It includes leaf beetles, that are small, brightly coloured. The larvae are soft, fleshy with short legs and usually brightly coloured. Many cause extensive damage to farm and garden crops. They pupate in soil. e.g. Colorado potato beetle (*Leptinotarsa decemlineata*) and the flea beetles.

Cibarium: The food pocket of the extraoral or preoral mouth cavity between the base of the hypopharynx and the under surface of the clypeus.

Cicadellidae: Formerly called Jassidae. Leaf hoppers of Homoptera order, which suck plant sap. They have small elongated bodies and can jump and fly with great speed. They spread many virus and fungus diseases among plants, e.g. hopper burn in rice.

Cicadidae: It includes Cicadas. Homopterous plant bugs of warm regions having a membranous wings and piercing mouth parts. The male can live only for a few days, can make a loude rattling noise with drum-like membranes on the sides of the body. The females are silent. They lay eggs in small holes after boring the twigs. The larvae fell to the ground and live for many years in the soil.

Cide: Suffix meaning "to kill".

Cimicidae: It includes bed bugs. Parasitic insects of the order Hemiptera, sucking blood from their pest time to time. The body is flattened and without wings, with suctorial proboscis.

Cinerins I and II: These are esters found in the pyrethrum extracts. However, they are of less insecticidal value than pyrethrum.

Circadian clock: Lat *circa* = about + *dies* = day; Exhibiting approximately 24 hour periodicity.

Circadian rhythm: A daily rhythmic activity cycle, based on 24 hour intervals, that is exhibited by many organisms.

Circular overlap: The phenomenon in which a chain of contiguous and intergrading populations curves back until the terminal links overlap geographically and behave like good species (noninterbreeding).

Circular System: See blood and blood circulation.

Circumfila: The elaborately looped sensilla found on cecidomyid antennae, which are essentially thinwalled chemoreceptors with the pores lying among a labyrinth of fine surface ridges.

Circumoesophageal connectives (Commissures): The connectives between brain and the ventral nerve cord embracing the stomodaeum. Primitively the connectives from the archicerebrum to the ventral ganglia; in insects the connectives between the tritocerebral ganglia and the mandibular ganglia.

Cistron: The genetic unit of biochemical function; the sequence of nucleotide pairs in DNA that specify the aminoacid sequence of a single polypeptide chain in protein synthesis.

Citrus caterpillar: Scientific name- *Papilio demoleus* Linn. (Lepidoptera: Papilionidae). It is a serious pest of citrus and found in Africa, Asia, Formosa and Japan. It has been a serious pest all over the India. The caterpillar cause damage by eating the leaves; which are green and has a horn like structure on the dorsal side of the last body segment. The adult butterfly had black head and thorax and the underside of the abdomen is creamy yellow.

Citrus leaf miner-*Phyllocnistis citrella* Stainton (Lepidoptera: Gracillariidae): This is a serious pest of citrus nurseries in Tamil Nadu, Madhya Pradesh, Assam, Uttar Pradesh, the Punjab. It also feeds on pomelo, willow, cinnamon. The larvae cause damage by making zig-zag silvery mines in the young leaves. The adult is tiny moth which has brown stripes and prominent black spots along the tips on the forewings.

Citrus mealybug: *Pseudococcus filamentosus* Cockerell, *P. citriculus* (Hemiptera: Pseudococcidae). This pest is more common in South Indian than in the north and it feeds on ferns, begonia, gardenia, poisettia. Other species, *viz. P. citriculus* Green found in China, the U.S.A., *P. gahani* Green in California and *Planococcum citri* (Risso) in Malaysia and Indonesia. Damage is caused by the nymphs and female adults. The adult female is wingless with a flattened body with piercing and sucking mouth parts. The male is winged, midge-like with long antennae and mouth parts. The nymphs are amber with a whitish waxy coating and filaments.

Citrus mite: *Paratetranychus citri* McGregor (Acarina: Tetranychidae). This pest has been recorded in the U.S.A., Indonesia, the Philippines, Israel, Iraq, Iran, UAR, Sri Lanka and India. It also infests sweet organe, lemon, grape fruit, and sour-lime. The adult is small, plump, and orange with thick deep brown patches on the dorsal side of its body. The newly hatched larva is light yellowish brown and has only three pairs of legs. The protonymph is orange brown and the deutonymph is orange brown with a greenish tinge.

Citrus nematode: *Tylenchulus semipenetrans* (Cobb.) (Tylenchidae: Tylenchuilidae). This pest is distributed in all the citrus growing regions of the world. In India it is a pest in Punjab and Delhi. Due to its attack, plants become unthrifty; their fruit-set is lowered and the incidence of fruit drop is increased. The female nematodes are found on thick, stunted rootlets.

Citrus psylla: *Diaphorina citri* Kuwayama (Hemiptera: Psyllidae). This pest is distributed throughout the Orient and has been reported from India, China, Formosa, Japan, Burma, Sri Lanka, the East Indies and New Guinea. Damage is caused by the nymphs and the adults by sucking the cell sap. The adults is small, active and rests on the lear surface with closed wings, the tail and of the body being turned upwards. The nymphs are flat, louse-like and orange yellow and seen congregated in large numbers on the young leaves and the buds.

Citrus red scale: *Aonidiella aurantii* (Maskell) (Hemiptera: Diaspididae). Three species of this pest cause damage to the citrus plant all over the world, viz. *A. aurantii, A. citrina, Chrysomphalus dictyospermi* Morgan. Of these, the red scale (*A. aurantii*) is the most destructive. Besides citrus it also feeds on fig, rose, willow, Shisham and many other plants. The insect cause damage in two fold-first, the scale insect feeds on the plant juice and devitalise the plants; second—they inject a toxic substance into the plant sap.

Citrus red spider mite: *Panonychus citri* (McG.) (Tetranychidae: Prostigmata). It is an important pest of citrus in India. The infested leaves are speckled with whitish spots, which later turn silvery or brown and drop prematurely. Fruits are also attacked which turn grey or yellow.

Citrus white fly: *Dialeurodes citri* (Ashmead) (Hemiptera: Aleyrodidae). The pest is distributed in the U.S.A., India, Pakistan, Vietnam and China. Damage is caused by the nymphs and the adults by sucking the cell-sap from the leaves which curl over and fall off. The adult is a minute insect, the males being smaller than the females.

Cladism: A taxonomic theory by which organisms are ordered and ranked entirely on the basis of "recency of common descent", that is, on the basis of the most recent branching point of the inferred phylogeny.

Cladistic: Based on the principles of Cladism.

Cladogram: A dendogram based on the principles of cladism; a strictly genealogical dendogram in which rates of evolutionary divergence are ignored.

Claspers: Any pair of processes on the hind and of the abdomen of a male insect serving to grasp other structures during mating. The claspers of male dragon flies grip the neck of the females, but in Diptera they grip the female abdomen.

Class: A subdivision of a phylum or subphylum, containing a group of related orders.

Classification: The delimitation, ordering and ranking of taxa. Zoological classification is the ordering of animals into groups (or sets) on the basis of their relationships. The process of assigning pesticides into groups according to common characteristics.

Claval suture: The suture of the front wing separating the clevus from the corium (Hemiptera).

Claval vein: A vein in the clavus (Hemiptera, Homoptera).

Clavate: Club-like or enlarged at the tip; clavate antenna.

Clavus: The oblong or triangular and portion of the front wing (Hemiptera and Homoptera).

Claws: Most insects have two horny claws derived from the last segment of the tarsus. Generally there is a median pad (arolium) between them but structures vary greatly in different orders. The protura and collembola, Hippoboscidae, and the larvae of few insects have only one claw.

Claw tuft: A dense tuft of hairs below the claws.

Clear-zone eye: The eye in which the light reaching a single ommatidium will have been refracted through several fecets.

Cleavage: The division of the egg or its nucleus and of the resulting cells forming the blastoderm.

Cleavage cells: The cells formed during cleavage.

Cleavage nuclei: The products of the division of the zygote nucleus are the *Cleavage nuclei.*

Cleft: Split or forked.

Cleptobiosis: A form of symbiosis in which one species of ant lives in or near the colony of another species and steals food from them.

Click beetles: Any member of the family Blateridae (Coleoptera). These are long, slim, shortlegged beetles having the habit of springing into the air with a clicking sound. The larvae called wireworms, that are agricultural pests living in the soil and attacking roots.

Climatograph: A polygonal diagram resulting from plotting temperature means against relative humidity.

Clinal: Varying gradually, of characters.

Cline: A gradual and nearly continuous change of a character in a series of contiguous populations; a character gradient.

Clinging leg: This type of leg is found in body house, in which the tarsus has a single curved claw.

Clistogastra: Apocrita. Hymenopterous insects with a deep constriction between the thorax and abdomen. They include ants, bees and wasps.

Clods: Hard lumps of soil formed by ploughing or cultivating soil.

Clone: All the offsprings derived by a sexual reproduction from a single sexually produced individual.

Closed cell: A wing cell bounded by on all sides by veins and not reaching the wing margin.

Closed coxal cavity: One bounded posteriorly by a sclerite of the thoracic segment (front coxal cavities in Coleoptera) or one completely sorrounded by sternal sclerites and not touched by any pleural sclerites (middle coxal cavities in Coleoptera).

Closing apparatus of a spiracle: The closing mechanism and opening, formed either of the lips of the atrium or by a valve at the inner end of the atrium.

Closing band: The movable valvular fold of the inner closing mechanism of a spiracle.

Closing bow: The rigid but elastic lip of the inner closing mechanism of an spiracle opposite the valve.

Clothing hairs: The hairs which invest the general surface of the body or its appendages and frequently exhibit various degrees of specialization.

Clove oil: It is a clearing agent, used in mounting of insects. Insects should not be left for long time in it, but mounted as soon as they are clear, otherwise specimen would brittle.

Clubbed: With the distal part (or segments) enlarged; clubbed antennae.

Clustering methods: Method of grouping related or similar species into species groups or higher taxa.

Clypeogenal sulci: Laterally, the clypeus may be delimited from the sides of the head by the *Clypeogenal sulci*.

Clypeus: The facial area of the cranium just above the labrum, usually separated from the frons by an epistomal suture and sometimes divided into anteclypeus and postclypeus; the dilator muscles of the cibarium are attached on its inner surface.

Coagulate: To change from a fluid to a jelly; to form a clot.

Coalescent: Running together, merging; as of wing base.

Coarctate larva: A larva some what similar to a dipterous puparium, in which the skin of the preceding instar is not completely shed but remains attached to the caudal end of the body; the sixth instar of a blister beetle; also called pseudopupa.

Coarctate pupa: A pupa enclosed in a hardened shell formed by the next to the last larval skin; pupae of the most dipterous insects.

Coccinellidae: It includes ladybird beetles (Coleoptera) that are brightly coloured, hemispherical beetles with well defined spots and other markings on the wing cases. The larvae are of *Campodeiform* type but differ those of

Coarctate pupa

ground beetles in having the appendages and sensory organs much reduced. They feed voraciously on aphids and on other small insects.

Cochineal: A red dye that is extracted from the bodies of scale insects *(Dactylopius coccus)*. About 500 gm of dye is produced from 70,000 insects. It is native of Mexico but now introduced in many other countries.

Cockroaches: *Periplaneta americana* Linn. and *Schelfordella tartara* Adel (Dictyoptera: Blattidae). This pest is world wide. The damage is done by the nymphs and adults. In northern India, *S. tartara* is the commonest species. *P. americana* is confined to south India. The other important species in India are *Blatella germanica* and *Blatta orientalis*.

Cocoon: A silken case inside which pupa is formed (Lepidoptera): a case of earth, wood fragments or other debris, protecting the pupa.

Code of Ethics: A set of recommendations on the propriety of taxonomic actions, to guide the taxonomist. Formulated in Appendix A of the International Code of Zoological Nomenclature.

Codling moth: *Carpocapsa pomonella.* A small grey or brown moth whose caterpillars are most destructive to apples and pears in most parts of the world. Eggs are laid on the leaves and the newly hatch larvae bore into the young fruits where they feed on the cores and eventually emerge to spin cocoons.

Codominace: A condition in heterozygotes where both members of an allelic pair contribute to phenotype in either homozygous condition.

Codon: A set of three nucleotides that is specific for a particular amino acid in protein synthesis.

Coeloconic receptors: Olfactory or other sense organs, more advanced than pasiconic receptors and differing from them in having the terminal processes of the sense-cell sank in a small pit.

Coelome: A body cavity formed of the coelomic sacs only.

Coelomic sacs: The paired segmental cavities of the mesoderm.

Coffee shot-hole borer: *Xylosandrus compactus* Eichhoff or *Xyleborus morstatti* Hugedorn (Coleoptera: Scolytidae). It is a serious borer pest of coffee in southern India. The adult is a cylindrical, dark brown beetle. Both the adult and larva produce a large number of pin holes in the bark, generally on the underside of the tertiary branches. Both the adults and the larvae tunnel through the bark and cause wilting.

Coffee stem borer: *Xylotrechus quadripes* Chevrolat (Coleoptera: Cerambycidae). This pest has been reported from Bruma, Sri Lanka, Thailand, Indonesia, the Philippines and India. In India, it is found in Assam and southern India. The white larvae of this beetle bore into the coffee stem, killing the young plants and making the older ones appear quite unhealthy. The adult is blackish brown beetle with characteristic pattern of yellowish bands on the elytra.

Cold-hardiness: The stage which is specially adopted by the insects to survice during exposure to extreme low temperature.

Coleoptera (Koleos, a sheath: pteron, a wing): It includes beetles and weevils. The beetles are minute to large and have their forewings modified to form elytra which are hard and lie flat on the body, the membranous hind wings being folded underneath. Mouth parts are biting type. Metemorphosis is complete and the larvae are campodeiform or eruciform.

Collateral: A lateral branch of an axon.

Collaterial glands: The accessory glands of the female reproductive system that secrets an adhesive substance for attaching eggs to support or for gluing egg in masses.

Collembola: Springtails; very small primitive wingless insects. They have biting mouth parts, short antennae and legs without tarsi. The abdomen is six segmented which are some times fused together. On the fourth segment is forcula (Spring organ) which, when not in use, is held beneath the abdomen by *retinaculum* (a catch), when this is released the forcula springs back, propelling the insect through the air.

Collophore: A tube-like structure located on the ventral side of the first abdominal segment (in Collembola).

Colon: The posterior part of the anterior infestine, between the ileum and the rectum; the large intestine.

Colouration: Colouring of chemical or material. Certain white or colourless pesticides are treated with a colouring agent which produce coloured product. The colour of the product does not change during normal conditions of marketing, storage and use. Seed treatment materials are examples of pesticides which have been coloured to indicate that seed has been treated.

Colulus: A slender pointed structure lying just anterior to the spinnerets (In spiders).

Comb: A group of stiff setae or spines, arranged like a comb.

Combustible: Able to catch fire and burn.

Commensalism: A living together of two or more species, none of which is injured thereby, and least one of which is benefited.

Commissure: A transverse tract of nerve fibre connecting the two ganglia of a segment or the lateral centers with in a median ganglion; a bridge structures.

Common name: Colloquial name; Vernacular name. Official common names of pesticides such as carbaryl, warfarin, phosphamidon, that when present must appear in the active ingredients list on container labels. Other official common names are given to a plant or animal. It is possible for two living things to have the same common name in different places. In addition, the same living thing may have several common names.

Community: All plants and animals living in a specific region.

Compatible: The ability to mix two or more pesticides without loss of their effectiveness.

Compatibility: The ability of a pesticide to mix with other pesticides without any undesirable effects, such as reduction in the pesticide activity, increase in the plant injury, promotion of blockage of the spray nozzles, and excessive runoff from the treated surfaces.

Competitive exclusion: The principle that no two species can exist at the same locality if they have identical ecological requirements.

Compensatory growth: Process by which the plant or crop mitigates or nullifies the effects of injury.

Complete metamorphosis: The process of insect development involving egg, larva, pupa and adult stages.

Complex metamorphosis: Metamorphosis in which the insect has four stages, namely egg, larva, pupa and adult; also known as complete metamorphosis; indirect metamorphosis. The insects involved in this type of metamorphosis are categorised as *holomometabola.*

Compound eye: An eye composed of many individual elements or ommatidia, each of which is represented externally by a facet; the external surface of such an eye consists of circular facets that are very close together or of facets that are in contact and more or less hexagonal in shape.

Compressed: Flattened from side to side.

Compressed air sprayer: Sprayer equipped with air pump to develop pressure, often with shoulder strap for carrying; not suitable for spraying at heights over 6 to 8 feet.

Compression sprayer: The sprayer which is fitted with an air pump is known as compression sprayer or pneumatic sprayer. The air pressure is employed in the sprayer for forcing a liquid through the nozzle for forming a spray.

Comstock-Needham system: The

most widely used system of nomenclature for the veins of the wings, intended by its authors to be applicable to insects of all orders and to be a 'natural' system related to the mode of formation of the principal veins from tracheae. It is least successful in Hymenoptera.

Concave and convex veins: The wing membrane is stiffened and made more effective, by a series of longitudinal pleats radiating from the base and, in principal, the major veins either run along a ridge (convex vein) or a trough (concave vein). This arrangement is adhered to most closely in fossil and most primitive living insects and in their most primitive living descendants (e.g. mayflies). But in most living insects, it is modified and blurred the relationship of the veins.

Concentrate: Opposite of dilute. A liquid or dry formulation containing a high percentage of toxicant to save shipping and storage charges and yet be of convenient strength and composition for dilution.

Concentrated solution: Commercial pesticide preparation before dilution for use.

Concentrate spraying: Direct application of the pesticide concentrate without dilution.

Concentration: Proportion of an active ingredient of a pesticide in formulation or spray/dust material.

Condemnation: The removal of a crop or product which does not meet the legal standards for tolerance on foods and is, therefore, not to be sold.

Conductivity: The property of nervous or other protoplasmic tissue by which changes in metabolic activity are propagated through it.

Condyle: A knob-like process forming an articulation.

Cone: The usually conical crystalltine body of a compound eye.

Congeneric: A term applied to species of the same genus.

Conglobate gland: A median gland of unknown function opening into the ejaculatory duct of the male reproductive system in certain insects.

Conjunctiva: See intersegmental membrane.

Connate: Fused together or immovably united.

Connective: A longitudinal cord of nerve fibres connecting successive ganglia.

Conspecific: A term applied to individuals or populations of the same species.

Constricted: Narrowed.

Contact poison: Pesticides that produce toxic effects by physical contact and, thereby, through absoroption by the body wall.

Contaminate: To alter or render a material unfit for a specified use, by the introduction of a foreign substance or a chemical.

Contiguous: Touching each other.

Contract: An agreement with someone to do a job or perform a service for him.

Control: To reduce pest population

to a level below the economic threshold.

Control (noun): Untreated subjects used for comparison with those given a particular crop protection treatment.

Control (verb): To reduce damage or pest density a level below economic threshold.

Control schedule: Time table for the application of control measures irrespective of presence or absence of the pest(s).

Conventional: Usual; Everyday; Customary.

Convergence: Morphological similarity in but distantly related forms.

Convergent: Becoming closer distally.

Cootie: Body louse and the head louse: Generally the term used in U.S.A.

Cope's rule: The generalization that there is a steady increase in size in phyletic series.

Copper butterflies: Butterflies of the family Lycaenidae of the order Lepidoptera. (See blue butterflies).

Copulation (Mating): The union of the two sexes for reproduction.

Copulatory bursa: See bursa copulatrix.

Corbicula: (Pl. Corbiculae); A smooth area on the outer surface of the hind tibia, bordered on each side with a fringe of long curbed hairs, which serve as "Pollen basket" (Bees).

Coreidae: A family of plant bugs of the order Hemiptera. The insects feed on melons, pumpkins, etc.

Corium: The elongate, usually thickened, basal portion of the frong wing (Hemiptera).

Cornea: The outer surface of the compound eye as a hole, and of each individual facet; cuticular lens.

Corneagen layer: Transparent epidermal cells beneath the cuticular lens (Cornea) in the compound eyes of insects and crustaceans. These lenses are shed with rest of the cuticle at each moult and are renewed by means of the corneagen cells.

Corneagenous cells: The epidormal cells that generate the cornea.

Corneous: Consisting of horns; horny.

Cornicle: (Pl. Cornicles); One of a pair of short dorsal tubular processes (horns) on the posterior part of the abdomen (Aphids). Contrary to the popular belief they do not secret the honeydew. The excess of sugars are excreted from anus. The cornicles produce the protective wax which is conspicuous in the wooly aphis.

Coronal suture: A longitudinal suture along the midline of the vertex, between the compound eyes.

Corpus allatum: (Pl. Corpora allata); A pair of small cellular bodies of ectodermal origin associated with the stomodael ganglia behind the brain; acts as a neurchemal and neurosecretory organs in insects. They secrete *juvenile hormone* which maintains the larval state at each moult. In absense of hormone metamorphosis takes place, i.e. larvae transforms to pupae.

Corpus luteum: The mass of degenerating follicle cells left in an egg chamber after the discharge of the egg.

Corrodentia: Psocoptera or booklice.

Corrosive: Having the power to eat away slowly. Some pesticides eat or wear away rubber hoses, nozzles, and other parts of spray machinery.

Cortical cytoplasm: The peripheral layer of cytoplasm in egg.

Cosmopolitan: A species occurring very widely throughout the major regions of the world.

Costa: The first vein of an insect's wing; a longitudinal wing vein, usually forming the enterior margin of the wing, connected basally with the humeral plate.

Costal area: The portion of the wing immediately behind the anterior margin.

Costal break: A point on the costa where the selecerotization is weak or lacking or the vein appears to the broken (In Diptera).

Costal cell: The wing space between the costa and subcosta.

Costal margin: The anterior margin of an insects' wing.

Costal vein: Seen costa.

Cotton boll weevil: *Anthonomus grandis*. A small north American weevil whose larvae feed on the bolls or seed pods of the cotton plants and cause wide spread damage.

Cotton grey weevil: *Myllocerus undecimpustulatus maculosus* Desbrocher (Coleoptera: Curculionidae). This insect is found throughout India and is a minor pest of cotton, pearlmillet, sorghum, soybean, pigeonpea, groundnut, etc. The plants are attacked both by weevils which are prominent above the ground and the grubs which feed on the underground parts. The weevils are grey and the grubs are white, legless and cylindrical.

Cotton jassid: *Amrasca devastans* (Distant) (Hemiptera: Cicadellidae). It is widely distributed in India and is the most destructive pest of American cotton, Okra, Potato, Brinjal. Damage to the crop is caused by the adults as well as by the nymphs, both of them very agile and move bristly forward and sideways. Adults (3 mm long) are greenish yellow during summer, acquiring a reddish tringe in the winter. Injury to plants is due to the loss of sap and also probably due to the injection of toxins.

Cotton leaf roller: *Sylepta derogata* Fabricious (Lepidoptera: Pyraustidae). It is an important sporadic pest of cotton in all the cotton growing tracts of India. The caterpillar which are greenish grey or pink, feed after rolling up the leaves. In years of outbreaks, cotton plants may be completely defoliated. Adult moths are yellowish white with black and brown spots on the head and throax.

There are series of dark brown wavy lines on the wings.

Cotton semilooper: *Trache notabilis* (Walker) (Lepidoptera; Noctuidae). This insect is found in Pakistan and India as a minor pest of cotton, but in certain years it assumes a serious form. The caterpillars are long and dark green with 6 pairs of black and bright yellow spot on the back cause damage to the crop by skeletonize it. In years of heavy infestation the plant may be completely denuded of leaves. The adult moth is white, stoutly built, with prominent grey and brown spots on the wings. Another species *Anomis flava* Fabr; occurs as a sporadic pest of cotton throughout India.

Cotton stainer: See seed cotton bug.

Cotton white fly: *Bemisia tabaci* Genn. (Hemiptera: Aleyrodidae). It is a very serious pest of cotton in northern and western parts of Indian subcontinent. It also cause damage to cabbage, cauliflower, mustard, toria, melon, potato, brinjal, okra, etc. The louse like nymphs suck the cell sap from underside of the leaves. The adults have two pairs of pure white wings and have prominent long hind wings. *B. tabaci* is known to transmit a number of virus diseases including leaf curl disease of tobacco, the vein clearing disease of okra and the leaf curl of sesame.

Cottony cushion scale: *Icerya purchasi*. A small scale insect originally from Australia but introduced in North America and other countries as a pest of Citrus trees. It feeds on large numbers of these trees and secrets; cotton like fibre.

Cotype: Suntype.

Coumafuryl: It is a rodenticide, now manufactured in India.

Coupling agent: A solvent that have the ability to solubilize or to increase the amount of solubility of one material in another.

Coverage: The amount of spread of pesticide over a surface.

Coxa: (Pl. Coxae); The basal (fist) segment of the leg articulating with the pleuron or also with the sternum.

Coxal Corium: The articular membrane sorrounding the base of the coxa.

Coxal gland: An excretory gland opening at the base of the leg in some insects.

Coxal plate: A plate-like extension of the hind legs in some water beetles.

Coxite (Coxopodite): The base of a reduced appendage, especially one of those the last two or three segments of the abdomen; refer to a region at the base of the limb, where coxa articulates with the segments from which it arises.

Coxomarginals: See basicoxite.

Coxopleurite: The sclerite of a generalized thoracic pleuron adjacent to the dorsal margin of the coxa, boating the dorsal coxal articulation, its anterior part become the definitive trochantin.

Coxopodite: The primary basal segment of appendage, representing the primitive limb basis.

Crabs: Crabs belong to class Crustacea of phylum Arthropoda. Crabs are mostly marine but there are fresh water and land crabs also. Many species of crabs cause economic loss to the crops.

Cramps: An involuntary and painful contraction of a voluntary muscle of group of muscles.

Crawler: The active first instar of a scale insect.

Cremastar: A spine-like or the hooked process at the posterior end of the pupae (chryalis) often used for attachment (Lepidoptera) or to help liberating pupae from the soil or other sorroundings just before emergence.

Crenulate: Wavy, or with small scallops.

Crepuscular: Active during the period of twilight either at dark or (more rarely) at day break.

Cribellum: A sieve-like structure (plate) lying just anterior to the spinneret opening near the upper surface of the mandibles in some insects. In spiders, it is situated on the hind end of the abodomen.

Cricket: Any member of the family cryllidae of the order Orthoptera. The chipping or singing of male cricket is brought about by rubbing of the bases of the fore wings together. They can be distinguished from the long-horned grasshoppers by brown colour and by the long, unsegmented anal corci. Like grasshoppers, they also have longback legs for jumping.

Critical areas: Places where pests are most likely to be troublesome.

Crochets (pronounced chroshag): Small, hooked, spines at the tip of the prolegs of Lepidopterous larvae.

Crop: Ingluvies; A thin walled dilation of oesophagus in certain insects for storing food for sometimes before passing it on. An enlargement of the oesophagus region of the stomodeum.

Crop hygiene: Phytosanitation; The removal and destruction of heavily infested or diseased plants from a crop so that they do not form sources of infestation.

Cross contamination: When one pesticide gets into or mixes with another pesticide accidently. This usually occurs in a pesticide container or in a poorly cleaned sprayer.

Cross pollination: Transferance of pollen from one plant to the stigma of the flower of another.

Cross resistant: When a pest population which has become resistant to one pesticide also becomes resistant to other chemically related pesticides.

Cross vein: A short vein connecting adjacent longitudinal veins and their branches; neumerous in net veined wings, in other generally few and located in definite positions.

Cruciate: Crossings; shapede-like a cross.

Cruciferous crops (Crucifers):

Plants such as Cabbage, Cauliflowers, Knol-knol, radish, turnip, mustard which belong to the mustard family.

Crustacea: A class of phylum Arthropoda. The crustaceans have two distinct regions of the body, *viz.*, the cephalothorax (head and thorax) fused and the abdomen. They have two pairs of antennae and atleast five pairs of legs, some of which are modified for swimming. It includes crabs, lobsters, shrimps, etc.

Cryptic: Hidden or concealed.

Cryptic colouration: Colouration of organism that blends with background.

Cryptic species: Sibling species.

Cryptocerata: Insects of the order Hemiptera and the suborder Heteroptera having very small inconspicuous antennae. They include aquatic bugs.

Cryptostigmatid mites: These mites are known as oribatid or soil mites. However, some of the species spend the whole or part of their life cycle on plants.

Crystalline body (Crystalline cone, Vitreous body): A transparent, refractive subcorneal part of the dioptic apparatus of an eye, formed of cells or of cell products, having an oval or conical shape. It acts as an additional lens between the cuticular lens and the retinal cells in each ommatidium.

Ctenidium (Pl. Ctenidia): A row of stout bristles like a teech of a comb on the tarsus of Psocopteran insects.

Cubitus (Cubital vein): The longitudinal vein immediately posterior to the media; the usual fifth vein of the wing between the median and anal veins, according to the comstock—Needham system. In other systems, the cubitus is not necessarily the same vein.

Cuckoo spit: A frothy spit produced by the nymphs of frog hoppers.

Cucurbits: It includes a group of plants among which the important vegetables are bitter gourd, bottle gourd, sponge gourd, pumpkin etc.

Culicidae: A family of two winged insects comprising most of the common grants and mosquitoes. They have long piercing mouth parts and long antennae with numerous whorls of bristles, especially in males.

Cultellus: The sharp knife-like proboscis of some blood sucking flies or one of its blade like lancets.

Cultural control: The control of pests by suitably altering or modifying the usual agricultural operations. It involves the use of ordinary farm practices and farm machinery.

Cumulative effect (of a pesticide): The effects of certain poisons depend not only on the dose received by the person at a particular time but on the doses of the poison previously received. This quantity of such a poison, innocuous in the first instalment may be poisonous in nature and is termed as "Cumulative effect".

Cuneus: A more or less triangular apical piece of corium, set off from the rest of the corium by a suture (in Hemiptera).

Curative: Having the power to heal or cure.

Curculonidae: A family of weevils belonging to the suborder Rhynchophora of the order Coleoptera. The head is prolonged into a snout which may be used for boring into grain or plant parts where the eggs are laid or perhaps to reach a deeper layer of the food material.

Current: New in use.

Currier: Inert material such as dusts, clays oils, water that is mixed with the pesticides and provides for more uniform dispersal of pesticide.

Cursoria: A name formerly used for cockroaches and mantids which were regarded as the running member of the order Orthoptera as distinct from the jumping members or saltatoria (grass hoppers). These names are no longer used and the cursoria is now known as Dictyoptera as a separate order.

Cursorial: Fitted for running; running in habit.

Cuticula (Cuticle): The outer non-cellular layers of the body wall of an arthropod. This is secreted by the epidermis. In insects it usually consists of following layers.

(i) Epicuticle: A thin outer waterproof layer of sclerotin often with minute hairs or microtrichia.

(ii) Exocuticula: A thicker middle layer composed of proteins and chitin together with melanin which colours the insects black or brown.

(iii)Endocuticula: A thick flexible laminated layer of protein and chitin. During mouldting endocuticle becomes liquidified by the action of enzymes secreted by the moulting glands. Thus the two outer layers detach and loosen and then cast-off.

Cutworm: The name appllies to the several larvae of noctuid moths which cut off the plants at the bottom and feed on foliage. The larvae and moths are nocturnal. These cutworms fall under genera *Agrotis* and *Euxoa*. Of the recorded species in India, only four are wide spread. (a) *Agrotis ijpsion* Hufn. is the abundant in India. The moths of these have the curious habit of hiding in sheltered spots or in thatched roofs of the house during the cold weather. *Euxoa segetis* Schiff and *E. spinifera* Hb. is also quite common in India being more in south (Also see *Agrotis flammatra, Agrotis ypsilon*).

Cyanide bottle: They are made in a strong bottle or tube with a specially well fitting cork. In the bottom is placed a layer of broken potassium cyanide about ¼" deep and powdered plaster of paris is used to fill the spaces between the lumps of cyanide and shaken down to level top. Some more plaster of paris is mixed with enough water to make a thick cream and this is poured over the previous mixture.

Cyclodiene insecticides: These are highly chlorinated cyclic hydrocarbon with endo-methylene-bridged structures. Such insecticides, e.g. chlordane, heptachlor, aldrin, etc. are prepared by the Diels-Alder reaction that was discovered by the chemists named Diels and Alder.

Cyclomorphosis: A seasonal non-genetic change of phenotype in species of planktonic fresh water organisms, particularly cladocerans and rotifers.

Cyclorrhapha: One of the three suborders of the Diptera, strictly Brachycera-cyclorrhapha, i.e. with short antennae and emerging from a pupa by a circular slit. The larvae are without head or legs and the pupae are coarctate (contained in a barrel shaped puparium), e.g. house fly.

Cylane: Dithiolane. Iminophosphate (Diethyl-N-1, 3, dithiolauryl-2-iminophosphate). It is a stomach and contact organophosphate insecticide with systemic activity in plants through roots and foliage absorption.

Cypermethrin: Cymbush, Ripcord. A synthetic pyrethroid insecticide, recommended for use against lepidopterous larvae, fruit and shoot borers, leaf folders, etc. of various crops.

Cyst: A sac; vesicle, or bladder-like structure.

Cyst forming nematodes: These nematodes form a cyst to protect their eggs. Golden nematodes of potato *(Heterodera rostochiensis)* soybean nematode *(Heterodera glycines)*.

Cystocytes: The cells enclosing the germ cells in a gonadial tube; the follicle cells of the ovary; the cyst cells of the testis.

Cytogenetics: The comparative study of chromosomal mechanism and behaviour in populations and taxa and their effects on inheritance and evolution.

Cytokinesis: The division of cytoplasm during the cell division.

Cytology: The study of the structure and physiology of the cell and its parts.

Cyton: See neurocyte.

Cytosine: A pyrimidine base occurring in DNA and RNA pairs with guanine in DNA.

D

Dactylopiidae: It includes Cochineal insects which are minute scale insects (Coccoidea) from which the red dye cochineal is extracted. Although, originally from Mexico, but now have been introduced into many other countries. The body of the insects is crimson but is usually covered with a white wooly secretion.

Dactylopodite: The terminal segment of a generalized appendage; typically claw-like in form with elevator and depressor muscles arising in the propodite; represented by a median claw of the pretarasus in some apterygote insects.

Dacus: A genus of tropical fruit fly of the family Trypetidae Order Diptera. The insects are commonly called 'fruit flies'. The larvae feed on fruits and flowers particularly of compositae, and cucurbitaceae, e.g. *Daccus cucurbitae.*

Damage: Injury resulting in measureable loss of yield or reduction in quality.

Damaging: Harmful; injurious.

Damsel flies: A general name for dragon flies of the group Zygoptera, usually smaller and more delicate than the true dragon flies (Anisoptera). Usually blue or green, with narrow based wings which are held vertically while at rest. The aquatic nymphs have legs as long as the body and have three elongated caudal gills at the hind end of the abdomen.

Denaidae: A family of the order Lepidoptera which includes milk weed or monarch butterflies. A wide spread family of large migratory butterflies of America, Southeast Asia, Pacific Islands and Australia. The general colour is orange brown with thick black lines outlining the veins. Many migrate long distances and sometimes even from one country to another, e.g. *Danaus plexippus.*

Dance of bees: A behaviour pattern of bees, which has recently been observed in details and interpreted by Von Frisch as a 'language'. With the behaviour the incoming worker-bees are able to communicate information regarding direction and distance of food availability. There are two types of movement known respectively, as a round dance (performed when food is close) and a 'wag tail' dance (when food is further away). In the latter case the movement resembles a figure of eight in which the two loops are separated by a straight run. The orientation of this and its frequency are said to give other bees the required information. The dancing of the bees has been studied in great detail but it is known also that some moths, blow flies and beetles perform similar activities.

Danger: Risk; hazard.

Darter: Any large dragon fly of the family Libellulidae. These are fat-bodied, broad winged dragon flies that dart out on brief flights with great speed.

Darwin principle: The taxonomic importance of characters that are not the result of a specific adhoc adaptation (in relation with insect taxonomy).

Dasyneura: A genus of midges (Cecidiomyiidae) whose larvae feed on leaves, buds, flowers and form galls, e.g. Linseed gall fly, *Dasyneura lini*.

Day fly: An alternative name for may flies (Ephemeroptera) given on account of the short duration of the adult life.

Days to harvest: The minimum number of days permitted by law between the final pesticide application and harvest of crops.

DDD: TDE, dichlorodiphenyldichloroethane, tetrachlorodiphenylethane: 1, 1-dichloro-2, 2-bis (4-chlorophenyl) ethane. It is similar to DDT but differs only in the absence of one chlorine atom in the ethane moiety.

DDE: (2, 2, bis (p-chlorophenyl) -1, 1-dichloroethylene). It is one of the detoxification products produced by insects from DDT.

DDT: Dichlorodiphenyl trichloroethane: (2, 2-bis (p-chlorophenyl)-1, 1, 1-trichloroethane); 1,1,1-trichloro-2, 2-bis (p-chlorophenyl ethane); pp DDT. It is synthetic chlorinated hydrocarbon insecticide first synthesised by Deidler in 1874. Its use in agriculture has been banned.

DDVP: See Dichlorvos.

Death's head moth: See till hawk moth.

Debris: Trash or unwanted plant parts or remains.

Death feigning: Many insects apparently feign death if disturbed. Many beetles, weevils, stick insects, caterpillars, etc. immediately become motionless when touched. Stick insects draw their legs in close to the body and their stillness makes them even more stick-like giving them a high degree of camouflage by protective resemblance to the plant to which they are attached.

Decamethrin: Decis. A synthetic pyrethroid insecticide recommended against various crop pests. Application rates vary from 7.5 to 25 g of active ingredient/ha. depending on the crop and pest stage. It is 100 times more powerful than DDT, 80 times than carbaryl, 50 times than malathion, 40 times than parathion. It has contact and stomach action.

Deciduous: Having a part or parts that may fall off or shed.

Decticous pupa: An exarate. The pupa that have free limbs and other appendages, having powerful mandibles which help the insect to escape from its cell or cocoon.

Decontaminate: To make safe by removing any pesticide from equipment or other surfaces as directed on a pesticide label.

Decumbent: Bent downward.

Deflexed: Bent downward.

Deguelin: It is a derivative of rotenone and has a benzene ring substituting with a pyron ring. These are two symmetrical carbon atoms and hence four optical isomers are possible.

Delation: Loss of wings by ants and termites; or other insects. In case of ants, rating take place during flight after which the fertilized queen descends and rubs or pulls her wings off. The loss of these appear to stimulate in her an instinct to go below ground and to start laying eggs.

Defensive secretion: Many insects produce and discharge caustic, irritant or evil smelling fluids through different organs as means of defence. They may be applied by sting (in bees and wasps), be a irritant secretions from hairs (in many ceterpillars), discharges of formic acid (in ants), cantharidin chemical (from beetle, lytta), butric acid (from carabid beetles), salicylaldehyde (from leaf beetles), a poisonous vapour (from Bombardier beetle).

Deflocculating agent: A substance that prevents rapid precipitation (setting out) of solids in the liquid in the spray tank.

Deflocculating material: The inert material added to a spray suspension to check or delay sedimentation.

Defoliant: Chemical sprays which induce premature leaf fall.

Defoliate: To lose or become stripped of leaves.

Defoliators: Insects that feed on leaves and denude trees, and other cultivated plants. This may be brought about by large number of insects feeding on the leaves (Bihar hairy caterpillar, *Spilosoma, obliqua*) or by spreading virus and fungus diseases which cause the leaves to curl up that eventually die.

Degradability: The ability of a chemical to decompose or break down into less complex compounds or elements.

Degradation. Break down of a complex chemical by the action of microbes, water, air, sunlight, or other agents.

Degrade: Decompose; breakdown.

Deleterious: Harmful; Injurious.

Delirium: A condition of extreme mental excitement, marked by a rapid succession of confused and unconnected ideas.

Delnav: Dioxathion; 1-4 dioxano-2, 3, 5, 5-bis (0-0-diethyl dithiophosphate). It is a general purpose persistent organophosphorus insecticide.

Deme: A local population of a species; the community of potentially inbreeding individuals at a given locality.

Demeton: Systox; Mercaptophos; 0-0-diethyl-2 ethylmercaptoethyl-thiophosphate). It is a systemic insecticide and acaricide. The technical material contains 70% thiono and 30% thiolo isomers.

Dendrogram: A diagramatic drawing in the form of a tree designed to indicate degrees of relationship.

Dendron: (Pl. Dendrites); Finely remifying branches given off from a nerve cell.

Density: The size of a population within a given unit of space.

Density dependent: Factor whose influence varies with the density of population.

Density independent: Factor whose influence does not vary with the number of individuals in a population.

Dentate: Toothed or with tooth-like projections.

Denticulate: With minute tooth-like projections.

Deoxygenation: Depletion of oxygen.

Deoxyribonucleic acid (DNA): A usually double stranded, helically coiled, nucleic acid molecule, composed of phosphate-dexyribose "backbones" which are connected by paired bases attached to the dexyribose sugar; the genetic material of many viruses, bacteria, plants and animals.

Depleted: Exhausted; reduced.

Deposit: Amount of pesticidal spray or dust deposited per unit area of the plant surface.

Depressed: Flattened dorsoventrally.

Derived character: A character that differs materially from the ancestral condition.

Dermal: Relating to the skin.

Dermal glands: Modified epidermal glands or group of cells on the outer surface of the body that secrete, wax, scent, irritant poisons, etc;, may also be applied to the hypodermal glands that secrete moulting fluid containing enzymes which liquify the endocuticle prior to moulting.

Dermal toxicity: How poisonous a pesticide is when it is absorbed through skin.

Dermatitis: Inflammation of skin.

Dermaptera: (derma, skin: pteron, a wing). Earwings. Insects have leathery forewings. The hindwings are large, membranous, and at rest lie folded underneath the fore wings. Mouth parts are biting type; one pair of foreceps present at the end of the body.

Dermat: Relating to the skin or outer covering.

Derris: The extract from *Derris elliptica* and *D. mallencensis* contain rotenone with certain rotenoid compounds of lesser insecticidal activity than rotenone.

Desert locust: *Schistocerca gregaria* found in Africa and Asia. It breeds in Punjab (India) during summer monsoon. Then in autumn it migrates through Iran and Arabia to the middle-east where it meets with others from the African continent. Together they spread through Egypt, Syria, Palestine and Soviet Asia. The migrants of this insect, have lately been shown to follow the movement of lowpressure systems, which provide the rain necessary for the survival of the egg-pods in the soil.

Desiccant: A compound that promotes drying or removal of moisture from plant tissues.

Desiccation: Excessive drying, either by natural loss of moisture, or by artificial means; drying.

Designated priority: In cases of simultaneous publication of several names, (scientific name of insect species), the priority established by the first reviser.

Detergent: A chemical (not soap) having the ability to remove soil or grime. Detergents can be used as sunfactants in some pesticide spray.

Deteriorate: To break down or bear away. To decay.

Detoxify: To make harmless.

Deutocerebrum: The part of the Arthropod brain containing the first antennal nerve centers. Also called the *Duterocerebrum* and *mesocerebrum*.

Deutonymph: The third instar of a mite.

Deutoplasm: The yolk, or nutritive materials of the egg enmeshed in the cytoplasm.

Deutro DDT: It is similar to DDT except that the one hydrogen of DDT is replaced to deuterium.

Development of host resistance: The manipulation of the morphology or physiology of a crop plant by selective breeding or hybridization so that pest can not become established.

Development stage: The stage of development is a general term for a section of ontogenetical development characterized by significant biochemical or morphological processes or by both, such as protopod, polypod, oligopod, and postoligopod stages in embryogenesis, the larval stage (comprising the first penaltimate larval instar); the metamorphosis stage (comprising the last larval and the pupal instars); the maturation stage; the stage of sexual maturity.

DFDT: Fix; 1, 1, 1-trichloro-2,-2-bis-p-fluorophenylethane. It is a chlorinated hydrocarbon. It is the fluorine analog of DDT.

Diacrisia: See Bihar hairy caterpillar.

Diagnosis: Identification of the nature and cause of problem. In taxonomy, a formal statement of the characters (or most important characters) which distinguish a taxon from other similar or closely related coordinate taxa.

Diamond back moth: *Plutella xylostella* (Linnaeus) (Lepidoptera: Plutellidae). It is a serious pest of cabbage and cauliflower and also feeds on cruciferous, solanaceous and liliaceous plants all over India. Damage is caused by the caterpillars which in earlier stages feed in mines on the lower side of the cabbage but in later stages feed exposed on the leaves.

Diapause: Temporary arrest of development at any stage of life cycle. It is an adaptive phenomenon or in born rhythm which usually facilitates survival during unfavourable periods. In some cases diapause lasts for several years. This is probably due to the fact that the insect fails to produce the normal harmones necessary for continued growth or development.

Diapause hormone: It is produced in some insects by the neurosecretory cells of the suboesophageal ganglion. It inhibits different body functions.

Diaphram: One of the horizontal partitions of the body cavity; Dorsal diaphram-the membranous and muscular sheets of tissue extending from the dorsal blood vessel to the laterodorsal parts of the body wall, separating the dorsal sinus from the perivisceral sinus; ventral diaphrom-a membranous and muscular sheet present in some insects stretched between the lateral edges of the abdominal sterna, sometimes extending into the thorax, separating the ventral sinus from the perivisceral sinus.

Diaspididae: A family of scale insects that attack many shrubs and trees, sucking the cap and secreting a waxy scale, e.g. San Jose scale, (*Quadraspidiotus perniciosus*) a pest of fruit trees.

Diazinon: 0, 0-diethyl-0-2-isopropyl-4-methyl-poyrimidyl-6-thiophosphate. It is a persistent general purpose insecticide used for controlling pests of plants and animals.

Dicapthon: 0, 0-diethyl-2-chloro-4-nitrophenylthiophosphate. It is a chlorinated analog of methyl parathion with cl in 2-position. It is a contact and stomach pesticide with fumigant action.

Dichlorovos: DDVP; 0-0-diethyl-0, 2, 2-dichlorovinylphosphate. It is a derivative of phosphoric acid and is the first representative of enol phosphate series.

Dichoptic: The eyes separated above (In Diptera).

Dichotomous: Divided or dividing into two parts.

Dicondylic joint: A joint with two points of articulation between the adjacent leg segments.

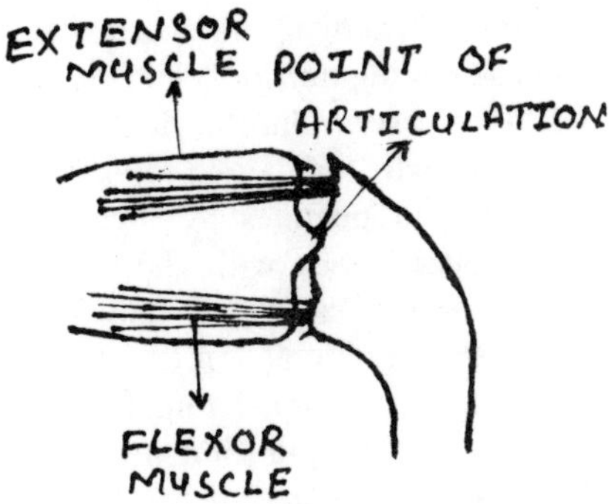

Dicondylic joint

Dictyoptera: An order of insects comprising the Cockroaches and the mantids. They were formerly included with grasshoppers and locusts in the order Orthoptera, but there are many points of difference, the chief being the shorter back legs of the Dictyoptera and their inability to jump.

Diecious: Having male and female organs in different individuals; any one individual being either male or female.

Dieldnin: HEOD; 1, 2, 3, 4, 10, 10-hexchloro-6, 7, epoxy-1, 4, 4a, 5,

6, 7, 8, 8a-octahydro-1, 4, 5, 8 endo, exodimethanonaphthene. It is a chlorinated hydrocarbon product of Diels-Alder reaction. It contains not less than 85% of the active ingredient HEOD. It is now banned in India for agricultural use.

Differential diagnosis (related with taxonomy): A formal statement of the characters which distinguish a given taxon from other specifically mentioned equivalent taxa.

Diffusion tracheae: Cylindrical tracheae having noncollapsible walls.

Diflubenzuron: Dimilin. It is an insect growth regulator produced in 1974, and operates by ingestion, and interference with the deposition of new cuticle at ecdysis. It has no systemic action. It is effective against a broad range of leaf eating insects, and some mites.

Digestion: Enzymatic hydrolysis of food into absorbable size molecules.

Digestive cells: The secretory and absorptive cells of the ventricular epithelium as distinguish from the regenerative cells.

Dihybrid: An individual heterozygous in two pairs of alleles.

Dihydrodeguelin: It is a optically active form of rotenone and is 10 times as active as the levo deguelin.

Dihydrorotenone: It is a natural organic insecticide produced by the catalytic hydrogenation of the side chain of rotenone with elimination of the double bond.

Dilan: It is a mixture of two insecticides viz. 53.3% Bulan (DNB) and 26.7% of Prolan (DNP).

Dilated: Expanded or widened.

Dilator muscles: Muscles extending form the body wall to the alimentary canal, also called suspensory muscles.

Diluent: Any material used to dilute a pesticide to the desired concentration.

Dilute: To make less concentrated. To mix a pesticide chemical with water, oil or other material before it can be safely and correctly used.

Dilution rate: The amount of diluent that must be added to a unit of a pesticide to obtain the desired dosage.

Dimecron: See Phosphamidon.

Dimefox: Pestox; Hnane; Terrasystam; Tetramethyl-phosphorodiamidic fluoride. It is a powerful systemic organophosphorus insecticides with very short residual activity.

Dimethoate: Rogor; Phosphamide; 0, 0-dimethyl-S-(N-Methylcarbamoyl methyl) dithiophosphate. It is systemic organophosphorus insecticide and acaricide.

Dimeton: (5, 5-dimethyl (dihydroresorcinol) N, N-diethyl-carbamate. It is a carbamate insecticide.

Dimorphism: Occurrence of two distinct morphological types (morphs, phena) in a single population.

Dinarmus spp.: It is a gregarious ectoparasite of the full grown larva of the girdle beetle (*Oberea brevis*) in India. The species of *Dinarmus* found in India are *D. coimbatorensis* Ferriere on the larvae of cotton boll weevil (*Pemphe-res affinis* Faust) and *D. sauteri* Masi on the larvae of amaranthus borer (*Lixus truncatulus* F.).

Dinocap: It is a good substitute for sulfur for controlling certain spores and mites.

Dioecious: Individuals producing either sperm or egg. In dioecious species, the sexes are separate.

Dioptric apparatus: The outer transparent part of an optic organ, consisting of the cornea and usually of a subcorneal crystalline body.

Diploid: Having a double set of chromosomes (2 n); the normal chromosome number of the cells (except for mature germ cells) of a particular organism derived from a fertilized egg.

Diplopoda: A class of the phylum Arthropoda. It includes millepedes or thousandlegged worms. Their body is rounded and the segments are internally fursed in twos.

Diplura: Bristle tails. Small, white, primitive wingless insects living in soil, having a forked tail formed from two long thin cerci and have biting jaws lying in sockets from which they are protruded when feeding.

Diptera: (Dis, two; pteron, a wing). It includes flies and their allies which have 2 wings; the hind pair is modified to form halteres used in balancing the body. Mouth parts modified for piercing and sucking or sponging liquid foods. The larva is wormlike, without legs and with an indistinct head, and has terrestrial, aquatic or parasitic mode of life. The pupa is obtect or exarate. Diptera is divided into 3 suborders: (i) Nematocora: Insects with filamentous antennae, e.g. midges, gnats, mosquitoes. (ii) Brachycera: It includes horse flies, beeflies (iii) Cyclorrhapha: It includes fruit flies, blow flies, etc.

Dipterex: See Trichlorofon.

Dip treatment: The application of a liquid chemical to a plant by immersing it, wholly or partially under the surface of the liquid, so as to coat the plant with chemical.

Directed application: An application to a restricted area such as a row, bed or at the base of the plants.

Direction finding: Most insects have a very good sense of direction and are able to find their way either by keeping a constant angle to the sun or to the polarized light of the sky (e.g. ants). Bees have much the same ability and also appear to be able to memorize the landmarks of their neighbourhood. The returning bees can communicate information about the distance and direction of food supplies to other bees of the colony by means of their dance. Certain male butterflies can fly several miles to the females by the scents released by the females.

Disc: The centre of any sclerite. away from its edges.

Discal: On the disc, or main surface of any part of the body as ossposed to 'marginal'; near its edges.

Discal cell: A more or less enlarged cell in the basal or central part of the wing, more particularly cell 1st M_2 in Diptera and cell R in Lepidoptera.

Discal cross vein: A cross vein behind the discal cell (In Diptera).

Discoidal cell: A cell near the middle of the wing (In Hymenoptera).

Discoidal vein: The vein forming a continuation of the median vein beyond the end of the transverse median vein, and extending along the posterior margin of the first discoidal cell (In Hymenoptera).

Discontinuous variation: Variation in which the individuals (insects, of a sample fall into definite classes which do not grade into each other.

Disinfectant: An agent that kills or inactivates organisms present on the surface of the plant or plant part in the immediate environment.

Dispersal: Movemen of individuals out of a population (emigration) or into a population (immigration).

Disperse: To spread out or scatter. To separate and move apart.

Dispersing agent: Spray agent which enables formation of stable suspension of solid particles in the liquid media.

Disposal: The process of discarding or throwing away unused spray material, surplus pesticides, and pesticide containers.

Dissipate: To spread out. Gases and vapours dissipate through the air.

Dissipation: Progressive loss of pesticides from the treated areas.

Dissolve: Usually refers to getting solids into suspension and or solution which is necessary for uniform application results.

Distad: A way from the body; towards the end; farthest from the body.

Distal: Near or towards the free end of an appendage; that part of a segment or appendage farthest from the body.

Distal process: The peripheral branch or one of several distal branches of a sensory nerve cell.

Distribution: The amount and way in which a pesicide chemical is spread when applied to a plant, animal or other surface. Also, part of the any country or world in which an insect, fungus or other pest is found.

Disyston: Dithiosystox; disulphotan; 0-0-diethyl-S-(2-ethyl mercaptoethyl) dithiophosphate. It is similar to those of thiometon but considerably more stable. It is also used as seed disinfectant.

Diurnal: Active during the hours of daylight.

Divaricate: Extending outward and then curving or toward each other distally (divaricate tarsal claws).

Divergent: Becoming more separated distally.

Dizziness: An imprecise term commonly used by patients in an attempt to describe various peculiar subjective symptoms, such as faintness, giddiness, lightheadness or unsteadiness.

Dollo's rule: The principle that evolution is irreversible to the extent that structures or functions once lost cannot be regained.

Dormancy: A state of quiscence or inactivity; see diapause also.

Dormant: Alive but not growing; quiescent; inactive; a resting stage.

Dorsad: Towards back or top.

Dorsal: Top or uppermost; pertaining to the back or upper side; opposed to ventral.

Dorsal blastoderm: The extraembryonic part of the blastoderm.

Dorsal organ: A mass of cells in the dorsal part of the embryo apparently produced by the invaginated serosa.

Dorsal sinus: The space of the definitive body cavity above the dorsal diaphram and the heart.

Dorsal trachea: The dorsal segmental trachea originating at a spiracle.

Dorsal vessel: The dorsal blood vessel, consisting of the pulsatile heart and nonpulsating aorta.

Dorsocentral bristles: A longitudinal row of bristles on the mesonotum, just laterad of the acroctichial bristles (In Diptera).

Dorselateral: At the top and to the side.

Dorsomesal: At the top and along the midline.

Dorso-pleural line: The line of separation between the dorsum and the pleural region of the body, often marked by fold or groove.

Dorsoscutellar bristles: A pair of bristles on the dorsal portion of the scutellum, one on each side of the midline (In Diptera).

Dorsoventral: From top to bottom, or from the upper to the lower surface.

Dorsum: The entire back of an animal above the pleural regions; or specifically when qualified by the designation of a segment, the back region of a segment.

Dosage: The amount of pesticide to be applied for given area or per unit of water.

Dose: Quantity of pesticide applied per individual, or per unit area, or per unit volume, or per unit weight.

Douse: Drench; soak.

Downwind: Direction towards which the wind is blowing.

Drench: Saturation (through soaking) of the soil with a pesticide.

Drench treatment: The application of a liquid chemical to an area until the area is completely soaked.

Dried current moth: See almond moth.

Drift: Movement of material out side the intended target area during or shortly after application; Spray or dust carried by natural air currents beyond the target area.

Drones: Male bees.

Duct: A tube for carrying secretion from a gland to the point of discharge.

Ductus ejaculatorius: The common outlet duct of the male reproductive system; Ejaculatory duct.

Dufours gland: See Alkaline gland.

Dulosis: Slavery among ants; e.g. the brown ants, living in the same nest but working for the red.

Dunnage: Materials, such as wooden plants, loose wood of any kind and sheet straw, which insulate the stored material against the seepage of moisture from the sides and floor.

Dursban: 0, 0-diethyl. 0, 3, 5, 6 trichloropyridyl thiophosphate. It is an active organophosphorus insecticide used as a stomach poison.

Dusky cotton bug: *Oxycarenus laetus* Kirby (Hemiptera: Lygaeidae). It is a pest of cotton, okra, hollyhock in India. The adults are 4-5 mm in length, dark brown and have dirty white transparent wings. The young nymphs have rotund abdomen.

Dusts: Dusts consist of a mechanical mixture of the active ingredients with or without an inert diluent pulverised to a particles size of 3 to 30 μ. Dusts can be classified into (i) undiluted toxic agents (sodium fluoride), (ii) Toxic agent with an active diluent (Rotenone with sulphur), (iii) Toxic agent with an inert diluent (DDT with prophylite), (iv) Inert dusts (Silica).

Dyar's law: An epirical law based on the observations that the head capsule and certain other parts of the caterpillars grow in geometrical progression, increasing in size by a constant ratio at each moult. Thus, if the logarith of the size of the head is plotted against the number of instar, the resulting gap will be a straight line.

Dynastodae: A family of Celeoptera order which includes large beetles, some of which are armed with large horns or protuberances extending forward from the thorax.

Dyscrasia: A general diseased state resulting from the presence of toxic matters in the blood.

Dysentery: A disease marked by frequent watery stools, often with blood and mucus, and characterized clinically by pain, fever, and dehydration. It may result from infection with *Entamoeba histolytica,* specific bacilli or virus.

Dytiscidae: A family of Coleoptera order. It includes large, aquatic, carnivorous beetles. They have large oval body and powerful back legs which are flattened and fringed with bristles for swimming. The males have broadened front tarsi which form cup-like suckers for gripping the females at mating.

E

Earias: See spotted bollworms.

Earwig: A member of the order Dermaptera. They are hemimetabolous insects in which the forewings are reduced to small thickened elytra and the hind wings are roughly semicircular. Each has a large forceps at hind end of the abdomen. Mouth parts are biting type and are generally omnivorous scavengers.

EC (Emulsifiable concentrate): Liquid formulation that immediately disperses as fine droplets in water to form an emulsion.

Ecdysial fluid: Moulting fluid. A liquid secreted by glands in the epidemis of an insect, having the function of dissolving away the endocuticle by enzyme action. In this way the exocuticle and the epicuticle, which together form the hard outer part of the exoskeleton, are loosened prior to being shed when insect moults.

Ecdysis (Pl. Ecdyses): Moulting; the process of shedding the exoskeleton (also see moulting). The stages between each moult is called instar. This process is initiated by a harmone (Ecdysone) secreted by the *Thoracic glands* which lie in the thorax behind brain and close to the oesophagus.

Ecdysone: Principal moulting hormone of insects. (See Ecdysis).

Ecdysoids: Ecdysones; The substances with moulting hormone activity.

Ecesis: Oikesis; The establishment of an organism in a new habitat; accidental dispersal and establishment in a new area.

Eclosion: The emergence of an insect from a pupa or cocoon or puparium. In the case of simple pupa, the cuticle is loosened by softening of the endocuticle, it then splits at the weakest part and the insect wriggles out. If a pupa is surrounded by a silken cocoon, the insect may gnaw its way out or may use spines to split the silk or in some cases may secrete a liquid which dissolve the silk. In housefly in which the pupa is surrounded by a capsule or puparium, made from the cast off skin of last larval stage, the insect uses *ptilinum* or bladder-like swelling on the head for pushing open one end.

Ecoclimate: Climate within the plant (crop) community.

Ecological isolation: A condition in which inter-breeding between two or more otherwise sympatric populations is believed to be prevented by mating in different ecologic niches.

Ecological race: A local race that owes its most conspicuous attributes to the selective effect of a specific environment.

Ecology: The study of organisms in relation to their environment, including both the physical and biotic components.

Economic control: The reduction

or maintenance of a pest density below the economic injury level.

Economic damage: The amount of crop injury that justified the cost of applied (artificial) control methods.

Economic injury level: The lowest population density of a pest that causes economic damage. The pest density that causes damage equal to the cost of preventing damage.

Economic pest: The pest which causes economic damage.

Economic threshold: The pest density at which control methods should be determined or undertaken to prevent an increasing pest population from reaching the economic injury level.

Ecophenotypic variation: (habitant variation). A nongenetic modification of the phenotype by specific ecological conditions, particularly those of a habitant.

Ecospecies: A group of populations so related that they are able to exchange genes freely without loss of fertility or vigou in the offsprings.

Ecosystem: It may be defined as "an area that includes living organisms and non-living substances interacting to produce an exchange of materials between the living and non-living parts". Agroecosystem refers to an ecosystem in an agricultural area.

Ectoderm: The outer embryonic layer from which the epidermis of the body wall is derived; the outer cell layer.

Ectognathous: Having the mouth-parts projecting outwards and well developed. Among Apterygote insects, Thysanura are ectognathous where as Diplura, Protura and Collembola are endognathous.

Ectoparasite: The parasites which feed outside their host or near the outside in the mouth or rectum. They include sucking lice (Anoplura), fleas (Siphonoptera), and bed bugs (Hemiptera). These insects have piercing and sucking type of mouth parts to suck blood or cell sap. Biting lice (Mallophaga) donot suck blood but feed on hairs, feathers, etc.

Ectoparasitic nematodes: These feed on roots but usually do not enter tissue. Variously known as sting, ring, awl, spiral, lance, pin, stylet nematodes. Their names are descriptive of their appearance and damage.

Ectophallus: The outer phallic wall in distinction to the endophallus.

EDCT Mixture: Ethylene dichloride-carbon tetrachloride mixture. It is used to control storage pests.

Edible: Safe to eat; food.

Eelworm: Nematode.

Effector: One of the organs of the body activated by nerve stimuli, principally a muscle or a gland.

Efferent nerve: A nerve that conducts from a nerve center toward the pariphery; the axon of a motor neurone.

Efficacy: Effectiveness; capacity for serving to produce effects.

Egg: Egg of an insect consists of a

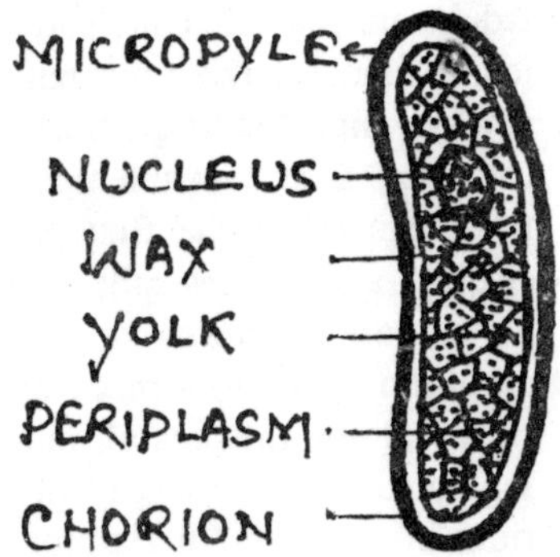

Insect Egg

nucleus and a mass of yolk, the latter being within the meshes of the central protoplasm, a state known as *Centrolecithal.* This is sorrounded by a thin vitelline membrane and outside this a relatively thick outer shell or chorion secreted by the ovarion follicle, sometimes with a wax layer in between. Small microphyles permit a sperm cell to enter. The flies has only a single micropyle at one end but locust and grass hoppers have 30-40. The size of the insect's egg varies from 0.15 to 30 mm. They are of various colours, shape and with chorion structures. The egg may be spherical, elliptical, conical or cylindrical, smooth, ridged or spiny. Some times these are sorrounded by a blue like substance for attachment, sometimes having silken threads (lace wing flies), sometimes fastened together to form a floating raft (Mosquitoes).

Egg-bursters: These are a set of hardspines or ridges located on the head or on other parts of the body of young larva. The young larva at the time of its emergence tears open egg-shell with these spines. The Colorado beetle larva has three pairs of spines and the embryo flea has a knife like projection.

Egg Chamber: One of the compartments, or follicle of an ovarial egg tube, formed of the follicle cells, containing an oocyte.

Egg guide: A median process of the subgenital plate behind the genital opening, particularly developed in Acrididae.

Egg plant: Brinjal plant.

Egg-pod: A sack-like capsule enclosing eggs of grasshopper or locust.

Egg tube: The tubular part of an ovariole containing the germ cells, the oocytes, the nurse cells, and the follicle cells.

Ejaculatory ducts: The ectodermal outlet duct of the male genital system, or the terminal portion of the male sperm duct.

Ekatin: Thiometon; Intruthiol; 0, 0 diethyl-S-2-ethyl mereapto ethyl dithiophosphate. It is a good organophosphate systemic insecticide with about the same duration of action of metasystox.

Elasmus brevicornis Gah. (Braconidae: Hymenoptera): It is a gregarious ectoparasite of the full grown larvae of *Aproraema modicella, Hedylepta indicata and*

Chrysodeixis sp.

Elateridae: A family contains click beetles. The family named on account of their habit of springing into the air with a clicking sound when placed on their backs. The larvae are known as wireworms, living in the soil and are pests of agricultural crops.

Elatiform larvae: These are elongate, smooth, heavily sclerotized, cylindrical, with short thoracic legs and with few body hairs, e.g. click beetles (wire worms) or a larva resembling wire worms or meal worms.

Elbowed antenna: An antenna with the first segment elongated and the remaining segments coming off the first segment at an angle.

Electrophoresis: A process of separating different molecules particularly polypeptides, owing to their differential rates of migration in an electric field.

Electrostatic charge: Electrostatic charges in dusts are produced between the particles themselves and between the particles and the dusting equipment.

Elytron (Pl. Elytra): The thickened leathery or horny front wings of beetles (Coleoptera), and earwigs (Dermaptera). These cover the membranous hind wings. In a few species of beetles, whose flying wings are reduced or absent, the elytra are cemented together down the middle line.

Embioptera: Insects of this order are commonly known as the web-spinners and they live in silken galleries made on the bark of trees. Adults are elongated and soft bodied, the males having two pairs of smoky wings. Mouth parts are biting type; tarsi 3 segmented; the first segment of the first pair is swollen and carries silk glands; females are wingless.

Embolium: A narrow piece of corium, along the costal margin separated from the rest of the corium by a suture (In Hemiptera); a part of hemielytron.

Emendation: In nomenclature, an international modification of the spelling of a previously published scientific name.

Emergence: The act of the adult insect leaving the pupal case or the last nymphal skin (also see Eclosion).

Emetic: A substance used to make humans or animals to vomit.

Emigration: The movement of individuals out of a population.

Empoasca: A genus of leaf hoppers of the family Cicadellidae. They suck the cell sap and also cause damage by their toxic saliva causing 'leaf burn'.

Empodium (Pl. Empodia): The single pad like structure between the paired pulvilli, or bristle like structure at the apex of the last tarsal segment between the two claws (In Diptera).

Emulsifier: Spray additive of colloidal substances such as alkali (sodium and potassium), soaps, cresols, phthalic glycerol, alkylresin, etc. which enables formation of a stable suspension of oil droplets in water.

Emulsify: To make into an emulsion. When small drops of one liquid are finally dispersed (distributed) in another liquid, an emulsion is formed. The drops are held in suspension by an emulsifying agent, which surrounds each drop and makes a coating around it.

Emulsifying agent: A material which facilitates the suspending of one liquid in another.

Emulsion: A suspension of fine droplets of one liquid in another. It is mostly milky in appearance.

Emulsifiable concentrate (EC): These are oil based liquid formulations concentrated to extra strength to save transport and storage costs. They contain solvents emulsifiers and wetting agent that make them easily soluble in water.

Encapsulation: A method of formulating pesticides, in which the active ingradient is encapsed in a material (often poly vinyl) resulting in sustained pesticidal release and decreased hazard.

Encephalitis: Inflammation of the brain. Virus caused encephalitides can affect man and horses. The viruses are transmitted by mosquitoes.

End Chamber: The germarium of a gonadial tube.

Endite: The basal segment of the spider pedipalp, which enlarged and functions as a crushing jaw; a mesal lobe of any limb segments.

Endochorion: The inner layer of the chorion or shell of an insect's egg.

Endocrine: A ductless gland producing hormones.

Endocuticle (Endocuticula): The flexible inner most soft layer of the cuticle, composed of chitin and proteins. During the moulting this layer is digested away by enzymes secreted by special epidermal glands. It thus leaves the outer layer of cuticle loose and ready to be cast off.

Endoderm: The innermost cell layer, forming the epithelium.

Endoparasite: The insect parasites which feed inside the host. A new host is affected either by the eggs being laid under its skin; by larvae entering through the skin and penetrating the tissues or by host swallowing the egg or larvae. These parasites are usually the larvae of small Hymenoptera, such as chalcids, wasps and ichneumaon flies and some Diptera, mainly Tachinidae. These parasites control agricultural pests in nature and some are also used as a tool of biological control.

Endoparasitic nematodes: These nematodes enter plant tissues or attach to it permanently, mainly on root, e.g. Root knot nematode, *Meloidogyne* spp.

Endophallus: The inner chamber of the phallus invaginated at the end of the adeagus, into which the ejaculatory duct opens; typically an eversible sac or tube but sometimes a permanently internal phallic structure.

Endopleurites: Y-shaped in growth of cuticle from the sides of the thorax, forming part of the internal or endophragmal skeleton of Arthropods. In the wing bearing segments of insects, these form the pleural ridges which support the wing above and articulate with the coxa of the leg below.

Endopodite: The mesal or internal branch of a biramous appendage; the main shaft of the limb beyond the basipodite.

Endopterygota: Insects in which the wings develop internally until the final moult; with complete metamorphosis. They are also called *Holometabola*.

Endoskeleton: Collectively the internal cuticular or apodemal ridges and processes of the body wall.

Endosternites: Apodeme or ingrowths from the ventral cuticular plates of an insect or other Arthropod. They form part of the endophragmal skeleton to which important muscles are attached.

Endosulfan: Thiodan; 6,7,8,9,10, 10-hexachloro-1,5,5a,6,9,9a-hexahydro-6, 9-methano,2,4,3-benzodioxathin-3-oxide. It is a chlorinated hydrocarbon insecticide containing 4 parts of £-cis isomer and 1 part of ß-trans isomer. It is contact poison and recommended for use against the insect pests.

Endotheca: The inner wall of the phallotheca.

Endothion: Phosphoryrone (0, 0-dimethyl-S-(5-methoxy-pyronyl-2-methyl thiophosphate). It is a systemic organophosphate insecticide.

Endrin: 1,2,3,4,10,10-hexachloro-6, 7-epoxy-1,4,4a,5,6,7,8,8a-octa hydro-1,4,5,8 endo, endodimethanonaphthalene. It is highly toxic chlorinated hydrocarbon insecticide of low persistance. It is sterio isomer of dieldrin. Its use in India is banned.

Ensifera: A suborder of Orthoptera that includes long horned grasshoppers, crickets, mole crickets having very long antennae.

Enteric: Relating to the digestive canal.

Enteric caecum (Gastric caecum): Sac-like diverticula at the anerior end of the midgut.

Enteromyiasis: Infection of the intestine by the larvae of flies.

Entire: Without teeth or notches, with a smooth outline.

Entognathous: Having the mouth parts in pockets from which they can be protruded when feeding.

Entomologist: One who collects and studies the insects.

Entomology: A branch of Zoology that deals with the insects only.

Entomophagous (Insectivorous): Animals that feed chiefly on insects.

Entomophily: Pollination of flowers by insects.

Enveloping cell: The intermediate cell of a sense organ or one of the components of sensory units of the organ, probably corresponding to the trichogen of a seta.

Environment: The totality of physical, chemical and biotic conditions sorrounding an organism.

Environmental manipulation: Skillfull and artful management or use of the environment in such a way as to assist man in pest control.

Environmental pressure: The adverse effects on a pest population induced by both living and nonliving environmental factors.

Enzyme: A protein catalyst produced within a living organism which regulates the rate of a specific biochemical reaction.

Ephemeroptera: (Ephemeros, living a day: Pteron, a wing). Mayflies are short lived, have soft bodies and large eyes; wings are membranous. Nymphs are aquatic with tracheal gills.

Ephestia: A genus of small moths (Microlepidoptera) whose larvae are common pest of food products (See Almond moth).

Epicranial plates: Two cuticular plates covering the dorsal surface of an insect head.

Epicranial suture: The dorsal-shaped (line of weakness on the head) suture of the cranium, including the median coronal suture of the vertex, and the divergent frontal sutures of the facial region. This is not a true suture, and without reference to segmentation. It is often miscalled the 'frontal suture'.

Epicranium: A term variously applied to the entire cranium: the cranium exclusive of the frons, or preferably to the upper part of the head, from the face to the neck.

Epiculticle: The non-chitinous film like covering of the exocuticle. It is usually very thin and is made of sclerotin which is impervious to water.

Epidemic: A sudden widespread increase in the incidence of a disease or organism.

Epidermis: The cellular layer of the body wall which secrets the cuticle.

Epididymis: A convoluted part of the vas deferens.

Epigastric furrow: A transverse ventral suture near the anterior end of the abdomen, along which lie the openings of the book lungs and the reproductive organs (In Spiders).

Epigynum: The external female genitalia of Spiders.

Epileptic: Convulsions owing to nervous disorder or mental disturbances.

Epimere: A dorsal process of the phaliobase.

Epimeron (Pl. Epimera): The area of the pleuron posterior to the pleural suture, sometimes divided horizontally into a supraepimeron or anepimeron, and an infraepimeron or katepimeron; the posterior sclerite of the pleuron of the thorax.

Epimorphic development: Immatures possessing same number of body segments as adults.

Epinasty: Increased growth on upper surface of a plant organ or part (especially of leaves) causing it bend downward.

Epineural sinus: The ventral part of the embroyonic haemocoel between the yolk and the ventral nerve cord.

Epiopticon: The middle section of the optic lobe of insects brain; also known as external medullary mass.

Epiphallus: A sclerite in some Orthoptera in the floor of the genital chamber proximal to the base of the phallus.

Epipharynx: A membranous mouth part structure on the inner surface of the labrum or clypeus; in chewing insects a maidan lobe on the posterior (ventral) surface of the labrum or clypeus. This is provided with the taste receptors. It is sometimes elongated to form part of the probosis, being associated with the labrum.

Epiphysis: A middle pad or lappet like structure on the inner aspect of the front tibia of some Lepidoptera.

Epiphyte: An air plant, one growing non-parasitically upon another plant or upon a non-living object.

Epipleura (Epipleuron): The deflex or inflex portion of the elytra; the bent down lateral edge of the elytra (Coleoptera).

Epipleurite: A small lateral sclerite in the membranous area between the thoracic pleura and the wing bases.

Epipodite: An exit of the coxopodite; often a gill bearing organ.

Epiproct: A process or appendage situated above the anus and appearing to arise from the tenth abdominal segment; actually the dorsal part of eleventh abdominal segment.

Episite: A predator; an organism that Feeds on other animals.

Episome: A genetic element that may be present in a given cell either on a chromosome or separately in the cytoplasm.

Episternum: The area of the pleuron before the pleural suture and above the trochantin, sometimes divided horizontally into a supraepisternum, or anepisernum, and an infraepisternum, or a katepisternum.

Episomal suture: The frontoclypeal suture; a suture between the frons and clypeus; a groove uniting the anterior ends of the subgenal suture across the face, forming internally a strong epistomal ridge typically straight, but often arched upward, sometimes absent.

Epistome: The part of the face just above the mouth; the oval margin (Diptera).

EPN (O-ethyl-0-(4 nitrophenyl) benzene thiophosphate): It is a powerful acaricide.

Equal weighting: The method which treats all taxonomic characters as equally important; a key assumption of phenetices.

Eradex (Quinoxaline): It is an active acaricide belonging to the group of derivatives of dithiocarbonic acid.

Eradicant: A chemical used to eliminate a pest from a plant or a place in the environment.

Eradication: The elimination of a pest or pathogen from an area where it has become established.

Ergatandromorph: An ant which combines some of the characteristics of both male and worker.

Ergataner: A male ant resembling the worker because it has lost wings.

Ergate: A worker ant.

Ergastogyne: A female ant without wings and looking like a worker.

Ergotism: Poisoning by a toxic substance contained in the sclerotia of the fungus, which grows on many grasses and on some grains.

Erinium: A growth of hairs in dense patches on plant leaves resulting from the attack of certain Eriophyidae (Acarina).

Eriophyid mites: The mites are elongate, vermiform, having only two pairs of legs both in the immature and adult stages. The body is differentiated into anterior short caphalothorax and annulated posterior abdomen. Pigeonpea sterlity mosaic is transmitted by two species of eriophyids. Sugarcane streak virus is transmitted by *Aceria sacchari*.

Eriosoma lanigerum: Wooly Aphis, A plant louse generally yellow or brown, originally from North America but now found in many parts of the world. It attacks apple trees, sucking the sap and producing a white wooly secretion, a mass of which may conceal many of these insects. They spend winter in cracks in the bark but in adverse conditions migrate down to roots and produce nodules resembling galls.

Erode: To bear away.

Error: In nomenclature, an unintentional misspelling of a scientific name, as a typographical error or an error of transcription.

Eruciform larva: Larvae or caterpillars with more or less cylindrical body, a well developed head, short thoracic legs, and with abdominal prolegs, e.g. moths and butterflies larvae.

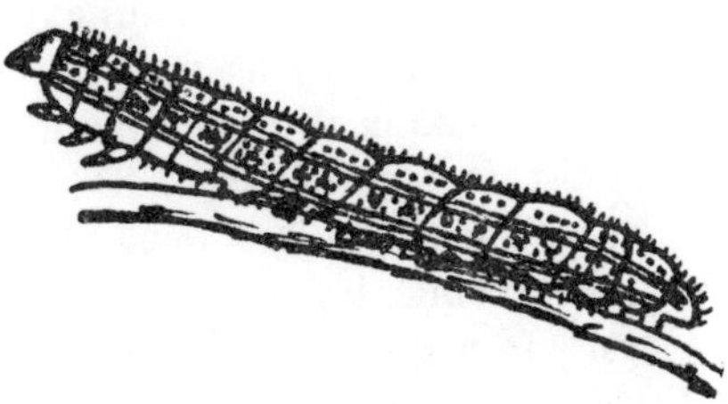

Eruciform (Polypod)larva

Escape: A plant in the treated area that either missed treatment or failed to respond to treatment in the same manner as other treated plants.

Escutcheon: An alternative name for the scutellum, particularly of beetles; a shield-like covering over the thorax.

Esophagus: The narrow portion of the elementary canal immediately posterior to the pharynx.

Essential: Necessary; a must.

Ester: Chemical compound formed by the elimination of water between a molecule of an alcohol and a molecule of an acid.

Estivation: See Aestivation.

Ethion: 0. 0, 0, 0-Tetraethyl-S, S-methylene bis dithiophosphate. It is an organophosphorus insecticide used for the control of mites whiteflies and aphids. It is marketed in India with the trade names of Fosmite and Dhanumit.

Ethology: The science of the comparative study of animal behaviour.

Ethopian region: This geographical region includes, all Africa, south of Sahara plus Madagascar, Southern Arabia and Islands connecting the two.

Ethyl acetate: An insect killing liquid, it acts quickly on most of the insects making them unconscious almost at once, though, it takes rather longer to kill them. It is clear, mobile liquid and escapes easily from corks and stoppers.

Ethylene dibromide: EDB. It is an insecticidal fumigant used against stored products pests and for treatment of fruit and vegetables, and for soil treatment against certain insects and nematodes. It is a colourless liquid, insoluble in water, but soluble in ethanol, ether, and most organic solvents.

Eucaryotes: Organisms with a well-defined nucleus and meiosis. All higher organisms above the level of procaryotes.

Eucone eye: The normal type of compound eye containing true crystalline cones.

Euparol: A compound mountant having a refractive index of 1.483 (canada balsam, 1.536). It gives better contrast with very transparent tissues. It also clears, so can be used straight from alcohol, without the need for clove oil or other clearing agent. It should be thinned with a special medium (euparal essence).

Euplantulae (Tarsal pulvilli): Pad-like structures on the ventral surface of the tarsal subsegments.

Eusternum: The intrasegmental ventral plate of a thoracic segment, exclusive of the spinasternum, but usually including the sternopleurites.

Eutrochantin: See Coxopleurite.

Evagination: An outpocketing or sac-like structure on the outside.

Evaporate: To form a gas and disappear into air; to vapourize.

Evaporation: The process of solid or liquid turning into a gas.

Eversible: Capable of being everted or turned down.

Evolution: Irreversible change in the characteristics of organisms, occurring in successive generations related by descent.

Excavated: Hollowed out.

Exarate pupa: This type of pupa has all the appendages free but folded and the abdomen is movable, e.g. pupa of most Coleoptera, Trichoptera, Neuroptera and some Lepidoptera and Diptera. In some cases such pupa is capable of limited locomotion.

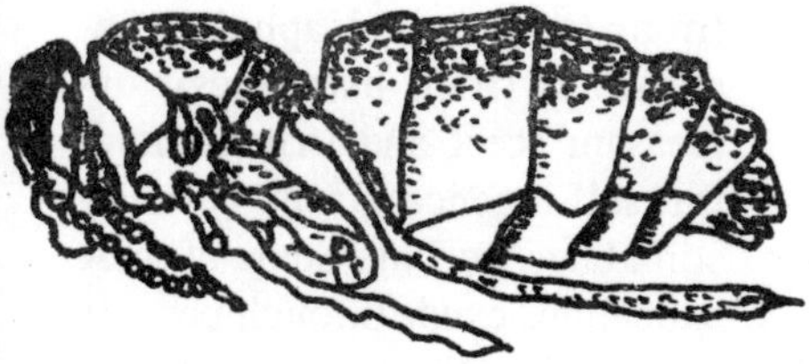

Exarate pupa

Exclusion: The prevention of the distribution and dispersal of a plant pest or pathogen into an uninfested area or the prevention of its establishment there.

Excretion: The elimination of waste products of metabolism. The principal excreting organs in insects are Malpighian tubules which absorbe nitrogenous products from the blood or hemolymph and discharge them into rectum. Some waste is re-absorbed during its passage down these tubes, so that most of the waste substances finally discharged consists of uric acid. The excretion is also carried out by fat-bodies present around and between other organs. Certain cells known as *nephrocytes* in this tissue can abosorb urates and other waste products which are then stored to be eliminated later when the insect pupates.

Exite: An outer lobe of any limb segment.

Exochorion: The thick outer layer of the shell or chorion of an insects egg. various shaped and sculptured and having a small hole or micropyle at one end, through which fertilization takes place.

Exocone eye: A ccmpound eye in which the cones are not true crystalline cones but are formed by invagination of the cornea, e.g. Lampyrid beetles.

Exocuticle: The middle cuticular layer of insect of Orthropoda, outside the endocuticle and beneath the epicuticle. It is very thick and is composed of chitin and proteins impregneted with sclerotin and sometimes melanin.

Exopodite: The external branch of a biramous appendage; an exite of the basipodite in Crustacea, often highly developed, giving the limb a biramous structure.

Exopterygota: (Hemimetabola). Insects in which the wings developing on the out side of the body and there is no pupal stage as in insects with simple metamorphosis. The youngs are known as nymphs which are similar to adults but are smaller and have external wing buds at an early stage.

Exoskeleton: A skeleton or supporting structure on the outside of the body; the external plates of the body wall. It is usually made of chitin impregnated with calcium compounds making it very hard, however, it is thinner and flexible at joints. Ingrowths from the exoskeleton known as *Apodeme* that serve as a point of attachment for muscles.

Exposed surfaces: Surfaces that can be successfully attacked by a pest. Surfaces that need to be protected by a pesticide chemical or other agent.

Exposure: When contact occurs with a pesticide through skin (dermal), mouth (oval), lungs (inhalation/respiratory) or eyes.

Exposure period: The length of time something has been under attack by a pest. The length of time a pest is in contact with a pesticide chemical.

Exserted: Producing or projecting from the body.

External: The outsider; that part away from the centre (midline) of the body.

External genitalia: The reproductive structures concerned with mating and deposition of eggs.

External respiration: The process of transferring the respiratory gases through the body wall; taking place in insect, through the thin areas of the ectoderm either at the body surface or in the wall of evaginations (gills) or invaginations (tracheae).

Exudate: A liquid discharge from diseased or healthy tissue.

Exudations: Secretions or modified excretory products from epidermal glands, including oil (from oil beetles), wax (from bees), silk (from caterpillar) sugary secretions (from termites and aphids), Sex attractant (from butterflies), unpleasant scent and irritant fluids for self-defence (from earwigs, stink bugs, bedbug. Some of these exudations exude through pores in the cuticle, others are discharged from special openings or tubes, e.g. wax in white fly.

Exuviae: (Always used in plural). The cast off skin of an Arhropod; the cuticular parts discarded at moult. Sometimes they remain to form the scales of a scale insect (Coccidae) or the puparium in the case of Diptera.

Exuvial fluid: See moulting fluid (See ecdysis).

Exuvial glands: Glands of the epidemis supposed to secrete the exuvial or moulting liquid.

Eye cap: A structure over hanging or capping the compound eye (Lepidoptera).

Eye compound: See compound eye.

Eyepiece micrometer: A linear scale in the field of vision of the eyepiece (or one of a pair of eyepieces) of a microscope for use as a measuring device.

Eye simple: See Ocellus.

F

Face: The part of the head below the antennae and above the mouth opening.

Face shield: A transparent piece of protective equipment used by pesticide applicator to protect his face from exposure to pesticides.

Facet: One of the lens-like divisions of the compound eye, formed by the cuticular lense covering the crystalline cone. A worker ant usually has a dozen, the house fly has about 4,000 and some dragonflies have 20,000 or more facets. Parasitic insects and those living in caves often have very few facets, or even a single one. When few, the facets are circular, but become hexagonal when tightly packed together.

Faeces: Excrement; The material passed from the alimentary tract through the anus.

Facilitate: Make it easier to do or to understand something.

Facultatism: Denotes a capacity to change their life habit at will; mode of life developed by organisms for their survival.

Facultative: Incidental; not necessarily compelled to live under one type of environment.

Fahrenheit (F): A thermometer scale that marks the freezing point of water at 32 degrees F. and the boiling point at 212 degrees F. Fahrenheit can be converted into celsius by -32, 5/9 x.

Falling reflex: A reflex action in insects by which a falling insect will, in a short distance, right itself so that it will either land on its feet or adopt a suitable position for flying. Legs and antennae are spread out so that the effect of air currents on these gives the insect the appropriate stimulus to adjust its position.

Fallow: Land which is allowed to lie idle during the growing seasons.

False legs: Prolegs or pseudopods. Fleshy pad-like outgrowth in larvae of insects from the abdomen. Caterpillars of butterflies and moths usually have five pairs, those of sawflies seven pairs. Prolegs are used for gripping. The prolegs of butterflies and moths bear rows of tiny hooks and those of sawflies are without these. Dipterous larvae have false abdominal legs, on ventral surface only and used for walking on surface, or in rings all round the body and used for pushing through narrow space (e.g. maggots).

False spider mites: These mites come under the family Tinuipalpidae. The mites are generally reddish, slow moving, dorso-ventrally flattened, minute usually varying from 200 μ to 300 μ in length. Many species are of economic importance.

Family: A taxonomic category including one genus or a group of genera or tribes of common phylogenetic origin, which is separated from related similar units (families) by a decided gap, the size of the gap being in inverse ratio to the size of the units (family).

Family name: The scientific designation of a taxon of family rank, recognised by the termination 'idae' which termination may not be used in names of other taxa.

Fastigium: The anterior dorsal surface of the vertex (grasshoppers).

Fatal: Deadly; able to cause death.

Fat-body: The masses of fatty tissues containing cells usually distributed throughout the body cavity. The food is stored in this tissue includes fat, protein and glycogen which form a reserve for hibernation or pupation.

Fat cells: One of the component cells of the fat body.

Fauna: The animal life a region.

Faunal work: A publication in which taxa are included on the basis of their occurrence in a specified area rather than on the basis of relationship.

Fecifork: A dung fork; A flattened fork-like appendages, also called an anal comb, attached to the hind end of the abdomen on the larvae of certain beetles and moths. It can be elevated and is used for shovelling up dung which is then deposited on the back of the larva to serve as camouflage and for protection from the sun.

Fecundity: Egg laying capacity of an insect. A queen bee may lay 2000 eggs a day; a queen termite lays 60 eggs a minute with a total of several millions. The parthenogenetic female aphids may produce many millions offsprings during a season.

Feelers: See Antennae and palps.

Female reproductive system: This system of insects consists of the two ovaries (sometimes of great size, complexities), these lead into oviducts which join to form a median oviduct or vagina, opening in the 9th abdominal segment. A pair of accessory glands or colleterial glands are open in vagina, and many secrete a silky substance with which the eggs are attached to a surface. However, in cockroaches, they secrete a cotheca in which the eggs are filled in. In most cases, a sac or spermatheca is present leading dorsally from the median oviduct or vagina which receives seminal fluid from the male, store it, and releases it over the eggs just before they are laid. Various modifications take place in different groups of insects.

Femur: Pl. Femora. The third segment of the telopodite, usually the principal and largest segment of the insect leg. It is situated between the trochanter and tibia.

Fenestrae: The degenerated compound eyes of the cockroach. They are represented by ocellar spot.

Fenitrothion: See Sumithion.

Fenthion: Baytex; Lebaycid; 0,0 dimethyl-0 (4-methyl mercapto-3 methyl phenyl) thiophosphate. It is a fast killing organophosphate insecticide with long residual action and it approaches chlorophos in its activity.

Fenvalerate: Sumicidin; Triumphcard. A synthetic pyrethroid insecticide recommended for use @ 25 to 100 g a.i./ha against the leaf defoliators, sucking pests, leaf folders, pod borers, shoot and fruit borers, etc. of various crops, waiting period is recommended to be 2 to 21 days.

Ferric oxide: It is used as a corrective in the formulations of arsenical insecticides to prevent arsenical injury.

Fertilization: The process of union of two germ cells.

Field cricket: A name for a number of crickets (Gryllidae) that live in burrows and feed on grasses and also on various agricultural crops. These are generally black with a band of yellow across the base of the elytra. The hind wings are shorter than those of the house cricket and are covered by the leathery forewings called elytra.

Field margin: Edge of a field.

Field Rat: *Rattus maltada*. The subspecies *R.m. pallidior* occurs in northern and eastern India, from Rajasthan to Nepal and from the Punjab to Gujarat, whereas *R.m. maltada* inhabits Bihar, Madhya Pradesh and southern India. The field rat usually damages irrigated crops.

Fig moth: See Almond moth.

Filament: A slender thread-like structure.

Filariasis: A disease caused by infection with certain round worms (nematodes) transmitted by mosquitoes. There is overgrowth of skin and tissues usually of leg and scrotum; also called elephantiasis.

File: A file-like ridge on the ventral side of the tegmen near the base; a part of the stridulating mechanism in crickets and long-horned grasshoppers.

Filiform: String-like or thread-like or hair-like with uniform segments; filiform antennae, e.g. ground beetle.

Filiform antenna

Filler: A diluent or innert material in powdered pesticide.

Filter: To screen out the unwanted material.

Filter apparatus: Finely branched processes of the atrial wall of some spiracles, forming often two thick but airpervious mats, just within the atrial orifice.

Filter chamber: A part of the alimentary canal in Homoptera in which the two ends of ventriculus and the beginning of the intestine are bound together in a membranous and muscular sheath.

Final treatment: Last pesticide chemical application before harvesting a crop or slaughtering an animal.

Fine dusts: Dusts having particle size less than 44 μm.

Finite tolerance: The maximum amount of pesticide which can legally remain on a food or feed crop at harvest after the pesticide has been directly applied to the crop.

Fire bug: A general name for the plant bugs of the family Pyrrhocoridae. They include red cotton bug or cotton stainer *(Dysdercus sp.)* which stains cotton fibres bright red by introducing fungi when it pierces the bolls. The fiery colours of most of these insects are probably a warning colouration, as they are distasteful to predators.

Fire fly: A name given to many luminescent insects but particularly to beetles of the family Lampyridae. These insects bear luminous organs on the 6th and 7th abdominal segments. Here a specialized fatty tissue is permeated by tracheae and a substance called luciferin is oxidised with the help of an enzyme, luciferase. This produces a cold light or chemo-luminescence. In some cases both sexes and in others only females are luminous.

First aid: Immediate assistance given in the case of injury or sudden illness by a by stander or other layman before the arrival of the physician.

First axillary: The anterior hinge plate of the wing base, associated with the base of the subcostal vein.

First maxillae: The second pair of appendages of the gnathal region of the head; in insects called simple maxillae.

First median plate: A small sclerite of variable shape lying in the angle between the second axillary and the distal arm of the third axillary at the base of the mediocubital field; accessory to the third axillary in function, and usually attached to it.

First thoracic spiracle: The spiracle of the mesothorax, often displaced into the posterior part of the prothorax.

Fishing fly: A name given to both may flies (Ephemeroptera) and Alder flies (Megaloptera).

Flabellate: With fan-like processes or projections; flabellate antennae has terminal segments sheet-like,

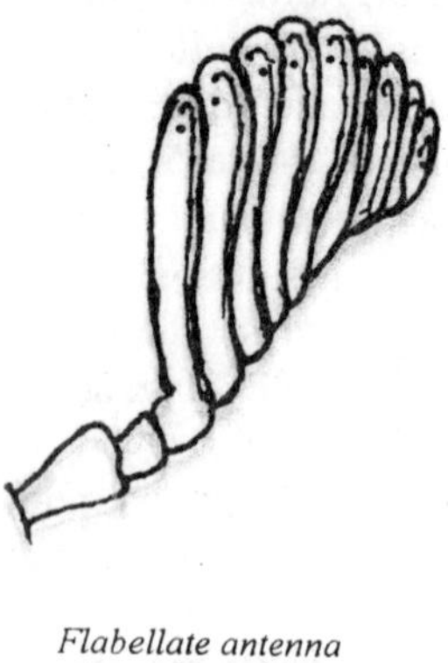

Flabellate antenna

tongue-like lobes, e.g. ceder beetle.

Flabellum (Pl. Flabella): A fan-like or leaf-like process (In Hymenoptera).

Flaccid: Plant tissues becoming limp.

Flagellum: The part of the antenna distal to the pedicel, typically filamentous, but of various forms, usually subsegmented or multiarticulate.

Flash point: It is the lowest temperature at which a combstible liquid will give off a flammable vapour which will burn momentarily.

Flea beetles: A group of small jumping beetles of the family Chysomelidae and the subfamily Halticinae, characterized by having thickened femora of the hind-legs. The larvae generally feed on the roots of plants and the adults attack young shoots and leaves.

Flexible: Easy to bend.

Flight: The movement of an insect wings during the flight is brought about by two sets of muscles (vertical and longitudinal muscles present with in the thorax). The vertical muscles pull the roof of the thorax and force the wings up and the longitudinal muscles drop the thorax and force the wings down. These muscles work automatically in turn by means of a special type of muscle fibers. In addition to these there are muscles attached to wing bases, giving a turning movement to the wings at each up and down stroke. These are under direct nervous control and thus enable the insect to move in any direction.

Flightless insects: Most insects bar ring few have two pairs of functional wings, but some are also wingless, they may categorized as wingless insects. (a) Springtails and bristle tails (primitive insects of aptereygota) have never evolved to the stage of having wings. These insects belong to the order Diptura, Thysanura, Protura and Collembola. (b) The ants and termites winged and wingless castes are found. Some aphids or stages of aphids are wingless. (c) The fleas, lice, certain bugs and flies are secondarily flightless, i.e. they have evolved from the winged insects but their mode of life has made the presence of wings a disadvantage. In some cases they may possess vestigial structures.

Flowability: It is an indicator of the feed rate of dusting equipment. The angle of slope is a measure of flowability.

Flowable pesticide: Very finely ground solid material which is suspended in a liquid and usually contains a high concentration or large amount of the active ingredient and must be mixed with water when applied.

Flucythrinate: Pay-off; fuching juji; cybolt. A synthetic pyrethroid insecticide, recommended for use against various insect pests of crops @ 30 to 50 g a.i./ha.

Fluid: In a liquid form.

Fluorescent tracer: Fluorescent material added to a spray to aid the assessment of spary deposits on plants.

Fluorine: Elemental fluorine as a gas at 100 to 1000 ppm is effective against a number of pests.

Fluted scale: An alternative name for the cottony cushion scale.

Fly: An insect of order Diptera, usually with one pair of wing.

Flying hairs: A name sometimes used for certain long hairs present on the wing bases of carabid beetles.

FMC (Marshal): 2, 3, dihydro-2, 2-dimethyl-7-benzofuranyl (di-n-butyl)-aminosulfenyl (methyl) carbamate. A carbamate insecticide, having broad spectrum systemic and contact action.

Foaming agent: A material which causes a pesticide mixture to form a thick foam. It is used to reduce drift.

Fog: Perticles between 0.1 and 0.5 m (Microns) in diameter which make a fine mist.

Fogger: An aerosol generator. A piece of pesticide equipment that breaks some pesticides into very fine droplets (aerosols or smokes) and blows or drifts the fog into the target area.

Fog treatment: The application of a pesticide as a fine mist for the control of pests.

Foliage: The leaves, needles and blades of plants.

Foliar spray: Droplets of a pesticide applied to leaves, needles, and blades of plants.

Folithion: See Sumithion.

Follicle: A minute cavity, sac or tube.

Follicle cells: The inner epithellial cells of an ovarian egg tube.

Fontanelle: A small depressed, pale spot on the frons of the head between the eyes (In Isoptera).

Foraging ants: A name for a number of genera of tropical ants that have no permanent nests but march in large numbers in search of food. Generally they travel in dense columns foraging on either side and driving out or killing all animal life. They drag or carry the queen, the eggs, the larvae and pupae as they go.

Foramen magnum: The opening on the posterior surface of the head to give passage to those structures that extend from the head into the thorax.

Forcipulae: See Forceps.

Forceps: Pincers or claspers at the hind end of the abdomen of certain insects, either for copulation or for defence, e.g. earwigs (Dermaptera).

Foregut: The anterior part of the alimentary canal lined by the ectoderm. It includes, buccal cavity, pharynx, cesophagus, crop and gizzard (See Alimentary canal also).

Forewings: The first pair of wings in insects are present on the mesothorax (middle segment of thorax). In Diptera, these are functional wings, the second pair being reduced to halters. In beetles and cockroaches, these wings are thickened and horny and called Elytra and in grasshoppes and locusts, these are called Tegmina.

Form: A neutral term for a single individual, phenon or taxon.

Formenkreis: A collective category of allopatric subspecies or species; in paleontology, a group of related species or variants.

Formothion: See Anthio.

Formula: A brief of writing a complicated idea by using abbreviations and symbols.

Formulation: The preparation of the active ingredient in a form suitable for use is referred as formulation.

Fossorial: Modified for digging; in the habit of digging or burrowing.

Frankliniella: A minute fringed winged insect related to the thrips. It is a pest of flowers in temperate regions.

Frass: Plant fragments made by a wood boring insects, usually mixed with excrement.

Free pupa: See Exarate pupa.

Frenate: A group of Lepidoptera comprising of the majority of moths. These insects have frenulum in wings.

Frenulum (Pl. Frenuli): A single spine (in male) or several (in female) arising from the humeral angle of the hind wings and projecting forward under the fore wings and often held here in frenulum hook. Thus frenulum interlocks hind wings with the forewings during flight.

Frequency of wing beat: Generally the large insects such as butterflies move their wings slowly and tiny insects such as gnats and midges very rapidly. The following are the estimated wing frequencies per second of some common insects: Butterflies-10; longer moth-50 - 70; bees and houseflies - 200; mosquitoes-600; small midges 1000, or more.

Frons: The facial area of the cranium between the frontal and epistomal sutures or the corresponding area when the sutures are absent, bearing the median ocellus and origins of the labral muscles. Generally loosely used between the ocelli and the antennae.

Front: That portion of the face between the antennae, eyes and ocelli; the frons.

Frontal bristles: Bristles above the antennae, away from the edge of the compound eye (Diptera).

Frontal ganglion: The median cerebral ganglion of the stomodaeal system, just in front of the brain and connected to the ventricular ganglion on the hind part of the oesophagus.

Frontal ganglion connectives: The connectives between the tritocerebral ganglion and the frontal ganglia.

Frontal Gland: A gland in the front of the head from which poison is discharged from certain insects, e.g. termites.

Frontal lunule: A small crescent shaped sclerite located just above the base of the antennae (Diptera).

Frontal pore: The pore of the frontal gland of a termite or other insects.

Frontal Sac: The ptilinum; a sac or vesicle on the head of a fly by which the insect pushes open the puparium as it emerges. Afterwards the sac is withdrawn into the head leaving a U shaped line called the *ptilinal suture,* enclosing a crescentric structure, the *lunule.*

Frontal suture: One or two sutures arising at the anterior end of the coronal suture and extending ventral towards the epistomal suture; a suture shaped like an inverted U, with the base of the U crossing the face above the bases of the antennae and the arms of the U extending downward on each side of the face (Diptera). The form is ambiguous and therefore be avoided.

Frontal vitta: An area on the head between the antennae and ocelli (Diptera).

Fronto-clypeus: A cuticular plate formed by fusion of the frons and the clypeus on the front next to the compound eye (Diptera).

Fossorial or Digging type leg: These legs are modified for digging purposes, being exceptionally powerful, broad and small in size. Each of the coxa, trochanter and femur is very strong and broad while the tibia is provided with four strong scraper-like prolongations at its broad distal end which are utilized for digging the ground or even cutting the roots of the plants. The tarsus is composed of two or three segments, each segment projecting like a tooth, and helps the tibia in digging, e.g. forelegs of mole cricket.

Fourth axillary: A postierior hinge plate of the wing base present in some insects.

Fruit borers: A name for a number of insects that lay their eggs on or in fruits and whose larvae bore into the fruits. They belong to many different orders and include fruit flies, sawflies, weevils, beetles, moths and butterflies.

Fruit sucking moth: *Ophideres* spp. (Lepidoptera: Noctuidae). It is a minor pest of the citrus, mango, grapes, and apple and is distributed throughout India. *O. conjuncta* Cramer, *O. materna* Cramer and *O. ancilla* Cramer are the commonest species found in Tamil Nadu, Madhya Pradesh, Western Uttar Pradesh and the Punjab. The moths are large and stout with piercing mouth parts.

Fuchsin: Acid or basic fuchsin are dyes brought in solid state, and are used in solution, generally in 20% alcohol.

Fulgoridae: It consists of large brilliantly coloured tropical plant bugs of the group Homoptera. The head is curiously elongated and rounded for which reason they are sometimes called 'Alligator bugs'.

Full bibliographical synonymy: A reasonably complete list of references to a given taxon arranged so as simultaneously to serve the needs of nomenclature (chronology of names) and zoology (pertinent taxonomic and biological sources).

Full coverage: The total amount of pesticide spray applied will thoroughly cover the crop being treated to the point of rub-off, or drop.

Fuller's earth: A silicate mineral used as an adsorbent base or dust diluent. It is flowable and possesses excellent grindability.

Fulturae: See Suspensorium of hypopharynx.

Fumes: A smoke; vapour; gas.

Fumigant: The substance used in the form of vapour and aerosol enters the body through the respiratory system used for destroying microbes and insects.

Fumigation: It is the use of a toxic chemical in gaseous form.

Funicle or Funiculuns: The antennal segment between the scape and the club.

Fundatrigeniae: Wingless parthenogenetic female aphids that later give rise to winged sexual forms.

Fundatrix: The first wingless female aphid of a season, later giving rise to numerous parthenogenetic females and eventually to winged sexual forms.

Fungus cultivation: Both ants and termites cultivate fungi as food; workers carry pieces of leaves, etc. into the nest, chew them and mix them with saliva. This forms a medium for the growth of certain specific fungi which have apparently evolved in associations with ants. A queen founding a new colony will take a pellet of this fungi into the new nest; in the case of termites, the leaves are taken in and fungi that grow on these are used as food, but this is an accidental phenomenon only.

Furadan: Carbofuran; 2, 3, -dihydro-2, 2, dimethyl-7-benzofuranylmethyl carbamate. It is systemic carbamate insecticide of broad activity.

Furca or Furcula: The forked endosternal process of higher insects, formed of the sternal apophyses supported on a median inflection of the thoracic sternum; any forked-like structure. The term furcula is generally used to the forked tails of collembola.

Furcasternum: A distant part of the sternum in some insects bearing the furca (This term is generally applied to the sternellum).

Furrow application: Placement of pesticides with seed in the furrow at the time of sowing.

G

Gadfly: The blood sucking flies which attack cattle, horses and other animals particularly belonging to Tabanidae family of Diptera order. The females suck the blood of horses and cattle causing sores. Some species are carriers of Trypanosomes and other parasites. The predatory larvae live mainly in wet situations, in mud or under shallow water.

Gaead: A young of Ametabola. The larvae of these insects undergo no metamorphose. The young ones are called larva, or nymphs, but these terms are inappropriate.

Galea: The hood-like outer endite lobe of a maxilla, provided with a muscle arising in the stipes.

Gall: An abnormal growth of plant tissues caused by the stimulus of various organisms which irritate the plants. The main gall causing insects belong to the Hymenoptera-(Gall wasps and some sawflies), Diptera-(Gall midges), Hemiptera-(Aphids, coccids), Lepidoptera-(The larvae of certain moths). These are also sometimes formed by the mites, elworms. The open galls are produced by the aphids, coccids, etc. due to the feeding from the outside causing leaf fold over and make a pocket in which it continues to feed. Closed galls are formed by the larvae of beetels, wasps, moths or flies which deposit their eggs beneath the epidermis of the host and the larvae bore into the tissues. The gall completely surrounds the larva and grow alongwith the size of the larva.

Galleriidae: A family of Lepidoptera order which includes Honeycomb moth *(Galleria mellonella)*. The moths are small, brown, or grey whose larvae infest beehives and feed on pollen and wax and causes considerable loss of honey. Some species also attack wasp's nests and others are scavengers.

Gall gnat: See Cecidomyiidae.

Gall midge: See Cecidomyiidae.

Gamete: A reproductive cell; an egg or sperm whose union in sexual reproduction initiates the development of new individuals.

Gamma BHC: It is one of the eight isomers of BHC. Lindane is a purified form of BHC containing 90% pure isomer.

Gamma taxonomy: The level of taxonomy dealing with various biological aspects of taxa, ranging from the study of inraspecific population to studies of speciation and of evolutionary rates and trends.

Ganglion: Pl. Ganglia. A knot like enlargement of a nerve, containing a coordinating mass of nerve cells.

Gardona: 2-chloro-1(2,4,5-trichlorophenyl-vinyl-dimethyl phosphate). It is a low toxicity organophosphorus insecticide with contact action and having moderate persistence.

Gas mask: A device which filters out chemicals in the spray, dust or gas form from air breathed by the bearer.

Gasolene: It is one of the lowest boiling petroleum fractions and is seldom used as insecticide because of its high volatility and fire hazards.

Gaster: The rounded part of the abdomen posterior to the node-like segment or segments (In ants).

Gastric: Pertaining to mid-gut.

Gastric caeca: See Enteric caeca.

Gastrocoele: The gastrulation cavity.

Gastroenteritis: The inflamation of the membrane of both stomach and intestine.

Gastrotheca: The part of an insect's pupal case covering the abdomen.

Gastrula: The embryo after gastrulation.

Gastrulation: The formation of endoderm, either by invagination of one wall of blastula or by internal proliferation of cells from blastoderm.

Gause's rule: The theory that no two species with identical ecological requirements can coexist in the same place.

Gelechiidae: A family of Lepidoptera order. The moths are small with prominent palps and with fringed hind wings. Some larvae make galls, others tie the terminal buds and leaves of plants by silken threads secreted by the labial glands, e.g. Angoumois grain moth (*Sitotroga cerealella*), Soybean leaf minor (*Aproaerema modicella*).

Gena: Pl. Genae. The areas of the head below and behind the eyes; between the frontal and occipital suture; the cheeks.

Genal comb: A row of strong spines borne on the anteroventral border of the head (In Siphonoptera).

Genathocephalon: The part of the head formed by the gnathal segments bearing the mandibles and the first and second mexillae.

Gene: A hereditary determiner; the unit of the inheritance, carried in a chromosome, transmitted from generation to generation by gametes, and controlling the development of the individuals; a particular segment of a DNA molecule generally located in the chromosome.

Gene frequency: The percentage of a given gene in a population.

Gene pool: The totality of the genes of a given population existing at a given time.

General equilibrium level: The average density of a population over a lengthy period of time in the absence of permanent environmental change.

Genetic drift: Genetic changes in populations caused by random phenomena rather than by selection.

Generation: A group of individuals of a given species that has been reproduced at approximately same time.

Genetic manipulation: The use of a pest species for its own destruction by introducing undesirable or incompatible genetic traits that reduce its fitness to survive and by mass sterilization techniques; e.g. autocide, chemosterilization.

Geniculate: An organ that is bent in the middle like a knee or elbow or abruptly bent; geniculate antenna of an insect.

Genital chamber: In the female, primarily an invagination cavity behind the eighth abdominal sternum containing the gonopore and spermathecal aperture, often coverted into a vagina or uterus, and in some insects opening secondarily in the ninth sternum containing the intromittent organ.

Genitalia: The external genital organs and associated structure, but usually applied only to the external genitalia (sexual organs). It is made up of following parts: (i) External male organs—These are on the ninth abdominal segment and oconsist of a pair of claspers or gonopods for grasping the female at mating; a pair of parameres, the penis or intromittent organ from which the gonopore opens. (ii) Internal male organs—Two testes made up of number of follicles, a vas defference or genital duct leading from each testis and both sides make single ejaculatory duct, a number of seminal vesicles for holding seminal fluid and a number of accessory glands. (iii) External female organs—An ovipositor with three pairs of valves from eighth and ninth abdominal segments. These may interlock or may be permanently united to form a tubular egg laying device. In bee and wasp workers the ovipositor is modified to form a sting. (iv) Internal female organs—Two ovaries each madeup of ovarioles, two oviducts, meeting inside the middle to form a main oviduct leading to vagina, several accessory glands secreting a glue-like substance, a spermatheca for storing and receiving a fluid. There may be many variations and adaptations in different insects.

Genital ridge: One of the embryonic gonadial rudiments, a ridge-like swelling of the splanchnopleure wall of the mesoderm containing the germ cells.

Genital segment: Specifically the ninth segment of the abdomen in the male though the other segments are frequently associated with the ninth in the genital complex.

Genotype: The genetic constitution of an individual or taxon.

Genovertical plate: An area on the head above the antenna, next to the compound eye (Diptera).

Genus: Pl. Genera. A taxonomic category including one species or a group of species, presumably of same phylogenetic origin which is separated from related similar units by a decided gap, the size of the gap being in inverse ratio to the size of the unit (genus); a group of related species; the first name in a binomial or trinomial scientific name of genera which are latinized, capitalized, and when printed are italized.

Geographical race: Subspecies.

Geographic isolate: The separation of a gene pool by geographic barriers; the prevention of gene exchange between a population and other by geographic barriers.

Geometridae: A family of Lepidoptera consisting of looper caterpillars. These loopers move by looping action owing to the absence of abdominal legs or prolegs, except at the hind end.

Geotaxis: A tendency to move towards the earth or into the earth (positive geotaxis) or away from the earth (negative geotaxis). The larvae of some insects go upward towards food and go downward or underground as thy have fed.

Germarium: The end chamber of an ovariole (ovary) or testicular tube (testis) containing the primary oogonia or spermatogonia, respectively. The ovary of an insect usually consists of a number of ovarioles, each containing a germarium and a vitellarium, the later produce the yolk. A testis consists of number of tubules of follicles with a germarium near the apex of each.

Germ band: The area of thickened cells on the ventral side of the blastoderm that becomes the embryo (Also called embryonic rudiment, germ disc, primitive streak, keimstreif, keimscheibe, bandelette primitive plague ventrale, paistra germinatriva).

Germ cell: Cells designed to become ova or spermatozoa, differentiated from the somatic cells during the cleavage; the early undifferentiated reproductive cells.

Germ tract: The cytoplasmic area of the blastula containing the germ cells.

Ghujhia weevil: *Tanymecus indicus* Faust (Coleoptera: Curculionidae). It is widely distributed in India and is a sporadic pest of considerable importants feeding on germinating rabi crops, particularly wheat, barley gram and mustard. The adult weevils cut the germinating seedlings at the ground level. Weevils are earthen grey and measure about 6-8 mm in length and 2-4 mm in width.

Giant African Snail: *Achatina fulica.* This is the largest known land snail with a maximum shell length of about 19.68 cm. It is a native of East Africa, but now has reached many parts of the world. It is the most important land snail pest known to man. The pest damages a large number of crops in main land and Andaman and Nicobar Islands. 5% metaldehyade pellate and 2% metaldehyde dust have been reported effective against the pest.

Giant silkworms: Insects that come under family Saturniidae of order Lepidoptera. These include some of the world's largest and most brightly coloured moths whose larvae spin siken cocoons, many of which are used commercially, e.g. *Philosemia cynthia.*

Gill: Evaginations of the body wall or hindgut, functioning in gaseous exchanges in aquatic animals. These are thin walled outgrowths of the abdomen in the form of lamellae or filaments. In May fly (Ephemeroptera) nymphs, paired lateral gills on nearly all abdominal segments; in the nymphs of Damselflies (Zygoptera), three elongated caudal gills; in Dragonfly nymphs (Anisoptera) a row of gills projecting into the rectum are present. These gills are supplied with tracheae and therefore are known as tracheal gills. In mosquito larvae, blood gills are present which absorb the dissolved salts and the respiration takes place over most of the body surface.

Gizzard (Proventriculus)**:** A part of the foregut immediately behind the crop. It is poorely developed in sucking insects. In chewing insects, it is equipped with powerful muscle and teeth with which the harder particles of food can be crushed.

Glabrous: Smooth; without hairs.

Gland: A cellular organ which secrets certain characteristic product such as saliva, silk; hormones.

Glandular hormones: They are phylogenetically the highest type of hormones. They originate in special ductless glands of which hormones secretion is the primary or sole function (e.g. moulting hormone and juvenile hormone secreted by the prothoracic glands and corpora allata, respectively).

Glia tissue: The cellular supporting tissue of the nervous system.

Globose (Globular): Spherical or nearly so.

Globuli cells: Specialized association cells of the brain, usually distinguished by their small size, poverty of cytoplasm and rich chromatic nuclei.

Globulin: One of the class of proteins in blood plasma, some of which (gamma-globulins) function as antibodies.

Glomerulus: A small compact mass of intermingled terminal arborizations of nerve fibers within a nerve center.

Glossa (Pl. Glossae): The two median lingular lobes of the labium between paraglossa, each provided with a muscle arising in the prementum.

Gnathal segments: The segments of the insect embryo, the appendages of which become the mandibles and first and second maxillae.

Gnathochilarium: A plate-like mouth part structure in Diptera, representing the fused maxillae and labium.

Gnorimoschema operculella: Commonly known as potato tuber moth, comes under family Gelechiidae and order Lepidoptera. The moths have narrow wings fringed with long hairs. The larvae known as 'tuber worm' are pest of potato, tobacco and also tunnel into potato tubers throughout word.

Golden nematode of potato:

Heterodera rostochiensis Wollenweber (Tylenchida: Heteroderidae). This nematode is wide spread in Europe, the America, Algeria, Israel and Japan. It has recently been recorded in India in the Nilgiri hills and near Simla. It is a serious pest of potato but also found feeding on tomato, brinjal, chillies and some other solanaceous weeds. The infested plants, when pulled out, show clusters of white and golden spherical bodies. These are the cyst of mature female bodies. The other eelworms attacking potatoes include *Meloidogyne incognita* and *M. javanica*.

Gonad: The ovary or testis, or the embryonic rudiment of either, formed of splanchnic mesoderm cells enveloping the germ cells.

Gonapophyses: Median proximal processes of the coxopodites of the gonopods, present in the male in some Thysanura.

Gonopore: In the female either one of the paired primitive openings of the lateral oviducts, or the main opening of the oviductus communis; in the male the external opening of the median ejaculatory duct usually concealed in the endophallus or one of the apertures of paired exit ducts.

Gonosomite: See genital segment.

Gonostyli: The styli of the ninth segment, when present, generally modified to form clasping organs.

Grade: In insect systematics a group of animals similar in level of organisation.

Gradient growth: In this type of growth, only certain parts of the body, the gradients, grow. Growth of the remaining parts is stopped and eventually they are wholly or partly removed either by histolysis or phagocytosis. This occurs, for example, in the embryogenesis of most animals and in the metamorphosis of insects.

Grain beetle: See saw-toothed grain beetle.

Gram: A metric weight measurement equal to 1/1000th of a kilogram.

Gram pod borer: *Heliothis armigera* (Hubner) (Lepidoptera: Noctuidae). It is cosmopolitan and widely distributed in India, and is serious pest of gram and red gram. It also feeds on cotton, sorghum, okra, maize. The moth is stoutly built and is yellowish brown. The caterpillars cause damage.

Granary: A storage area for threshed grain.

Granular pesticide: A pesticide chemical mixed with or coating small pellets or a sand-like material. They are generally applied with seeders, spreaders, or special equipment.

Granules: These consist of free flowing granular basic materials impregnated with an insecticide. Basic materials commonly used are clays, bentonite, etc.

Grapevine beetle: *Sinoxylon anale* Lesne (Coleoptera: Bostrychidae). It is distributed in France, Italy, U.S.S.R., Japan, China and India.

It also feeds on sal, teak, shisham, wooden packing cases. The adult is sturdy and a very lively beetle that walks slowly and flies rarely. The diagnostic three bladed antennae and a pair of spines, on the posterior elytral extremety are present. The full grown larvae of the beetles are thick, yellowish, white curbed grubs, often found feeding alongwith the adult beetles.

Grapevine girdler: *Stenias grisator* Fabricius (Coleoptera: Cerambycidae). It is a serious pest of grapevine throughout India. Besides grapes, it feeds on rose-bushes, mulberry, various garden shrubs, creepers, crotons, mango, almond etc. Girdling by the adults is followed by the drying-up of the wood.

Grapevine leafhopper: *Erythroneura* sp. (Hemiptera: Cicadellidae). The adults and the nymphs of this pest suck the cell sap from the lower surface of the leaves. In heavy infestation, the leaves turn yellow and fall from the vines and the quality and quantity of the fruits are greatly affected.

Grapevine leafroller: *Sylepta lunalis* Guenee (Lepidoptera: Pyralidae). It is a serious pest of grapevine during August to October. Only the caterpillars cause damage by defoliating the leaves. They are green, and the adults are dirty brown with white spots on the fore and hind wings.

Grapevine thrips: *Rhipiphoroth-rips cruentatus* Hood (Thysanop-tera: Heliothripidae). This is most destructive pest of grapevine in India and is also found feeding on rose, Jamun, and aak. Both adults and nymphs cause damage by sucking the cell-sap from the leaves. The adults are minute 1.4 mm long, blackish brown, with yellowish wings. The nymphs are yellowish brown.

Grasshopper: Any of the jumping members of the order Orthoptera. These are straight winged hemimetabolous insects with biting mouth parts and having the head bent at right angles to the body. The forewings are thickened and called Tegmina which cover the membranous hindwings. Some species are wingless throughout their life. They produce noise either by rubbing the hind femur against the tegmen or by rubbing the two tegmina together. The grasshopper which are non-migrating are called true grasshoppers but the migrating grasshoppers are called locusts.

Greater wax moth: *Galleria mellonella* Linn. (Lepidoptera: Galleriidae). The moths of this pest attack honey bee comb or beewax in bee hives or in stores. Two species, viz. wax moth, *G. mellonella* and lesser wax moth, *Achiora grisella* Fab. are common. Of the two, the former one is more important all over the world and in India, it is a serious threat to bee keeping industry in plains and lower hills. Damage is caused by the caterpillars which feed on old combs, propolis, pollen, larval skins, and other proteinaceous matter. A full grown larva is dirty grey with a brownish head. The moth is brownish grey.

Gregarious: Living in groups.

Gregarious phase: A name given to the stage in the life of a locust when millions of them swarm together and migrate. This may occur after several generations of solitary phase during which they behave like grasshoppers. Perhaps due to overcrowding, there comes a change in their appearance and habits. In the solitary phase these are light green, grey or brown but in gregarious phase they turn to boldly pattern in yellow and black. In gregarious phase they fly off in huge swarm to find new feeding and/or breeding grounds, e.g. Desert locust.

Groundnut aphid: *Aphis craccivora* Koch (Hemiptera: Aphididae). This is most important pest of groundnut and leguminous crops such as lentil, lathyrus, lab-lab, etc. It has been recorded from Africa, Argentina, Chile, USA, Europe and Australia. The colonies of the aphid present underside of leaves and on tender stem suck the cell sap resulting into wilting of the crop. It also spreads virus disease known as the rossette of groundnut.

Groundnut leaf miner: *Aproaerema (Stomopteryx) modicella, A. nertaria* Meyrick (Lepidoptera: Gelechiidae). It is a serious pest of groundnut and soybean in various states of India and also found in Pakistan, Sri Lanka, Burma, and south Africa. The tiny caterpillars mine into the leaves, skeletonize them and web them together. The adult is small bronze moth with a wing expanse of about 1 cm.

Groundnut stem borer: *Sphenoptera perotetti* G. (Coleoptera: Buprestidae). It infests groundnut crop in Andhra Pradesh, Bihar, Delhi, Gujarat, Kerala, Tamil Nadu, Madhya Pradesh, Maharashtra and Karnataka. Besides groundnut it also attacks sesamum, gram and some pulses. The grub of this beetle bore into the stem and the root. The adult is a small jewel-like beetle with a striking metallic shine over dark brown background. The full grown grub is whitish in appearance.

Group: In systematics, a neutral term for a number of related taxa, expecially an assemblage of closely related species within a genus.

Growth: It involves increase in size by increase in cell size, cell numbers or both.

Growth inhibitor: A substance that inhibits the growth.

Grub: A scarabaeiform larva; a thick bodied larva with a well developed head and thoracic legs; without abdominal prolegs, and usually slugish. The body is fleshy and rounded with little difference

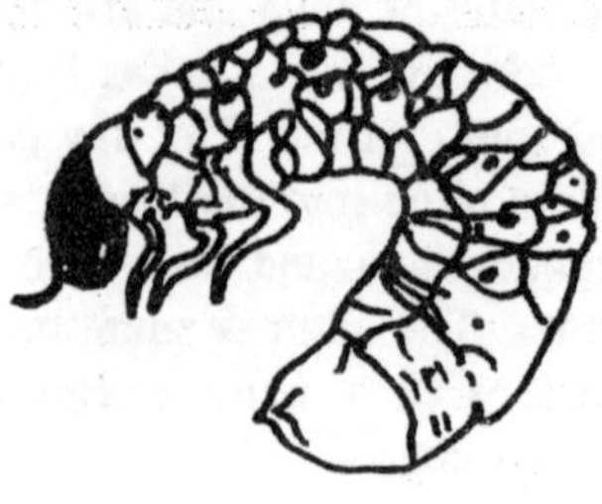

Grub

between the two ends. They may be distinguished from the more pointed legless maggots of the fly.

Gryllidae: It includes crickets, members of the order Orthoptera resembling grasshopers but distinguished from them by their brown colour, by long unsegmented anal cerci, and by their less predominently vegetarian habits. *Gryllus domesticus* is a common house cricket whose chirping can be heard in houses and warm places.

Gryllotalpidae: It includes mole crikets and comes under order Orthoptera. These insects have large flattened forelegs modified for burrowing. They live generally underground, feeding on worms, beetles and other insects but also sometimes gnawing roots. At night they may fly clumsily with their transparent wings fluttering rapidly and their heavy body hanging vertically.

Gryllus: The principal genus of crickets. (See also Gryllidae).

Guanine: A purine base occuring in DNA and RNA. Pairs normally with cytosine in DNA.

Gula: A median ventral cuticular plate of the head in some prognathous insects formed by sclerotization of the neck region proximal to the posterior tentorial pit, continuous with the postmentum or submentum.

Gular sutures: The ventral ends of the post occipital sutures extended forward on the underside of the head in some prognathous insects.

Gum chloral (de Faure's medium): Distilled water 50cc; gum arabic 30 g; glycerine 20 cc; chloral hydrate 50 g.

Gurdaspur borer: *Bissetia steniellus* (Hampson) (Lepidoptera: Crambidae). This is most destructive pest of sugarcane in Haryana, the Punjab, Uttar Pradesh, and Rajasthan in India and in Pakistan and Vietnam. The damage is caused by the larvae which feed inside the canes by making spiral gallaries. The tops of infested canes dry up and break off easily. The caterpillar is creamy white with orange brown head having four wide longitudinal violet stripes on the body. The moth is dull brown and has a number of dark spots along the outer margins of the forewings, and hind wings are white.

Gustatory organs: Relating to the sense of taste.

Gymnocerata: Insects of the group Hemiptera-Heteoroptera comprising most of the terrestrial forms of bug. They are distinguished from the aquatic forms

(Cryptocerata) by their conspicuous antennae. Notably among the Gymnocerata are the bedbugs (Cimicidae) and the plant eating shieldbugs (Pentatomidae).

Gynandromorph (Gynander): An individual in which one part of the body is masculine, the other feminine; most frequent are bilateral cynandromorphs, in which the left and right halves are of different sex. In some butterflies the wings on one side show the female pattern and those on the other the male. This is because of variation in chromosome content of cell in different parts of the body.

Guthion: Gusathion. Azinophos methyl (0, 0-dimethyl-S (3, 4-dichloro-4-ketro-1, 2, 3 benzotriziny 1-3-methyl) dithiophosphae. It is a broad spectrum organophosphorus insecticide.

H

Habitat: Physical place where an organism lives.

Habrobracon: Microbracon. A long-legged braconid wasp. The larvae parasitize the caterpillars of *Ephestia* and also the Lepidopterous crop pests. The adult wasp search out its victims for egg laying by using its sense organs. Before egg laying, the adult wasp paralyses its host by sting, lay eggs on it and larvae eat away the body of the host.

Hadda beetles: *Epilachna duodecastigna* (Mulsant), *E. vigintioctopunctata* (Fabricious) and *E. demurilli* (Coleoptera: Coccinellidae). Two species viz. *E. Duodecastigma* and *E. Vigintioctopunctata* attack different solanaceous vegetables. *E. demurilli* attacks cucurbitaceous vegetables exclusively. Damage is caused by the grubs and also by the adult beetles by feeding on the upper surface of the leaves. E. *vigintioctopunctata* beetles are deep-red and have 7-14 black spots on each elytron; *E. deodecastigma* are deep copper-coloured and have 6 black spots on each elytron; *E. demurili* beetles have dull appearance with 6 black spots which are surrounded by yellowish rings.

Haemocoel: The body cavity filled with blood or the blood cavity or the cavities of the embryo between the mesoderm and the other germ layers, probably a remnant of the blastocoel. Blood from the dorsal tubular heart is pumped forwards, but it does not go into the capillaries but goes directly into the haemocoel and the every organ is bathed in it. Finally it reaches the pericardial sinus from which it re-enters the heart through small openings called ostia.

Haemocoelous viviparity: A phenomenon that occurs in certain insects which have no oviducts, e.g. in Strepsiptera, and in some Cecidomyiidae which develop parthenogenetically. Eggs escape into haemocoel, develop into the larvae and finally escape through secondary opening in the body wall.

Haemocytes: Blood corpuscles. These are colourless and of various kinds found in insects. Important being: (i) Prohaemocytes (Proleucocytes)—young cells (ii) Plasmocytes (Phagocytes)—Amoeboid cells that digest bacteria, etc. (iii) Granular haemocytes: These are also phagocytic in nature. (iv) Oenocytoids - Rounded acidophil cells whose function is obscure. (v) Cystocytes - Specialized granular haemocytes. (vi) Spherule cells: Containing large round inclusions. (vii) Adipohae-mocytes - Containing fat granules.

Haemoglobin: An iron-protein pigment of blood, functioning in oxygen-carbon dioxide exchange of living cells. It is absent in most of the insects. The best known species containing this substance are the small red 'bloodworms', aquatic larvae of chironomids.

Haemolymph: The colourless blood of insects; the blood plasma; liquid part of the blood. It fills up all the body cavities and combining the functions of blood and lymph.

Haemophilia: A metabolic disorder characterised by free bleeding from even slight wounds because of the lack of formation of clotting substances. It is associated with a sex linked recessive gene.

Hair: Insect hairs originate from the cuticle and may be of following types: (i) Microtrichia - Minute non-movable hairs formed from the cuticle, (ii) Spines - Similar but thicker and stronger, (iii) Macrotrichia - Single or branched hairs, sometimes called setae. These are hollow and longer than microtrichia, and arise from a pit or socket. They may be fined end thickened, (iv) Glandular hairs - The hairs of caterpillars that secrete an irritant fluid, (v) Sensory setae: They act as tactile, or sometimes chemical receptors with nerve cells leading from their base, (vi) Scales - These are flattened and coloured but otherwise similar to setae in structure as in butterflies and mosquitoes.

Haltere: (Pl. Halteres), Balancer; The reduced second pair of wings in flies (Diptera), shaped-like short drumsticks. They vibrate and act as gyroscopic stabilizers.

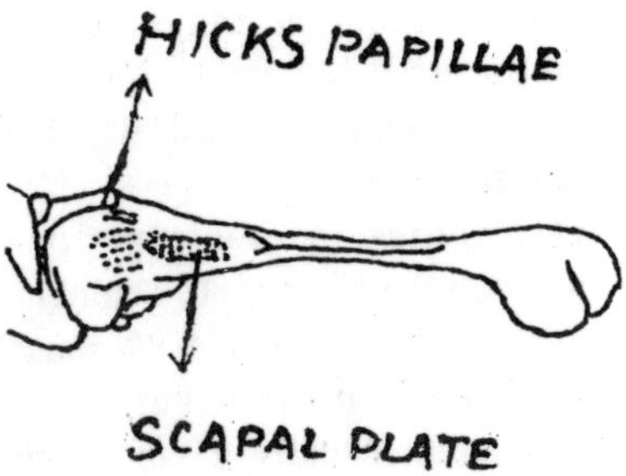

Haltere

Halticinae: It includes flea beetles of Chrysomelidae family of Coleoptera order. The insects are minute and jumping and have long thick hind legs. The larvae feed on roots of various crops.

Hamula (Retinaculum): A hook-like structure which with furcula enables insects of the order Collembola (Springtails) to leap forward into the air. This is formed from a pair of appendages on the ventral side of the abdomen. Before leaping, the caudal furcula is bent forwards and lock to the hamula. When this catch is released the furcula springs back propelling the insect violently forwards.

Hamulus (Pl. Hamuli): A series of minute hooks on the anterior margin of the hind wings which catch into the uprolled hind margin of the forewings (In Hymenoptera) when the insect is in flight.

Haploid: An individual or cell having a single set of chromosomes.

Hard Water: Water with minerals such as calcium, iron and magnesium dissolved in it. Some pesticides added to hard water will curdle or settle out.

Harpagonos: Movable periphallic processes of the ninth segment individually provided with muscles, probably derivatives of the gonostyli, usually having a clasping function.

Harpes: Claspers formed from modified stylets at the hind end of the abdomen of male butterflies and moths.

Hatching membrane: A membranous sheath investing the young insect at the time of hatching, probably the embryonic exuvial cuticula, shed during hatching or shortly after (not the amnion).

Hatching spines: A spine or series of spines on the head of an embryo insect, used for breaking open the egg shell when hatching; also called an egg-burster.

Haustellate: Mouth parts formed for sucking, the mandibles not for chewing.

Haustellum: The distal portion of the proboscis of some dipterous insects bearing at its extremity two large oral hooks or labella containing numerous tubes or pseudotracheae through which food is sucked.

Hawk-moths: Moths of the family Sphingidae. The moths are large, stoutly built, fast flying moths whose forewings are much larger than the hind ones. The proboscis is extremely long and the antennae end in hooks. The larvae have 10 prolegs and usually bear a horn or spine at the hind end.

Hazard: It refers to exposure to risk associated with the handling of pesticides. It is function of two primary variables, toxicity and exposure.

Hazardous: Dangerous; risky. Pesticide chemicals that may cause injury or death if not used as directed on the label.

HCCP (Hexa chlorocyclopentadiene): It is the base produced by the Indian Petrochemical plants and used for the manufacture of all chlorinated hydrocarbons in India.

Head: The first division of an insect body which bears the eyes, ocelli, antennae and mouth parts. It is formed from the 5 embryonic segments fused together and contained in a cuticular capsule. In cockroach and locust, the head is bent at right angle to the body so that the mouth parts are ventral, arrangement is known as *hypognathous*. Beetles have mouth parts in front and this pattern is called *Prognathous*. The back is perforated by the occipital foramen which facilitates nerves and oesophagus to pass into the thorax.

Head louse: *Pediculus humanus capitis*; a small, wingless blood-sucking insect of the group Siphunculeta. It is a carrier of typhus and other diseases. This and body louse are generally considered to be biological races of one species, *Pediculus humanus*.

Heart: The heart of an insect is a tubular structure that lies dorsally with a muscular wall and with a series of lateral openings or ostia by which blood enters. In some insects it is divided into compartments roughtly corresponding to the segments of the body, but often it is a continuous straight vessel. With the contraction of the heart, blood is pumped forwards into the aorta. Due to "open blood circulation" the blood eventually finds its way to the pericardial sinus from which it again enters the heart through the ostia.

Heart chamber: One of the segmental swellings of the heart.

Heavy oils: Oils having a viscosity greater than 55 seconds saybolt and are usually classified as heavy and used for insecticidal sprays, especially for dormant sprays.

Hectare: In the metric system, a land measure to 10,000 square meters. One hectare is equivalent to 2.471 acres.

Heliotropism: A term used to denote the attraction of insects to the sun.

Hellabore: *Veratrum album* or V. viridae. It has been found to be toxic to certain species of insects.

Hemicephalous: The insect larvae in which the head is incomplete posteriorly and is partly embedded in the prothorax.

Hemielytron: (Pl. Hemielytra); The front wings of Hemiptera; the basal part of the wing is thickened and apical part is thin and membranous, e.g. shield bugs.

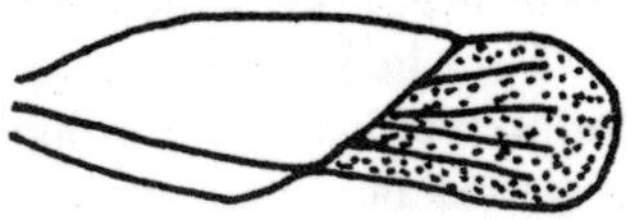

Hemielytron

Hemimetabolous (Exopterygotous): Insects in which the metamorphosis is gradual without pupal stage and the wings develop externally. The young is known as nymph and are generally similar to the adult, but are smaller and have external wing pads in immature stage.

Hemipneustic (Hypopneustic): Insects with some of the spiracles functionally suppressed or insects in which one or more pairs of spiracles are non-functional.

Hemiptera (Rhynchota): A large order of insects that includes bedbugs, cicadas, aphids, and many kinds of plant bugs. The mouth parts are piercing and sucking type with maxillae and mandibles in the form of long stylets lying in a trough-like labium, palps are reduced or absent. This order is divided into two suborders namely; Homoptera and Heteroptera. The Homopterous insects have forewings of uniform consistency, either leathery or membranous, and the Heteropterous insects have half membranous and half-leathery. These wings are called Hemielytra. Some insects are wingless.

Hemipteroid insects: A group name suggested in system of classification proposed by Wardle (1936) to include Hemiptera, Psocoptera, Anoplura and Thysanoptera.

Hepatic cells: See nephrocytes.

Heptachlor: 1, 4, 5, 6, 7, 8, 8-heptachloro-3a, 4, 7, 7a-tetrahydro-1, 7-methano-indene. It is a chlorinated hydrocarbon insecticide, produced by substantive chlorination of chlordane at low temperature like 0 to 5°C in carbon tetrachloride solutions.

Hermaphrodite: An individual, such as conttony cushion scale, *Icerya purchasi*, which contains both the male and female reproductive organs. The individual is self-fertilizing. True males are rare and true females unknown, the apparent females being functional hermaphrodite.

Hermaphroditism: A state characterised by the presence of both male and female sex organs in the same species.

Hermatic storage: An airtight storage.

Herbivorous: Insects that feed on plants.

Heterodynamic life cycle: A life cycle in which there is a period of dormancy.

Heterogamy: Alternation of bisexual reproduction with parthenogenetic reproduction.

Heterogony (Heterogonic growth): See Allometric growth.

Heteromerous: The 3 pairs of tarsi differing in the number of segments (Coleoptera; for example, with a tarsal formula of 5-5-4).

Heteromatabola: Insects which donot go a complete metamorphosis. It is an alternative name for Hemimetabola or Exopterygota. In some classification the Odonata and Ephemeroptera are not included in this group.

Heteroneura: Lepidopterous insects, whose veins of forewings are differently arranged from those of hind wings. The group comprise of all the butterflies and the majority of moths with the exception of Micropterygidae and Hepialidae.

Heteroptera: It is one of the suborders of the Hemiptera. Insect with the basal half of the forewing is of a much harder consistency than the rest of it and known as Hemielytra. It includes plant bugs which are medium insects and the important crop pests such as the rice bug, *Leptocorisa* acuta (Thumb).

Heterozygous: Having different alleles at homologous loci of the two parental chromosomes.

Hexapoda: Insects are also called hexapods (Hexa-six; podous= legs), owing to the presence of six legs.

Hibernating aggregation: Some times many thousands insects congregate together before bibernating in a suitably sheltered place. This phenomenon is called hibernating aggregation. This gives them a greater degree of protection from extreme winter ecological conditions.

Hibernation: Dormancy during the winter. During this the metabolism is slowed down in winter. This is due to the increase or decrease in the secretion of certain hormones. In some cases the shortage of food is stimulus and in other cases a shortening of the period of daylight may give the required stimulus. Hibernation may occur in egg, larval, pupal or in adult stages.

Hierarchy: In classification, the system of ranks which indicates the categorial level of various taxa (i.e. kingdom to species).

Higher category: A taxonomic category of rank higher than the species (i.e., from subgenus to kingdom).

Higher taxon: A taxon ranked in one of the higher categories.

Highly volatile: A liquid that quickly forms a gas or vapour at room temperature.

High-volume spraying: Full cover spraying; plants are sprayed thoroughly to cover them to the point of run-off or drip. The foliage is thus, completely wetted. It requires 500-1000 litres/ha of the spray fluid in the case of field crops.

Hindgut: The posterior portion of the alimentary tract, between the midgut and anus.

Holoblastic division: The type of cleavage in which the entire egg is divided.

Holocephalous: A term used to denote insect larvae in which the head is fully developed.

Holometabolous (Endopterygotous): Insects with complete metamorphosis and wings develop internally. The insects' larva is very different from the adult and does not become more like adult, but transforms suddenly by means of a pupal stage.

Holopneustic: Insects having the usual (generally 10, 8 on abdomen and 2 on thorax) pairs of open spiracles. However, few primitive insects of the order Diplura have more than 10 pairs of spiracles. *Japy* has 11 pairs of spiracles, 4 of them on the thorax.

Holoptic: The eyes contiguous above (In Diptera).

Holotype: The single specimen designated or indicated as "the type" by the original author at the time of the publication of the original description.

Homodynamic life cycle: A life cycle in which there is a continous development, without a period of dormancy (hibernation, diapause or other interruptions to their regular development).

Homoeosis (Homoosis): The replacement of a damaged appendage by a different one. If, for instance, the antenna is removed from the stick insect (*Dixippus*) the regeneration that follows may not produce another antenna but sometimes give rise to an extra leg in place of it.

Homologous: In systematics, a feature in two or more taxa which can be traced back to the same feature in the common ancestor of these taxa.

Homoneura: A name formerly used to denote moths having almost identical wing venation in the two pairs of wings. This group included small biting moths of Microterygidae and the much larger but some what primitive moths. But now this classification does not exists.

Homonyms: If the same name is used for more than one genus or species.

Homoptera: It is the suborder of the Homiptera. It includes insects with forewings, when present, are of uniform consistency. Some insects are crop pests, as leaf hoppers (Ciadellidae), aphids (Aphididae).

Homozygous: Having identical alleles at the two homologous loci of a diploid chromosome set.

Honey dew: Certain insects such as aphids, coccids, etc. feed continuously on the juice of plant and take more sugar and water than need. The excess excreted as sweetish excretion is called "honey dew".

Honey stomach: The crop of a honey bee in which nectar is mixed with enzymes from the salivary glands and converted to honey before being disgorged.

Honey stopper: A valve at the posterior end of a bee's crop specialized for preventing the passage of nectar into the stomach.

Hopper: (i) Small plant bugs of the group Homoptera, closely related to the aphids, very varied but all having powerful hindlegs for jumping; they include leaf hoppers (Cicadellidae), tree hoppers (Membraçidae) and frog hoppers (Cercopidae). (ii) Nymphs of locusts are also called hoppers when they march in bands.

Hopper burn: A disease or burning resulting from the feeding of a leafhopper.

Horizontal classification: Classification which stresses grouping together species in a similar stage of evolution, rather than location on the same phyletic line.

Hornworm: A caterpillar (larva of Sphingidae) with a dorsal spine or horn on the last abdominal segment.

Horny or leathery flies: *Hippolosca maculata* Leach (Diptera: Hippoboscidae). These flattened horny or leathery flies are the permanent parasites of horse, goat, sheep, dog, etc. They have developed tarsal claws adapted for clinging to the hair of their hosts.

Horny: Thickened or hardened.

Horse flies or Gad flies: *Tabanus* spp. (Diptera: Tabanidae). More than 2000 species of blood sucking flies in the Tabanidae family, of which 150 occur in India. The female fly bites and sucks blood from its host's which include horse, camel, cattle, elephant and sometimes man. The male feeds on nectar, honey dew, oozing tree sap, etc. The flies have thickest body, large head and big eyes.

Hoses: These are tubes that are used in the sprayers.

Host: The organism in or on which a parasite (or parasitoid) lives; the plant on which an insect feeds.

Host races: Different genetic races of the same species in oligophagous food specialists or parasites occuring on different hosts.

House crickets: *Gryllodes sigillatis* Linn., *G. domesticus, Acheta domestica* Linn. (Orthoptera: Gryllidae). Various species of crickets are found in the world. *G. sigillatis* is the commonest. Both the nymphs and the adults damage cloths, starchy materials, and pollute food stuff. Both *G. sigillatis* and *A. domestica* are common in India.

House fly: *Musca nebulo* Wied., *M. vicina* and *M. domestica* Linn. (Diptera: Muscidae). The house fly is distributed world over. In India about 3 species are common. It transmits diarrhoea, dysentry, cholera, typhoid, and other enteric fever. It also serves as an intermediate host for Helminths, of which three species of tapeworm are parasitic on poultry and three species of nematodes are parasitic on horses, mules, and donkeys. The larvae are legless, without head and pupae are coarctate, i.e. contained in a barrel shaped puparium. The mouth parts are specialized and the method of feeding is to regurgitate saliva or gastric juice over the food, this partly digests and softens the food so that it can be sucked up.

House rat: *Rattus rattus*. About 16 sub species of this rat have been recognized from the Indian subcontinent, but *R. rufescens* is the most important one from economic point of view.

Human louse: *Pediculus humanus* Linn. (Anoptura: Pediculidae). The pest is world wide in distribution, especially in overcrowded slum tenements, military barracks, prisons, etc. These are blood sucking insects which cause irritating lesions and also transmit certain rickettsial diseases, like typhus fever. *P. humanus capitis* lives on the skin and on the scalp. *P. humanus corporis* de Geer lives on garments adjacent to the body and the *Phthirus pubis* Linn. infests human hair in the pubic region. The adult louse is a small, greyish white, flattened, wingless, six legged insect.

Humectant: Material added to a spray to delay evaporation of the water carrier.

Humeral: Pertaining to the shoulder, located in the anterior basal portion of the wing.

Humeral angle: The basal anterior angle or the portion of the wing.

Humeral bristles: Bristles on the humeral callus (Diptera).

Humeral Callus: One of the anterior lateral angles of the thoracic notum, usually more or less rounded (Diptera).

Humeral cross vein: A cross vein

in the humeral portion of the wing between the costa and the subcosta, e.g. in the wings of a mosquito.

Humeral plate: The anterior preaxillary sclerite of the wing base supporting the costal vein, very large in Odonata.

Humeral suture: The mesopleural suture (In Odonata).

Humeral vein: A branch of the subcosta that serves to strengthen the humeral angle of the wing (Neuroptera).

Humerus (Pl. Humeri): The shoulder; the late angle of the pronotum (Hemiptera).

Humidity: Amount of moisture in the air.

Hyaline: Like glass; transparent; colourless.

Hybrid belt: A zone of interbreeding between two species, subspecies or other unlike population; zone of secondary integration.

Hybrid index: See character index.

Hybridization: The crossing of individuals belonging to two unlike natural population, principlly species.

Hybrid sterility: It refers to infertile progeny that is produced by crossing different species.

Hydrated: The addition of water.

Hydraulic: Pertaining to water or other liquids.

Hydraulic agitation: It refers to the agitation of the spray fluid caused by the return flow of some spray fluid from the pump.

Hydraulic agitator: A device which keeps the tank mix from settling out by means of water flow under pressure.

Hydaulic sprayer: A machine which applies pesticides by using water at high pressure and volume to deliver the pesticide to the target. Same as high pressure sprayer.

Hydrocarbon: Compounds that contain carbon and hydrogen.

Hydrofuse: Hydrophobe; Repelling water; Unwettable; The insects have waxy surface that repels water and is not wetted. The cuticle and cuticular hairs are responsible for this. This property enables many aquatic insects to breath under water by taking down a bubble of air trapped by the hairs under the body.

Hydrolysis: The splitting of a substance into the smaller units of which it is composed by the addition of water elements.

Hydrophil: A substance or system which attracts or is attracted to water is hydrophobic in nature.

Hydrophobe: A substance or system which repels or is repelled by water is hydrophobic in nature.

Hydrophobe hairs: See hydrofuge.

Hydrophyle: A specialized region of the shell of an insect's egg through which water can enter.

Hygroscopic: Substances capable of absorbing water form the atmosphere under normal conditions of temperature, pressure and humidity, are called hydroscopic substances.

Hymenoptera: (Hymen, a membrance; pteron, a wing). This order includes ants, bees, wasps, sawflies, which have membranous wings that can be hooked together by small hooks when flying. The hind pair wings are being smaller than the forewings. Mouth parts are biting type and sometimes also of the lapping or sucking type. The first segment of the abdomen is fused with the thorax, often like a narrow petiole. The pupae are generally enclosed in cocoons. The ovipositor may be adopted for sawing, piercing or stinging. The order is divided into two suborders: (i) Apocrita - with a narrow waist and (ii) Symphyta - without narrow waist.

Hyperestrogenic Syndrome: A condition in which the esterogenic hormone is present in undue high proportion.

Hypermetabola: Insects having several distinct and different forms of larva.

Hypermetamorphosis: A type of complete metamorphosis in which the different larval instar represent two or more different types of larvae, e.g. some blister beetles, pass through at least four larval stages including Triungulins, Caraboids, Scarabaeoids and Co-arctate.

Hyperosmotic: Having a salt concentration greater than that of the sorrounding medium.

Hyperparasite: When a parasite develops (lives) on another parasite, e.g. *Bracon hebetor* is parasitized by *Brachynuria* spp.

Hyperpneustic: With supernumarary spiracles, as in thorax of Diptera.

Hypertrophy: Abnormal growth and hypoplasia; underdevelopment; stunting.

Hypodermis: The single cellular layer underlying the cuticula.

Hyposmotic: Having salt concentration less than that of the sorrounding medium.

Hypodigm: The entire material of species that is available to taxonomist.

Hythergraph: A polygonal diagram resulting from plotting temperature means against rainfall.

Hypoglottis: A sclerite or cuticular element sometimes present between the mentum and the labium of an insect, i.e. at the base of the tongue in some Coleoptera.

Hypoglycemic: An abnormally small concentration of glucose in the circulatory blood, i.e. less than the minimum of the normal range.

Hypognathous: With the head vertical and the mouth parts located ventrally.

Hypomere: A ventral process of the phallobase.

Hypopharynx: A median chitinous sclerite of mouth part structure anterior to the labium; the ducts from the salivary glands are usually associated (open) with the hypopharynx, and in some sucking insects, the hypopharynx is the mouth part structure containing the salivary channels. In mosquitoes, it is enormously elongated and forms the part of the proboscis down which saliva flows. In robberflies (Asilidae), it becomes the principal piercing organ.

Hypopleural bristles: A more or less vertical row of bristles on the hypopleuron, usually directly above the hind coxae (In Diptera).

Hypopleuron (Pl. Hypopleura): The lower part of the mesepimeron; a sclerite on the thorax located just above the hind coxae (In Diptera).

Hypopneustic (Oligopneustic): Having a reduced number of functional spiracles, e.g. In Coccidae and Thysanoptera.

Hypostoma: The part of the subgenal margin of the cranium posterior to the mandibles, usually narrow, but sometimes extended mesally as a hypostomal plate or as a hypostomal bridge in the ventral wall of the head.

Hypostomal suture: The part of the subgenal suture posterior to the mandibles, often obsolete or suppressed.

I

Iapygidae: An alternative spelling of Japygidae.

Icerya: See cottony cushion scale.

Ichneumonoidea: A superfamily of order Hymenoptera, comprising Ichneumon flies, and the Brachonids which are parasitic. The insects have narrow waist and large, sharp, piercing ovipositor with which they lay their eggs in the bodies of other insects. The developing larvae live for sometime in the body of the host, eventually killing it and emerging to pupate in a silky cocoon. Many ichneumons and braconids parasitize agricultural pests.

Identification: The determination of the taxonomic identity of an individual insect.

IGR: Insect Growth Regulator.

Illegal residue: Residue that is in excess of a pre-established government enforced safe level.

Ileum: The anterior part of the anterior intestine, between the ventriculus or pylorus and colon.

Imaginal disc.: Clusters of cells which undergo rapid division to form the rudiments of future organs during metamorphosis of an insect.

Imaginal moulting or Emergence: Moulting which results in the adult form.

Imago: Pl. Imagoes or Imagines. The adult or a reproductive stage of an insect.

Immigration: The movement of individuals into a population.

Imminent hazard: A situation which exists when the continous use of a pesticide would likely result in unreasonable, harmful effects on the environment.

Immune: A state of not being affected by poison or disease or insect. A zero relaionship beween plant and a causal agent.

Imidan: Phthalophos 0, 0-dimethyl-S-(naphthalimide-methyl) dithiophosphate. It is an organophosphorus insecticide.

Impermeable: Not capable of being penetrated.

Impervious: Hard to penetrate or soak through.

Inchworm: A colloquial american name for the Geometrid caterpillars.

Inclinate: Bent towards the midline of the body.

Inclined eye piece: The tubes carrying the eye pieces may be tilted towards the operator so that he can sit comfortably erect, without having to lean forward. These eye pieces prevent the use of a Camera lucida, an aid to drawing.

Incompatibility: The inability of a pesticide to mix with other pesticide.

Indian meal moth: *Plodia interpunctella* (Hubner) (Lepidoptera: Phycitidae). It is distributed world wide feeding on grains, meals, soybean, dried fruits, nuts, seeds, dried roots, herbs and dead insects. The larva causes damage which is small whitish caterpillar, often tinged with green or pink with a light brown head and prothoracic shield. The adult moth have coppery lusture on the outer two thirds, and is whitish grey on the inner portions and on the end of the body.

Indicator: Marker.

Indirect metamorphosis: Complete metamorphosis in which there are 4 stages, viz. egg, larva, pupa, and adult.

Industrial melanism: The evolution of a darkened population owing to sooty sorrounding of an industrial area.

Inert: A material having no biological activity, such as diatomaceous earth, silica, aerogels, magnesium oxide, aluminium oxide, activated charcoal, cowdung ash and soap stone.

Infect: To enter and establish a pathogenic relationship with a plant (host); to enter and persist in a carrier; to occupy and cause injury to either a plant, soil or stored products.

Inferior appendage: The lower one (Anisoptera) or two (Zygoptera) of the terminal abdominal appendages used in grasping the female at the time of copulation (In male Odonata).

Infraepisternum: A ventral subdivision of an episternum.

Infraorder: An optional category below the suborder.

Infrasubspecific form: Individual and seasonal variations in a single interbreeding population.

Ingluvial ganglion: A paired ganglion of the stomodaeal nervous system in some insects, situated on the side of the crop.

Ingluvies: See crop.

Inhalation toxicity: How poisonous a pesticide is after a single exposure? It is measured as acute, oral, acute dermal, and acute inhalation toxicity.

Injector: A device for ejecting a pesticide below the soil surface, or into a transport system of a tree.

Injury: The effects of insect feeding or other activities on the growth or appearance of crop plants.

Inactive: Not involved in the pesticide action. Not reacting chemically with anything.

Incinerator: A special high-heat furnace or burner which reduces everything to a non-toxic ash or vapour. This is used for disposing of some highly toxic and moderately to slightly toxic pesticides.

Incisors: A pair of teeth on the upper and lower jaw of rodents.

Incompatible: Not capable of being mixed or used together. In case of pesticides, the effectiveness of one or more is reduced or they cause injury to plants or animals.

Incorporate: The mixing of pesticide into the soil by mechanical means.

Incubation period: The time which passes from the deposition to the hatching of eggs or time between penetration of a host by a pathogen and the first appearance of disease symptoms on the host.

Indian Flying Fox: Fruit bat. *Pteropus giganteus* Brunnich. It is distributed all over the Indian Subcontinent, and is essentially frugivorous mammal. The bats inflict severe damage to orchards.

Indian Hare: *Lepus nigricallis* Cuvier. It is found in almost every habitat throughout India. It feeds mainly on green grass, wheat, barley and mustard crops.

Indicator: Marker.

Individual variation: Variation within a population.

Induce vomiting: To make a person or animal throw up.

Inert ingredient: A inactive ingredient. An ingredient in a formulation which has no pesticide action.

Infection: The establishment of a parasite with a host plant.

Infest: To be present in number (e.g., inscets, mites, nematodes, bacteria, fungi) on surface matter or in the soil. Do not confuse with infect (infection) that applies only to living, diseased plants or animals.

Infestation: Pests that are found in an area or location where they are not wanted.

Infraspecific: Within the species, usually applied to categories (subspecies) and phena (varieties).

Infra-sub-specific name: A name given to an infrasubspecific form.

In-furrow: An application to or in a furrow in which crop is planted.

Ingested: Taken into the digestive system.

Ingredients: The simplest constituents of the economic poison which can reasonably be determined and reported.

Inhalation: To take air into the lungs; to breathe in.

Inject: To force a pesticide chemical into a plant, animal, building or other enclosure, or the soil.

Injurious: Harmful; can or will cause damage.

In lien of: Instead of; in place of.

Inner vertical bristles: The more mesally located of the large bristles on the vertex, between the ocelli and compound eyes (In Diptera).

Innocuous: Harmless; having no adverse effect.

Inoculation: The introduction of an infective agent (the inoculum) into living tissues producing specific disease.

Inorganic compound: The compounds lacking carbon.

Insect: Any of numerous small invertebrate animals of the class *Insecta* or *Hexapoda*, with an adult stage marked by three pairs of legs, a segmented body with three major divisions, and two pairs of wings.

Insecta: Hexapoda; A class of the phylum Arthropoda. It includes insects whose body is divided into three parts, viz., head, thorax, and abdomen. They posses one pair of compound eyes, two pairs of wings, one pair of antennae and three pairs of legs.

Insect growth regular: A synthetic, organic pesticide which mimics insect hormonal action so that exposed insect can not complete its normal development cycle, and dies without becoming an adult.

Insectary: A place where living insects are kept or bred.

Insecticide: A chemical that kills the insects.

Insectivore: An organism feeding on insects.

Insect mouth parts: Typically constitute two mandibles, one labrum, two maxillae, one labium and a hypopharynx. These are variously modified according to the feeding habits.

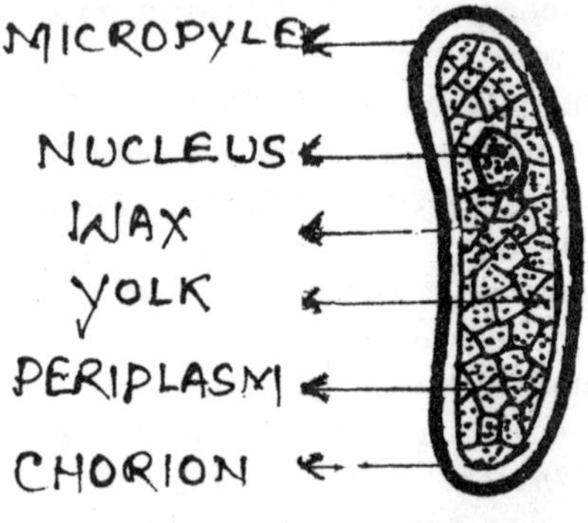

Insect Egg

Insectorubin: A brown pigment usually incorporated with melanin in the epidermis and the cuticle of insects formerly considered to be a mixture of a group of pigments called Ommochromes.

Insectoverdin: A green pigment in the bodies of some insects; usually a mixture of the blue-green waste substance, *biliverdin* with yellow carotinoid substance derived from the food.

Instar: The stage of an insect between successive moults, the first instar being the stage between hatching and first moult; the form assumed by an insect during the stadium or stage between the two moults. The number of larval instars varies from three to about thirty.

Instinctive behaviour: Unlearned sterotyped behaviour in which the nerve pathways involved are hereditary.

Integradation: Gradually merging of population through a continuous series of intermediate forms or populations.

Integrated control: The Control of pest(s) by selecting and utilizing more than one control techniques so that the economic damage is avoided and adverse side effects on the agro-ecosystem are minimized. Also see pest management.

Integument: The outer covering of the body.

Intensity: Measurement of incidence of pests.

Interantennal suture: A suture extending between the bases of the two antennae (Siphonoptera).

Intercalary vein: An extra longitudinal vein that develops from a thickened fold in the wing, more or less midway between two preexisting veins (Ephemeroptera).

Intercastes: Types that are intermediate between the recognised castes, especially in social insects such as bees, ants and termites.

Such variations may be due to the variations in diet or to invasion by parasites. Either of these factors may have an effect on the relative dominance of certain genes or may cause arrested allometric development.

Intercostal vein: The subcosta (Hymenoptera).

Intermediate mesenteron rudiment: A median strand of cells of the ventral endoderm remnant taking part in the regeneration of the mesentron in some insects.

Intermediate phase: A stage between the solitarius and migratorius phases in locust.

Intermittently: Starting and stopping from time to time.

Internal parasite: See Endoparasites.

Internally: Inside.

Internal respiration: The process of oxidation accompanying metabolism in the cells of the body tissues.

International Codes of Zoological Nomenclature: The official set of regulations dealing with zoological nomenclature.

Internucial neurone: A neurone lying within the central system that intermediates between sensory and motor neurons, or between other association neurons.

Intersex: An individual which shows both male and female features due to an upsetting of the balance between the sex producing factors associated with sex chromosomes. A strong female crossed with a weak male may produce some offsprings with female characteristics male, e.g. Gipsy moth (*Lymantria* sp).

Intersternite: An intersegmental sclerite on the ventral side of the thorax; the spinasternum.

Intersternite: Primary intersegmental sclerites of the ventar, becoming the spinasterna of the thorax.

Interstitial: Situated between two segments (interstitital trochanter of Coleoptera); coincident (the end of two veins meeting).

Intertidal: Occurring on the beach between high water and low water levels.

Interval: Period of time; the time between two applications; the distance between two points.

Intervalvulae: Sternal sclerites in the venter of the ninth abdominal segment between the second valvifers.

Intestinal myiasis: Disease of man or other vertebrates caused by the larvae of certain Diptera living as internal parasites in the intestine (also see myiasis).

Intestine: The proctoderm, or the part of the proctodaeum beyond the pylorus (the term applied also to the entire alimentary canal).

Intima: The cuticular lining of the foregut and hindgut.

Intra alar bristles: A row of two or three bristles situated on the mesonotum above the wings base between the dorsocentral and the super-alar bristles (Diptera).

Intradermal: Within the skin.

Intracellular: Within cells.

Intraspecific: Within the species; usually applied to categories (subspecies) and phena (varieties).

Intrasubspecific name: A name given to an intrasubspecific form.

Invagination of the embryo: The direct infolding of the embryo, into the egg.

Invertebrate: Animals without a spinal column or a back bone.

In vitro: In culture or outside the host plant.

In vivo: On the host plant or in the field.

Involution of the embryo: Invagination of the embryo accompanied by a revolution and final reversal of position in the egg.

Iodine value: It is a measure of the unsaturated aliphatic compounds in oil.

Iris: Dark pigment sorrounding the dioptric apparatus of an eye.

Iris cell: Pigmented cells sorrounding each ommatidium or optical unit of a compound eye. In bright light they spread a round each crystalline cone isolating it from its neighbours so that the vision is mosaic; in dim light, they may retract so that the light entering the each ommatidium merges with that of its neighbours causing continuous but rather ill defined vision (also see compound eyes).

Iristapetum: A network of fine tracheae sorrounding the ommatidia or optical units in some compound eyes. They are particularly well developed in nocturnal insects such as the Noctuid moths and serve to reflect light, so giving the retinal sensitive cells an increased stimulation.

Irreversibility rule: See Dollo's rule.

Irritating: Annoying. Making a person or animal uncomfortable by burning, stinging, tickling, making the eyes water, etc.

Ischiopodite: The third segment of a generalised limb or second segment of the telopodite (second trochanter; prefemur).

Isle of weight disease: A common disease of honey bee caused by the minute parasitic mite *Acarapis woodi*, which infests blocks their tracheal tubes.

Isodrin: 1,2,3,4,10,10-hexachloro-1,4,4a,5,8,8a-hexahydro-1,4,5,8-endo-dimethanonaphthalene. It is a chlorinated hydrocarbon, remiscent of aldrin in properties. However, it is used mainly for preparation of endrin.

Isolating mechanism: In ecology, properties of individuals that prevent successful interbreeding with individuals that belong to different populations.

Isomer: Any of the two or more chemical compounds having like constituents atoms, but differing in physical or chemical properties owing to differences in the arrangement of the atoms.

Isometric growth: Hormonic or proportionate. The type of growth where there is an equal rate of increase in all parts of the body, i.e. there is no morphological changes. The shape of the body at the end of an isometric growth period is homothallic figure of that at the beginning.

Isophene: A line connecting points of equal expression of a character; lines at right angle to a cline on a map.

Isopneustic: A term used to denote a primitive arrangement of the spiracles, two thoracic and eight abdominal pairs, all situate in the intersegmental membranes. The present day insects have some degree of specialisation or reduction of spiracles as well as migration on to one of the adjacent segment. The isopneustic arrangement persists only in some embryos.

Isopropyl alcohol: It is sometimes used instead of ethyl alcohol, because there is no excise trouble. Mixed in all proportions with xylol, cedar wood oil, and water.

Isoptera: (Iso: equal, pteron: wing). It includes termites or white ants. These are social insects, and live in large colonies, each colony having single queen. Body is soft; live in earthen gallaries or mounds; mouth parts are biting type and have two pairs of similar elongated wings placed in sockets and are easily shed at first flight. Cast system is well developed and have worker, soldier, both are sterile.

Itching: A peculiar irritating sensation in the skin that arouses the desire to scratch.

Iteroparous: A type of life cycle in which the animal reproduces two or more times during its life time.

Itonidiadae: Gall midges; an alternative name for Cecldomyidae.

J

Jackal: *Canis aureus* Linnaeus is the pest species, which causes serious losses to maize, sugarcane, melon, pineapple and sweet potato crops.

Japygidae: A family of Diplura order. It consists of bristle tails which are small, wingless insects having a forked tail formed from two long thin cerci and having biting mouth parts lying in pockets from which they are protruded when feeding. These insects are primitive and may perhaps be considered as an evolutionary link between insects and millipedes.

Jasmids: A common genus of small Asiatic butterflies of the family Lycaenidae, bright blue with broad black margins round the forewings and border of black spots round the hind wings.

Jasmolin II: It is one of the alkaloid compounds found in pyrethrum extracts.

Jassidae: A alternative name for cicadellidae or leaf hoppers; a family of agile plant bugs which infests plants and leap from leaf to leaf with great speed.

Jaws: Mandibles and maxillae; a typical insect has two pairs of horizontal moving jaws, viz. upper pair or mandibles and lower pair or maxillae. The maxillae bear a pair of palps. Both these jaws are modified in different insects according to their feeding habits. In cockroaches and beetles, which have biting type mouth parts, the mandibles are stout, thickly sclerotized and toothed, and the maxillae have two movable lobes (i.e. lacinia and galea). In dragon flies, the mandibles, maxillae and labium all bear teeth. In butterflies, the mandibles are reduced or absent but the galeae of the maxillae are extended to form a tubular proboscis. In Hemiptera, both mandibles and maxillae are extended to form long piercing stylets; when not in use these are bent back under the head and thorax. In female mosquitoes, the arrangement is similar to bugs but the stylets are very thin, sharp and sometimes toothed for piercing the skin of their victims. The male dipterous donot use mandibles to obtain food except tsetse flies, stable flies, in which the both sexes suck the blood by the stiffened labium as a piercing organ.

Jet: Liquid pesticide emitted from a nozzle orifice.

Jet agitator: A device that keeps a tank mix from settling out of suspension by means of water flowing under pressure.

Jewal wasps: A name given to brightly coloured wasps of several families, such as Pteromalidae, Chrysididae, etc

Johnstonian organ: An organ of the scolopophorous type located in the second segment, or pedicel of the antennae of nearly all insects.

Johnston's organ: A sense organ similar to a chordotonal organ, located in the second antennal segment of most insects; functions in sound reception in some Diptera.

Joint: An articulation of two successive segments or parts.

Joint worm: The larva of the chalcid wasp, *Harmolita tritici* which tunnels into stems of wheat and is a pest in North America.

Jugal lobe: A lobe at the base of the wing, on the posterior side, proximad of the vannal lobe (In Hymenoptera).

Jugal region: A posterior basal lobe or area of the wing set off from the vannal region by the plica jugalis containing the vena arcuata and vena cardinalis when these veins are present.

Jugum: A small lobe at the base of the forewings which catches under the anterior edge of the hind wing and so locks the wings together during flight; present in moths of the group Homoneura and caddis flies (Trichoptera)

Jumping: Insects use jumping for locomotion or for sudden escape. In grasshoppers, fleas, flea beetles, leaf hoppers, the femora and the tibia of hind legs are specially strong and lengthened and extremely powerful.

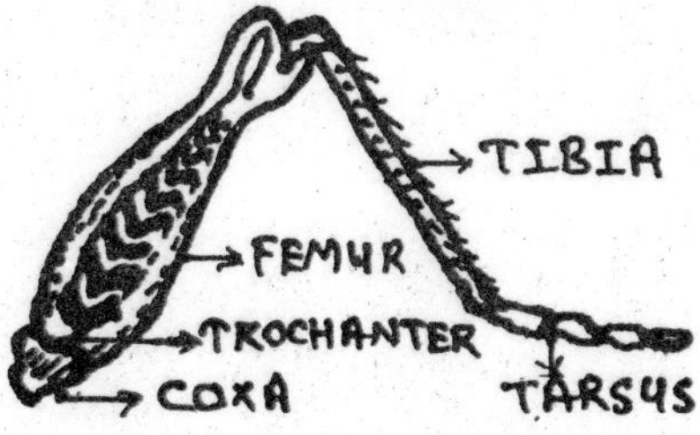

Jumping leg

Junior homonym: The more recently published of two or more identical names for the same or different taxa.

Junior synonym: The more recently published of two or more available synonyms for the same taxon.

Jute mealy bug: *Phenacoccus hirsutus* Cr. (Hemiptera; Coccidae). It attacks roselle fibre crops *(Hibiscus sabderiffa L.)* an important fibre allied to jute and responsible for an appreciable decrease in the yield of fibre. The nymphs and females feed on the apical parts of a plant which becomes stunted.

Jute semilooper: *Anomis sabulifera* Guenee (Lepidoptera: Noctuidae). It is a specific pest of jute and reported from India, Bangladesh, Burma, Sri Lanka and in parts of Africa. The caterpillars cause damage by feeding on the foliage.

Jute stem girdler: *Nupserha bicolor postbrunnea* Dutt (Coleoptera: Lamiidae). It is serious pest of jute, mesta in India and Bangladesh. The main damage is caused by the adult which makes girdles on the stem for egg laying resulting into the breakage of the fibre length at various places.

Jute stem weevil: *Apion corchori* Marshall (Coleoptera: Apionidae). It is a pest of jute in India and Bangladesh. Adult weevil is small, brown or dullblack having small whitish setae on the body. It bores holes for oviposition with its rostrum in the jute stem and the larvae feed inside the stem.

Juvenile hormore: A specific group of complex organic substances which as hormones regulate the development of larval characteristics in insects. These are internally connected by the endocrine glands in insects. However, synthetic chemical analogs are being developed as insecticides.

Juvenoids: (Juvenile hormone analogues): The substances of different chemical compositions with biological effects similar to those of the true juvenile hormone of the insect's corpus allatum.

K

Karyological character: A character involving chromosome structure or number.

Katepimeron: The lower part of the epimeron of an insect in those cases in which it is divided by a transverse suture (See also epimeron and anepimeron).

Katepistemum: The lower part of the episternum or anterior sclerite of the subcoxa at the base of an insect's leg (See Episternum).

Katydids: A colloquial name used in the United States for a number of the larger and more noisy nocturnal long-horned grasshoppers (Tettigoniidae).

Keeled: With an elevated ridge or carina.

Key: A tabulation of diagnostic characters of species in dichotomous couplets facilitating rapid identification.

Key character: In taxonomy a character of special utility in a key.

Key pests: Have been defined as serious, perennially occurring pests that dominate control practices because in the absence of control their population inflict severe economic damage.

Khapra beetle: *Trogoderma granarium* Everts (Coeloptera: Dermestidae). It is a native of India but now has gone abroad and has been reported from England, Germany, Israel and the USA. It damages wheat, sorghum, rice, barley, gram, maize, poppy, pulses, walnut and other dried fruits. The grubs eat the grain near embryo or on any other weak point and from their proceed inwards. The adult is small and brown beetle, with a retractile head and clubbed antennae. The entire body is clothed in fine hairs.

Killing liquids: Following chemicals are used as killing liquids: ethyl acetate ($CH_3COOC_2H_5$); ammonia (NH_4OH); benzene (C_6H_6); Chloroform ($CHCl_3$); carbon tetrachloride (CCl_4); trichloro-ethylene (C_2HCl_3).

Kinetopause: An arrestation of activity without necessarily an arrestation of development. It is therefore, not the same as a diapause but may accompany it. Rest, sleep and death feigning are examples. Many insects show long periods of motionless inactivity interupted occasionally by bursts of energy.

King: Fully sexually developed male among termites.

Kissing bug: A colloquial name used in the United States for *Reduvius personatus*, a predatory insect that enters houses and hunts for bedbugs and other insects. Some related species also suck human blood.

Klinokinesis: An 'avoiding action' manifested by some insects. The insects fly in a straight line when the environment is favourable but perform rapid turning movements as soon as there is any unfavourable stimulus. The turning appears to be a random movement and takes place when the stimulus is an undirected one such as a change in temperature or humidity.

Klinotaxis: A movement of an insect in a definite direction in relation to a stimulus, e.g. towards or away from light.

Knapsack duster: A duster carried on the back.

Knapsack sprayer: A sprayer that can be strapped on the back and used to apply pesticide chemicals.

Knee: The joint between the femur and the tibia of an insect; named by analogy with the knee of a vertebrate.

Knock-down effect: The rapid falling down of the insects treated with an insecticide.

Kynurenine: An oxidation product of the aminoacid tryptophane that gives rise to some of the red and brown pigments in insects.

L

Labellum: The expanded tip of the labium (In Diptera); one of a pair of oral lobes or pads present at the distal end of the proboscis (In Diptera). In mosquitoes they are less developed but in blowflies they are greately enlarged and partly covered by a membrane containing a series of channels (Pseudotracheae) with which liquid is sucked.

Labial: Of or pertaining to the labium.

Labial glands: The usual 'salivary glands' of insects, opening by a median duct between the base of the hypopharynx and the labium, or on the hypopharynx.

Labial hooks: Hooks or spines on the labial palps of the nymphs of dragon flies (Odonata).

Libial kidneys: Excretory glands in the head with tubules opening near the base of the labium (In some collembola and Thysanura insects).

Labial palps: A pair of small segmented, feelerlike sensory structures arising from the labium of and insect.

Labial suture: The suture of the labium between the prementum and the postmentum, always distal to the mentum when the latter is present.

Labile: Easily destroyed.

Labiostipites: The prementum, or that part of the labium formed by the stipites of the component labial (Secondary maxillae) appendages.

Labium: The posterior median appendage of the insect's mouth part or a lower lip of insect. It is composed of the union of the second maxillae. It consists of the fixed mentum and the movable prementum to which one attached laterally two labial palps and distally the ligula.

Labrum: The preoral lobe of the head suspended from the clypeus, with muscle arising on the frons; the upper lip of the mouth parts and lies infront of the mandibles.

Lace bug: Small plant bugs of the family Tingidae of suborder Heteroptera. These insects have lace-like network of veins on the forewings and a lace-like embossed pattern on the head and prothorax.

Lacerate: To tear.

Lacinia (Pl. Laciniae)**:** The inner lobe of maxilla borne on the stipes, provided with a muscle arising in the stipes, and often with a second muscle arising on the cranial wall.

Lacewing flies: A group of predatory Neuroptera having a narrow brown or green body and large gauge-like wings. Both the adults and the larvae prey on aphids.

Lac insect: A scale insect similar to the coccids and cochineal insects found in India and southeast Asia. The females crowd together on the stems of fig and other trees and they secrete a resinous substance from which lac is obtained, e.g. *Laccifer lacca.*

Lacciferidae: A family of scale insects including the lac insects which secrete resinous substance.

Lactating animal: Any animal that is producing milk.

Lactation: The period during which an animal produces and secretes milk.

Ladybird beetles: Small, brightly coloured beetles come under family Coccinellidae of order Coleoptera. These beeltes have well defined spots and other markings on the wings. Larvae are Campodeiform type but differ from those of ground beetles in having the appendages and sensory organs much reduced. The adults and the grubs feed voraciously on aphids, e.g. *Coccinella septempunctata.*

Lamella (Pl. Lamellae): A leaf-like plate.

Lamellate: With plate-like structures or segments; Lamellate antenna.

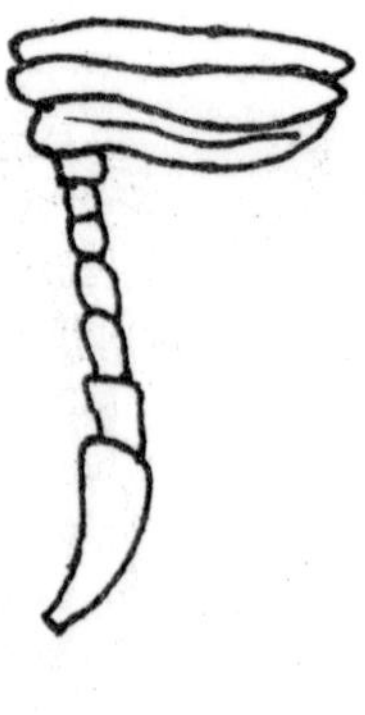

Lamellate antenna

Lamellicornia: The beetles having the last few segments of the antennae expanded into movable leaflets. These comes under super family Scarabaeidae of order Coleoptera. Also called leaf horned beetles.

Lamina lingualis: One of the two median distal plates in the millipede gnathochilarium.

Lammert's cycle: A daily cycle of temperature changes observed in beehives during winter. The temperature falls to a maximum of 13°C. This stimulates bees to greater muscular activity and the temperature rises rapidly to about 25°C. At this temperature the bees cease their activity and the temperature again falls gradually to 13°C. In an average hive about 20 g of sugar is consumed during this cycle.

Lampides boeticus: (Lycaenidae: Lepidoptera), commonly called blue butterfly having thin black tails on the hind wings. It breeds in Asia and migrates across Europe where it is common in gardens. In India, the larvae feed on leguminous plants.

Lampyridae: A family of order Coleoptera. It includes predatory beetles including glow-worms and fireflies and also many non-luminous species. These insects have elongated soft body and leathery wing cases.

Lanceolate: Spear shaped, tapering at each end.

Lancets: The first valvalae.

Laphygma: One of the genus of noctuid moths. The larvae of these moths are pests of cultivated crops, grasses in many parts of the world, e.g. *Spodoptera (Laphygma) exigua*, commonly called linseed caterpillar in India.

Lapsus calami: In nomenclature, a slip or the pen, especially an error in spelling.

Larentiinae: A subfamily of the Hydriomenidae. The geometrid moths which rest with their wings spreadout exposing the hind wings and whose looper caterpillars rest in an arched position.

Larridae: A family of solitary hunting wasps that burrow into the ground and stock their nests with other insects that have been stung and paralysed to provide food for the developing larvae.

Larva: The free living immature stage, differing in structure and habit from the adult, with holometabolous development. The larvae of insects may be classified as below: (a) Protopod: Primitive parasitic larvae with barely incipient limb-buds and with no abdomen segmentation, e.g. larvae of some parasitic Hymenoptera. (b) Polypod or Eruciform: Caterpillars with 6 legs on the thorax and a number of cushion feet or prolegs on the abdomen, e.g. larvae of Lepidoptera and sawflies. (c) Oligopod or Campodeiform: Being predatory these have efficient sense organs and long legs but devoid of prolegs, e.g. beetle larvae. (d) Apodous or legless: Larvae without any kind of legs, but with segmented body and minute head with few sense organs, e.g. larvae of house fly.

Larval moulting: Any moulting which results in larval form (See moulting also).

Larvarium: A nest or shelter made by insect larvae, sometimes a silken hammock or tube but often made from pieces of leaf, pine needles, soil particle, etc. wooven together.

Larvicide: An insecticide used to kill larvae of insects.

Larviparous: An insect giving birth to larvae instead of eggs, e.g. Tsetse fly (*Glossina* sp.).

Larviposition: The deposition of larvae.

Larvula: The first larval instars of Odonata.

Lasioderma: A genus of anobiid beetles whose small, hairy larvae are pests of tobacco and also some stored products.

Latent: Dormant.

Lateral: Referring to sides (left or right).

Lateral movement: Chemical movement in a plant or in the soil to the side or horizontal movement in the roots or soil layers.

Lateral plates: The lateral areas of the germ band after differentiation of the middle plate.

Lateral Ocellus: Simple eye in larvae.

Lateral tracheal trunk: The longitudinal tracheal trunk on each side of the body closely connected with the lateral spiracles.

Lateral oviduct: See oviductus

lateralis.

Laterosternite: The lateral part of a definitive thoracic sternum apparently derived from the ventral area (sternopleurite) of the subcoxa.

Laterotergite: A lateral sclerotization of the dorsum distinct from a principal median tergite (paratergite).

Lateroventral: To the side (away from the midline of the body) and below.

Lay-by treatment: Application of pesticides after the last cultivation.

LC 50: A means of expressing the toxicity of a compound present in air as a dust, mist, gas or vapour. It is generally expressed as micrograms per litre as a dust or mist but in the case of a gas or vapour as ppm. The LC 50 is the statistical estimate of the dosage necessary to kill 50 per cent of a population of the test species, through toxicity or inhalation under stated condition.

L.D. 50: Amount (lethal dose) of toxicant or active ingredient that will kill orally or dermally 50% of the tested population. The lower the L.D. 50 the more poisonous the pesticide. It is usually given in milligrams per kilogram (mg/kg) of body weight or parts per million (ppm).

Leaf beetles: The beetles of family Chrysomelidae of order Coleoptera, are voracious leaf feeders. These are generally small, oval, and sometimes brightly coloured or with a metallic sheen. Antennae and legs are usually short and the head is often sunk into the thorax. The adults feed on foliage and the larvae generally on leaves, root nodules and stems.

Leaf bug: A name generally given to insects that fed on or suck the plant juices from leaves. They have piercing-sucking type mouth parts. More strictly these belong to family Miridae or Capsidae of order Heteroptera.

Leaf crumplers: Moths whose larvae live in twisted conical silken tubes, usually covered and mixed with pelletts of faeces. Leaves are drawn in and tied together in crumpled masses on which the insect feed.

Leaf cutting ants: These insects cut pieces of leaf, carry them above the head like a parasol and take them down into their underground nests and here they smother them with saliva and faeces and use them as medium on which to cultivate fungi as food for the colony.

Leaf folders: Also called leaf sewers. These insects fold over the edges of leaves, fasten them together with silk secreted by labial glands and make their nest inside. These are larvae of moths of families, Tortricidae and Pyralidae of Lepidoptera order, but few midges of the family Cecidomyiidae also behave in this way. The larvae of soybeen leaf folder (*Hedylepta indicata*) folds the leaves.

Leaf footed bugs: The plant bugs of the family Coreidae of suborder Heteroptera. These insects have leaf like dilations on the hind legs and mostly live in tropics.

Leaf hoppers: These insects belong to the family Cicadellidae or Jassidae of suborder Homoptera. These feed on plants' juices by piercing-sucking mouth parts. They are small with elongated bodies and can jump and fly with great speed. They also spread many virus and fungus diseases among crop plants.

Leaf insects: Insects of the group Phasmida, also called stick insects. These have greatly flattened bodies which resembles to a leaf. The hind-wings are atrophied and the wingcases are expanded giving a perfect appearance of a dead leaf. The legs also have a flattened leaf like outgrowths.

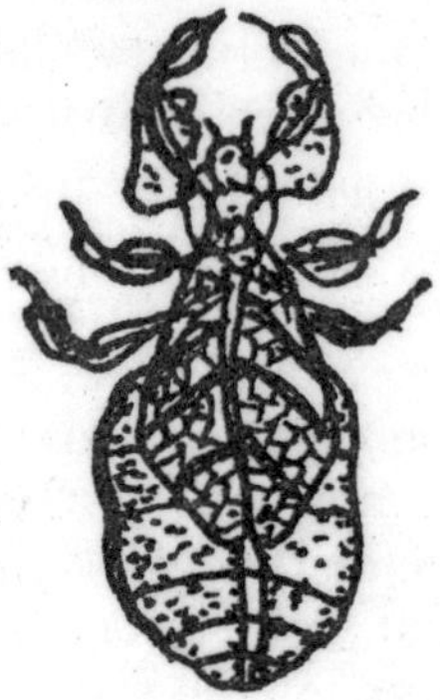

Leaf insect

Leaf miner: An insect that lives in and feed upon the leaf cells between the upper and lower surface of a leaf, e.g. leaf miner (*Phytom-yza* sp.). Some larvae produce linear or serpentine mines other feed around a centre and produce blotch mines.

Leaf rollers: Insects whose larvae make nests for living or for feeding with rolled leaves held together with silk, secreted by the labial glands. They include large number of moths of the family Tortricidae and few beetles and sawflies.

Leaf webbers: The insects whose larvae use silken threads to bind together a large number of leaves for feeding or for nest making, in which they live gregariously or solitary.

Lebaycid: See Fenthion.

Lecaniidae: Hemispherical scale insects of the Coccidae. The plant bugs in which the wingless females attach themselves to a plant and remain protected by a scale like covering formed from cast-off skin.

Lectotype: One of a series of syntypes which, subsequent to the publication of the original description is selected and designated through publication to serve as 'the type'.

Leg: When applied to insect the term only refers to anticulated organs, and does not include various prolegs or pseudolegs. The true legs are always on the thorax and an insect has 3 pairs. A great modifications in true legs takes place in different insects. The bee's legs have pollen baskets on the tibia and the pollen brushes on the first tarsal segment. Water beetles have fringes of long hairs usually on hind legs which serve as oars. The grasshoppers have greatly enlarged femora for jumping and these often bear a row of bristles for the production of sound by rubbing them against the thickened wing cases. The mantids and water scorpions have different legs modified for grasping their prey. The mole crickets have front legs flattended and spade-like for burrowing. The head lice have short legs with a single claw by which they can cling to their hosts.

A typical insect leg consist of 6 segments. The subcoxa is constituted by the pleurites and is not adistinct segment, the coxa is the base, the trochanter articulates with the coxa but fixed with next segment, the femur is the largest part of the leg, the tibia is only a slender shaft, the tarsus is divided into 2-5 segments and the terminal pretarsus.

Legal control: The control of pests through the enactment of legislation that forces control measures or imposes regulations, such as quarantine, to prevent the introduction or spread of pests; regulatory control.

Legal residue: Residue that is within safe levels according to the regulations.

Length of life: The length of life of insects ranges from a few hours to many years. Often the insect spent its greater life in the larval stage, as in some cicadas which takes many years to develop. The aquatic nymph of the mayfly lives for 2 years or more but the adult which can not feed, dies after a few hours. A very active life may cause an insect to 'bear itself out; a worker bee lives on an average for about six weeks but a queen bee lives for 4-5 years. A queen ant may live 15 years and some inactive scale insects have been known to live for atleast 17 years.

Lens: A lenticular outer part of the eye, generally formed of the cornea.

Lentil pod borer: *Etiella zinckenella* Treitschke (Lepidoptera: Phyciti-dae). It is a serious pest of lentil and pea in northern India and is also found on various other pulses in various parts of the country and in Burma and Sri Lanka. The tiny greenish caterpillars enter the pods and eat away the grains. The moths are grey and the full grown larvae are rosy, with a purplish tinge.

Lepidoptera: (Lepidos, a scale; pteron, a wing). It includes moths and butter flies, whose bodies are covered with scales. Galeae are modified to form a spirally coiled suctorial proboscis used for sucking nectar and in some cases juice. Pupae are obtect and larvae have well developed head, 3 thoracic and 10 abdominal segments. Abdominal prolegs are present on segments 3 to 6 and 10, however, lopers and semi-lopers have reduced number of legs.

Lepidosceles: A genus of small parasitic wasps whose females ride on the back of grasshoppers and lay their eggs on these grasshoppers.

Lepismatidae: A family of primitive insects of the order Thysanura. It includes silver fishes, bristle tails. They have scaly body without wings, long antennae, biting mouth parts and triple tail fork formed from two long cerci and a median filament. *Lepisma saccharina* is a very common and world wide species inhabiting houses and feeding on starch in old books, wall papers, clothers, etc.

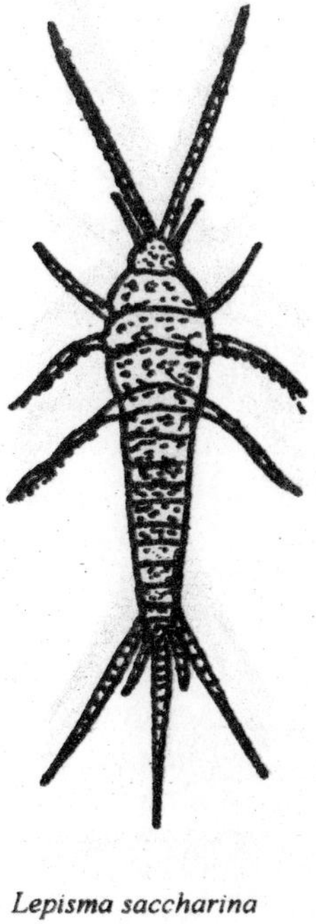

Lepisma saccharina

Leptocorisa: See rice gandhi bug.

Lesion: An injury.

Lesser grain borer: *Rhizopertha dominica* Fabicious (Coleoptera: Bostrichidae). It is a native of India but now has been reported from Algeria, Greece, United States, Australia, Japan and China. It is a destructive pest of wheat, rice, maize, sorghum, barley, lentil, stored grians, dried potatoes, corn-flour, beans, pumpkin seeds, tamarind seeds. The adult is small cylindrical beetle with shining brown or black.

Lethal: Deadly or fatal.

Leucocytosis: An abnormally large number of white blood cells.

Leucopterin: A white pigment in the scales of the cabbage and other pierid butterflies; a derivative of uric acid.

Life cycle: The complete succession of developmental stages in the life of an organism.

Life history: The development of an insect from egg to adult stage and changes therein. It has series of instars due to ecdysis. The instar means the form assumed by an insect during any particular stadium (stage). Many authors, however use instar as synonym to stadium.

Light oil: Oils with a viscosity between 40 and 65 seconds saybolt are light oils. These are highly refined and contain 90% insulfonatable residues.

Light producing organs: The organs in insects which produce light. The colour and brightness of light and the position and number off light organs varies considerably from insect to insect. *Pyro Phosphorus* gives a green light; Photinus gives orange; Photuris gives yellow. Some insects emit a steady ight which can, however, be switched off, others such as the *Luciola* sp. emits flashes at regular intervals. In some insects the eggs and larvae glow and in others only adults glow. In some insects only male produces light and in others only female. The luminous organs may be distributed over the whole body or may be concentrated in particular areas such as on the head, or on the tip of the abdomen. The light produced by the luminous insects (glow worm and fire fly) is not strictly phosphorescence but chemo-luminescence, a process of oxidation of the substance luciferin in the presence of enzymes luciferase. This is a chemical change under the nervous system control. During the luminescence a considerable amount of oxygen is used and so the organs concerned are well supplied with respiratory tracheae.

Ligula: The terminal lobe (distal part) of the labium collectively, or a terminal part of the labium formed by the union of a pair of inner lobes (glossae) and a pair of outer lobes (paraglossae). In some insects paraglossae are reduced or absent and in bees the glossae are greatly lengthened to form a tubular proboscics.

Limaciform larva: Legg-less, slug-like larvae, as in some saw flies and moths.

Limacodidae: Also called Heteroge-neidae and Cochlidiidae. A family of moths whose caterpillars are legless and slug-like.

Limb bases: The primary basal segment of an appendage implanted in the pleural area of the bodywall (coxopodite) supporting the telopodite; sometimes subdivided into a proximal sobcoxa (pleuropodite or pleuron) and a distal coxa.

Limitation: Restriction.

Lindane: See Gamma BHC.

Line precedence: In systematics (Nomenclature), occurrence of a name on an earlier line of the same page than another name for the same taxon.

Linnaean hierarchy: A structure of categorical ranks for taxa where each category except the lowest includes one or more subordinate categories.

Linear: Line-like; long and very narrow.

Linguate: Shaped like a tongue.

Linseed caterpillar: *Spodoptera exigua* (Hubner) (Lepidoptera: Noctuidae). This is a polyphagus pest generally feeds on cotton, jute, linseed, soybean, gram, lintil, etc. It is found in Europe, South Africa, America and Orient. It is also called lucerne caterpillar being noted on lucerne crop in India, USA and Canada. The caterpillars feed on leaves, making webs, within they feed gregariously for 2-3 days.

Linseed Gall midge: *Desyneura lini* Barnes (Diptera: Cecidomyidae). It is a serious pest of linseed in Andhra Pradesh, Madhya Pradesh, Bihar, Uttar Pradesh, Delhi and the Punjab. The adults is a small orange fly. The tiny maggots destroy floral buds and thus prevent pod formation.

Liparidae: An alternative name for Lymantriidae or Tussock moths. The larvae of these moths are with heavy hairs over the body.

Lipoid: A fat or fatlike substance.

Lipophilic: Fat loving, used to describe a molecule or group of atoms tending to promote oil, rather than water solubility.

Litter: Straw that is spread specially for animals to lie upon. Also refers to a brood of an animal.

Littoral: Shore dwelling, either the sea shore or the edge of fresh water.

Local: This term refers to the site of action of a poison and means that the action takes place at the point of area of contact. The site may be skin or mucous membrane.

Localized: Limited to a given area or part. Something that occurs in a small area or on a certain part of a plant or animal.

Local population: The individuals of a given locality which potentially form a single interbreeding community.

Locus: The position of a gene in a chromosome.

Locust: Three species of locusts are found:

(a) Bombay locust—*Patanga succincta* Linnaeus
(b) Migratory lucust—*Locusta migratoria* Linn.
(c) Desert locust—*Schistocerca gregaria* Forskall.

Of these, the desert locust is most important. It is distributed in India, Afganistan, Iran, Iraq, Arabia, and northern Africa (also see Locustidae).

Locustidae: It includes locusts, migratory orthopteran of family Acrididae or short horned grass hoppers. These are hemimetabolous insects with biting mouth parts, with head bent at right angles to body and with very short antennae. The forewings are thickened and called horny tegmina, which cover membranous hindwings; the hind femora is greatly enlarged for jumping. Locust differ grasshoppers in being gragarious, migratory and extremely voracious. This is, however, only a phase in their life. They may live for many years in a solitary or nonmigrating phase and at times they become greatly overcrowded and then enter the gregarious phase. The colours changes from green to reddish brown and become greatly excited, collect together in large number and then start to fly in a swarm of millions, often travelling hundred of miles. When they settle, they denude the plants completely.

Long horned beetles: See Cerambycidae.

Long horned grasshoppers: Members of the family Tettigonidae, a subgroup of Orthoptera, commonly called bush crickets. They differ from true grass hoppers and locusts by having very long antennae, and differing from crickets by the absence of long cerci. They are generally carnivorous and some are crop pests also. The male can produce a loud song by stridulation by rubbing a rough part of one fore wing against a thickened part of the other.

Longicornia: A large and world wide family of beetles (Coleoptera) having very long antennae. Also called Cerambycidae. Most of the insects of this family are serious crop and timber pests.

Longitudinal: Length wise of the body or of an appendage.

Loopers: Larvae possessing alongwith the thoracic legs, two pairs of abdominal legs which are located on the sixth and tenth abdominal segments. The larvae crawl by bringing the tail to the thorax and forming a loop of the intervening segments. Such larvae are found in the family Geometridae of the order Lepideptera.

Looper Larva

Lorum (pl. Lora)**:** The cheek; A sclerite on cuticular plate on the side of the head (Homiptera; Homoptera); the submentum in bees which is V-shaped plate supporting the elongated proboscis.

Louse: A common name for a number of wingless ectoparasitic insects belonging to the Anopleura and Mallophaga. The body is flattened and the legs have strong claws for clinging to the host. Antennae, eyes, and cerci are reduced or absent. The human body louse, *Pediculus humanus* is a carrier of typhus and other diseases. Biting lice (Mallaophaga) are chiefly parasitic on birds but some lives on mammals also. Unlike the sucking lice they donot pierce the skin but bite off small particles of feathers, hairs, etc.

Luminescent: Producing light.

Lumper: A taxonomist who emphasizes the demonstration of relationship in the delimitation of taxa and who tends to recognise large taxa.

Lung-books: The repiratory cavities of the mites, ticks, spiders and scorpions. They are so called because their walls are produced like the leaves of book.

Lunate: Any moon-shpaed structure but particularly the frontal lunule of many Diptera; A crescent shaped sclerite (scar) between the eyes and above the bases of the antennae.

Lunule frontal: A small crescent shaped sclerite located just above the base of the antennae. (In Diptera).

Lycaenidae: A family of the order Lepidoptera. It includes blue butterflies, coppers, and hairstreaks. The upper surface of the wings, particularly of the males, are usually bright blue, green or copper coloured. The undersides are generally paler and are decorated with a pattern of dots, circles, or lines by which the species can be identified. The caterpillars are generally slug-like. Some butterflies are serious pests of leguminous crops and some larvae are predaceous.

Lygaeidae: A family of the order Hemiptera. It includes ground bugs, and chinchbugs. The body of insects is elongated, soft. Plant bugs are brightly coloured. Some insects are with short wings and flightless, other have long wings.

Lymantriidae: A family of the order Lepidoptera, whose caterpillars are brightly coloured and have a number of conspicuous tufts or tussocks of hairs which exude an irritant poison when handled. Many of these are destructive crop pests.

Lyochromes: Yellow and red pigments of insects; flavine derivatives possibly originating from the anthocyanin pigments of plants.

Lysolecithin: One of the substances in the venom of bees and wasps, which is very toxic substance as it breaks down the cells of its victim and set free histamine.

M

Machilidae: Bristle tails; a primitive group of order Thysanura. The insects have scaly body, a pair of longitudinal abdominal cerci, a median tail filament, and stylets on nearly all the abdominal segments and found under the stones.

Macraner: An usually large male ant.

Macrergate: An usually large worker ant.

Macro: A combining form meaning large.

Macropterous: Large or long-winged.

Macroscopic: Visible to the naked eye without the aid of a microscope.

Macrotaxonomy: The classification of the higher taxa.

Macrotermers: A genus of tropical termites. They build very large nests called termitaria above ground which are made up of earth and sand cemented together with salive. Each termitaria contains several galleries and ventilation shafts and as high as 3 m and it contains more than 2 million workers.

Macrotrichia: Large hairs or setae consisting of extended epidermal cells sorrounded by the cuticle. They may be of following types: (a) Simple or branching hairs covering most of the body including also larger hairs on wings. (b) Sensory setae with numerous connections which respond to touch, taste, smell, etc. (c) Setae which secrete (exude) irritant secretions (e.g. caterpillars). (d) Pigmented sclaes, e.g. scales on butterflies wings and on the bodies of mosquitoes.

Maggot: A vermiform larva; a small legless larva with a pointed head and a truncated rear but without a well developed head capsule (In Diptera).

Maize stem borer: *Chilo partellus* (Swinhoe) (Lepidoptera: Crambidae). It is the most destructive pest of maize and sorghum in Sri Lanka, India, Pakistan, Afganistan, Uganda, Central and East Africa. Larvae also found feeding on pearl millet, sugarcane, Sudan grass, and on some grasses. The caterpillars cause damage by boring into stems, bobs or ear. The grown up caterpillars are about 20-25 mm long and dirty greyish white with, black head and four brownish longitudinal stripes on the back. The adults are yellowish grey moth.

Malaria: An infectious disease of human beings caused by protozoa of the genus *Plasmodium* which are transmitted by the mosquitoes of the genus *Anopheles.*

Malathion: Carbphos; 0,0-dimethyl S-1, 2-dicarboethoxy ethyldithiphosphate. It is organophosphorus insecticide, highly toxic to mites and is generally non-phytotoxic.

Malaxation: Softening by chewing with the mandibles, e.g. as inwasps.

Male reproductive sytem: See Genitalia.

Mallophaga: (Mallus, Hair: phagein, to eat). This order includes bird lice or the biting lice. Insects have biting mouth parts and are ecto-parasite of birds and also of few mammals; thoracic segments more or less free and the tarsi may be 1-or-2-segmented. They attach eggs to the feathers of their hosts and complete life cycle without dropping off.

Malpighian glands: Glands lining the malpighian tubules of an insect, normally functioning as secretory organs but in the larvae of some Coleoptera and Neuroptera they are able to spin the silk threads.

Malpighian tubules: The chief excretory tubes (organs) that arises near the anterior end of the hind gut and extend into the body cavity. These are narrow glandulor tubes, sometimes branched and in large numbers. Waste products from the insects blood are excreted by the glandular lining of these tubes and pass into the rectum where they mix into the faces.

Malpighian Tubules

Mammalian toxicity: The toxicity of a chemical to mammals.

Mammals: Warm blooded animals that nourish their young with milk; their skin is more or less covered with hair.

Mandibles: The first pair of appendages of the gnathal region of the head in the mandibulata; biting jaw like organs in their generalized form (also see jaws).

Mandibular glands: A pair of glands often present in insects opening mesally at the bases of the mandibles.

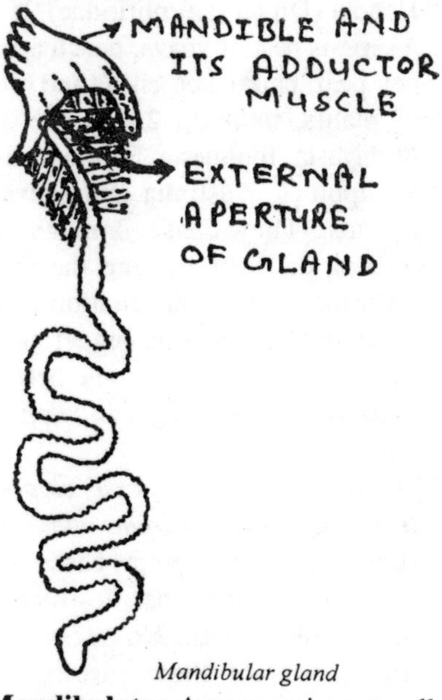

Mandibular gland

Mandibulata: A name given to all the arthropods that have maxillae, mandibles and antennae, i.e. insects, crustacea, millipedes and centipedes, but not spiders and other Arachnids which are classed Chelicerata or insects with biting mouth parts as distinct form those with a sucking or haustellate proboscis.

Manage: A group of contagious skin diseases caused by certain mites. In such diseases, the hair coat is lost and the skin becomes thickened and wrinkled; scab.

Mango bud mite: *Acaria mangiferae* Sayad (Acarina; Eriophyidae). It is a pest of mango inflorescence in India, Pakistan and the USA. The bud mites suck the sap from inside the buds and cause necrosis of the tender tissue.

Mango fruit fly: *Dacus dorsalis* Hendal (Diptera: Tephritidae). It is a serious pest of guava, peach apricot, pear, chiku, ber, citrus and other plants, totalling 250 plants in Malayasia, Indonesia, Formosa, the Philippines, Australia and Hawaii islands. They cause damage by feeding on pulp, turning the fruit unfit for human consumption. The adult is stout, a little bigger than the ordinary housefly, is brown with transparent wings and yellow legs.

Mango hoppers: *Idioscopus clypealis* (Lethierry); *Amritodus atkinsoni* (Lethierry); (Hemiptera: Cicadellidae). These are most destructive pest of the mango. No alternative host plants of these insects are known. The nymphs and adults suck the cell sap from the inflorescence and tender shoots. Adults of *I. clypealis* measured 6.3 mm in length are greyish. The *A. atkinsoni* adult is about 5.1mm long.

Mango malformation mite: *Aceria mangiferae* (Sayed). This is an eriophyid mite and causes malformation in mango throughout India. Drying and death of terminal and auxilliary buds, crowding of short twigs at the axils and inflorescenses with crowded blossoms turning green and becoming sterile are the main symptoms.

Mango mealy bug: *Drosicha mangiferae* Green (Hemiptera: Coccidae). It is widely distributed in India, especially in Indo-gangetic plains from the Punjab to Assam. It also attacks 63 other host plants including banyan, guava, papya, citrus spp., Jamun etc. The nymphs and wingless females which are oval and flattened and body covered with a white mealy powder cause damage. The males have one pair of black wings and are crimson red in colour.

Mango stem borer: *Batocera rufomaculata* Dajean and *B. rubus* Linn. (Coleoptera: Cerambycidae). These species are the serious pest of mango, fig. and other trees in north-western parts of the Indian subcontinent. Grub causes damage, killing a branch of the tree by boring it. The full grown larva is stout, yellowish-white and fleshy grub with dark head and strongly developed Mandibles. The adults had long antennae, well built large and are pale greyish.

Mango stone weevil: *Sternochaetus mangiferae* Fabricious (Coleoptera: Curculionidae). It is widely distributed throughout the tropics. The adult is a stout, ovoid, dark brown weevil which is found inside the stone of the mango or burrows into its pulp.

Mania: An emotional disorder characterized by greater psychometer activity; excitement; a rapid passing of ideas; exaltation and unstable attention.

Mantidae: A family of order Dictyoptera, suborder Mantodea, formerly classed with Orthoptera. The insects are medium to large with elongated thorax and flexible head with biting mouth parts. The fore legs are adopted for catching the pray, being greatly enlarged and can be folded back like a jack knife and the edges are sharply toothed. Only male can fly, e.g. Preying mantids.

Manuscript name: In nomenclature, an unpublished scientific name.

Marginal cells: A cell in the distal part of the wing bordering the cotal margin (In Diptera, Hymenoptera).

Marginal veins: A vein on or just within the wing margin; the vein forming the posterior side of the marginal cell (Hymenoptera).

Margined: With sharp of keel-like lateral edges.

Marine: Of the ocean or sea. The plants and animals which live in, on, or around an ocean or sea.

Mask: The elongate, enlarged prehensile labium which can be thrust forward to seize prey (of dragon fly nymph).

Mastigium: A telescopic anal filament such as on the caterpillar of the Puss moth (*Cerura vinula*).

Mastotermes darwiniensis: A termite which is a pest of sugarcane and trees in North Australia and other tropical countries.

Material: In taxonomy, the sample available for taxonomic study.

Mating (Copulation): Mating of male and female insects for internal fertilization, may occur in flight or on ground or under water. In many cases male is equipped with claspers with which the female is gripped during the process. In some cases the legs and antennae are modified to hold on to the female. The sperm cells may be released into the female by the spermatop-hore, a gelatinous case enclosing the spermatozoa. In most cases, the eggs are not fertilized immediately but the semen stored in a pocket (spermatheca) present adjecent to oviduct. The sperm enters the egg through small hole or micropyle. Dragonflies have unique mating. The male by flexing the abdomen first transfer some semen to a special receptacle beneath second abdominal segment. He then uses his terminal claspers to grasp the female round the neck while she bends her abdomen forward to bring the genital aperture opposite the second segment of the male. Transfer of sperm takes place in this position sometimes while flying but more usually while resting on a plant. The male may still be grasping the neck of the female while she is laying her eggs under water.

Maxilla (Pl. maxillae): One of the paired mouth parts structures immediately posterior to the mandibles; the second pair jaws in insects (See jaws also).

Maxillary: On or pertaining to the maxilla.

Maxillary glands: Glands present in insects opening mesally at the bases of the maxillae.

Maxillary palp: A small jointed sensory, feeler like structure arising from the maxillae. The number of segments may be upto 7. These are sometimes enlarged and specialized in various ways.

Maxillipeds (Maxillipede): The three pairs of appendages in Crustacea following the second maxillae. The first pair sometimes (Amphipoda) united to form a labium like structure attached to the head.

Maxillulae: The first maxillae of Crustacea.

Maximum Allowance Concentration (MAC): The concentration of a poision or toxic substance in the air expressed in suitable units, above which its presence becomes harmful.

Maximum Dosage: The largest amount of a pesticide that is safe to use without resulting in excess residues or damage to whatever is being protected.

Maximum permissible dose: For an individual is that dose accumulated over a long period of time or resulting from a single exposure, which in the light of present day knowledge carries a negligible possibilities of severe somatic organic injuries.

Mayflies: See Ephimeroptera.

Mayfly

Meal beetle: *Tenebria molitor.* A medium sized, dark brown or black beetle with elongated grooved elytra. This is pest of flour and grain on which both adults and grubs feed.

Meal moth: *Plodia interpunctella.* A small pyralid moth whose larvae are common pests in flour mills, grainaries etc. The wings are dark brown with broad waxy transvers bands of light brown edged with thin white line. Whitish larvae feed on flour and on organic matter, may live for about 2 years protected by long silken tubes.

Mealy bugs: Coccids; Scale insects. The body of the insects is covered by a white flour-like or waxy secretions. These are destructive pests of many agricultural crops (See Coccidae).

Measuring worm: An American name for geometrid or looper caterpillars.

Mecarbam: 0, 0-diethyl-S-(N-methyl-n-carboethoxy carbomoylemethyl) dithiophosphate. It is an organophosphorus insecticide similar to dimethoate in action.

Mechanical agitation: It refers to the agitation of the spray fluid with the help of mechanical devices, such as paddles or propellers.

Mechanical control: To control the pests by the use of mechanical devices or by manual operations.

Meconium: A dark red or brown fluid excreted as a butterfly emerges from its chrysalis and consisting of the waste products of pupal metabalism.

Mecoptera: A family of small or medium sized carnivorous insects livings in damp places and have large biting mouth parts. The upturned abdomen of the male giving it the appearance of a scorpion, e.g. scorpion fly.

Media: The longitudinal vein between the radius and cubitus; usual fourth vein of the wing; when not united with the radius, associated with the median plates of the wing bases along the fold of the plica basalis.

Median: In the middle; lying along the midline of the body.

Median cell: A cell between the subcostal and median veins at the base of the wing (Hymenoptera).

Median cross vein: A cross vein connecting two branches of the media of a wing.

Median nerve cord: The median strand of nerve tissue produced from the ventral neuroplasts.

Median nerves: Unpaired nerves arising from the ganglia of the ventral nerve cord between the roots of two connectives.

Median oviduct: The single duct formed by the merger of the paird lateral oviducts. This duct opens posteriorly into a genital chamber or vagina.

Medio cubital cross vein: A cross vein connecting the media and cubitus.

Medullary substance: The dense fibrous mass of nerve terminals forming the interior of ganglia.

Megaloptera: A suborder of neuroptera having primitive wing venation, eg. snake flies.

Meganeura: A fossil dragonfly of the Carboniferous and Permian periods, having a wing span of 70mm, the largest known insect.

Melanagromyza: A small fly of the family Trypetidae order Diptera. The larvae bore into the stems pods and leaves (See Soybean stemfly and tur pod fly).

Melanin: An insouble organic pigment which produces black-amber and dark brown colours by deposition in the cuticle of insects. It is formed by the oxidation of amino acid tyrosine in the presence of specific melanogenic enzymes.

Melanism: An usual darkening of colour owing to increased amounts of black pigments, sometimes a racial characters, sometimes a case of polymorphism, restricted to certain percentage of individuals within a population.

Melittin: A poisonous protein like substance present in the venom of bees.

Meloidae: A family of order Coleoptera. The beetles are elongated, greenish or black often short winged. They exude an oily, irritant secretion which is the chief source of drug Cantharidin. It has three or four kinds of larvae of a genus, e.g. oil and blister beetles.

Melon fruit fly: *Dacus cuccurbitae* (Coquillett) (Diptera: Tephritidae). This is most common pest of musk melon and other cucurbits throughout India. It is also found in Pakistan, Malayasia, China, Formosa, Japan, East Africa, Australia and the Hawaiian islands. It also feeds on tomato, chillies, guava, citrus, pear, fig, cauliflower, etc. Two othere species—*D. cillatis* Loew and *D. dorsalis* are also common in India. Only maggots cause damage by feeding on the near ripe fruits, riddling them and polluting the pulp. These are legless, dirty white, wriggled creature thicker at one and tapering to a point at other. The adult flies are reddish brown with lemon yellow markings.

Membracidae: Plant bugs of the group Homoptera; found mostly in tropics; characterized by a large hood like extension of the pronotum. Insect are generally small, many of them have protective colouration and often have bizarre shapes.

Membrane: A thin film of tissue, usually transparent; that part of the wing surface between the veins; thin apical part of a hemialytra (Hemiptera).

Membranous: Like a membrane; thin and more or less transparent (wings); thin and pliable (cuticle).

Menazon: Sayfos; 0-0-Diethyl-S-4, 6-diamino-1, 3, 5,-tiazinyl-2-methyl dithiophosphate. It is a good systemic organophosphorus insecticide with low mammalian toxicity for the control of vectors of viruses.

Mendelian population: A population with unrestricted interbreeding of individuals and free reasortment of genes.

Menotaxis: Movement in a direction that makes a constant angle with a source of light, as for instance in some ants. This is to keep a constant visual pattern.

Mental setae: Setae on the mentum (Odonata).

Mentum: The distal part (plate) of the labium between the prementum and submentum, which bears the palps and ligula; a median more or less triangular plate in the millipede gnathochilarium.

Meosis: Process of reducing the number of choromosomes from diploid to haploid in gamet formation.

Mercury vapour lamp: An electric discharge lamp containing mercury vapour is widely used in light traps because it gives a light with a high content of ultra violet rays.

Meristic variation: Variations in characters that can be counted, like number of vertebrae, scales, fin rays and so forth.

Meroblastic division: The type of egg cleavages in which the only nucleus and the nuclear cytoplasm are divided.

Meroistic egg tube: Ovariole; A type of egg tube containing both oocytes and trophocytes, including the acrotrophic and polytrophic types. Polytrophic types—when the food cells are contained in all ovarian follicles, e.g. Dermaptera, majority of Holometabola, Acrotrophic or Telotrophic—when the food cells are only at the apex of the ovariole, e.g. Siphonoptera and Heteroptera.

Meron: The lateral postarticular area of the base of the coxa, in some insects greatly enlarged and extended distally in the posterior part of the coxa, but always proximal to the basicostal suture.

Meropleuron (Pl. Meropleura): A sclerite consisting of the meron (basal) of the coxa and the lower part of the epimeron.

Meropodite (Femur): The fourth segment of a generalized limb.

Mesad: Toward the medline of the body.

Mesadenia: Accessory glands arising from the mesodermal lining of the seminal vessicles; evaginatinations of the vasa deferentia.

Mesal: At or near the midline of the body.

Mesenchyme: Mesoblastic tissue formed of loosely connected or scattered cells.

Mesenteron: The midgut or the middle portion of the alimenary canal; the endodermal stomach, or ventriculus. This part of the alimenary canal is lined by the endodermal tissue and here most of the digestion and absorption of food normally takes place.

Mesepimeron (Pl. Mesepimera): The epimeron of the mesothorax.

Mesepisternum (Pl. Mesepisterna): The episternum of the mesothorax.

Mesh: It is the number of open spaces in a screen per unit area.

Mesinfraepisternum: A ventral subdivision of the mesepisternum.

Mesoblast: The middle cell of the embryo, including mesenchyme and mesoderm.

Mesocerebrum: Also called the deutocerebrum; the middle part of the brain of an insect including the antennary lobes.

Mesoderm: The mesoblastic tissue that takes the form of a definite middle layer.

Meson: The midline of the body, or an imaginary plane dividing the body into right and left halves.

Mesonotum: The dorsal sclerite (notum) covering the middle segment of an insect's thorax.

Mesopleural bristles: Bristles on the mesopleuron (In Diptera).

Mesopleuron (Pl. Mesopleura): The lateral sclerite of the mesothorax; the upper part of the episternum of the mesothorax (In Diptera).

Mesoscutellum: The scutellum of the mesothorax; usually called simple scutellum. It covers the hind part of the mesothorax of an insect.

Mesoscutum: The scutum of the mesothorax; the middle and largest of three sclerites which typically form the mesonotum or cuticuller covering of the middle segment of an insect's thorax.

Mesosternum: The sternum or ventral sclerite (plate) of the mesothorax.

Mesothorax: The middle segment of the thorax.

Mesyraol: Metamercapturon (3, 5-dimethyl-4-methyl-mercapto phenyl-N-methyl carbamate). It is an active carbamate insecticide and acaricide with wide spectrum and long residual activity.

Metabola: All those insects that undergo metamorphosis (see Hemimetabola and Holometabola).

Metabolism: The whole process of a series of changes of food into tissues and cell substances and of these later into waste products; the first of these change being anabolic or building up and the second catabolic (catabolism) or breaking down.

Metaldehyde: It is molluscicide, now manufactured in India.

Metamere: One of the body divisions of the embryo; an embryonic somite or primary body segment.

Metamorphosis: A drastic change of forms during growth and development from larval stage to imago. It may be of following types. (a) Ametabola—In some primitive insects (springtails) there is no difference between larva and adults. (b) Hemimetabola—In cockroaches, there are only slight changes at each mould, the main difference being gradual increase in size of the wing buds. (c) Holometabola—In bees, butterflies, beetles, etc. there are several changes from caterpillar to caterpillar, from caterpillar to pupa and from pupa to adult.

Metanotum: The dorsal sclerite or cuticular covering of the metathorax.

Metapleuron: (Pl. Metapleura): The lateral cuticular sclerite or plate of the metathorax.

Metapneustic: The term applies to insects which have only last pair of spiracles open or having functional spiracles only on the posterior segments of the abdomen, e.g. mosquitoes larvae.

Metascutellum: The scutellum or the dorsal cuticular shield covering the hind part of the metathorax or third segment of an insect's thorax.

Metascutum: The largest part of the dorsal cuticular shield covering the metathorax or-third segment of an insect's thorax. In most insect's each segment is covered by a single plate (notum) but in some cases this is made up of three parts, viz., prescutum, scutum and scutellum.

Metasternum: The sternum or ventral sclerite of the metathorax.

Metasystox: Methyl demeton; Methylmercaptophos (0, 0-dimethyl-2-ethyl merceptoethylthiophosphate). It is a rapidly acting systemic and contact insecticide containing only the thiolo isomer of methyl demeton.

Metatarsal spinners: Insects such as, of Embioptera have numerous silk glands opening on the enlarged metatarsus (basitarsus) or first tarsal segment of each front leg. Both sexes of all ages insects of this order have such glands with which they make communal underground silk lined nests. Similar glands occur in males of the genus *Hilara* (Diptera of family Empididae).

Metatarsus (Pl Metatarsi): A term often used for the basal (first) segment of the tarsus of an insect's leg. According to some taxonomists an erroneous term for the first segment of the tarsus; see basitarsus. Better terms are basitarsus or first tarsomere or first tarsal segment.

Metathetely: Delay in appearance of adult characters as a result of two much hormone, which may be due to an upsetting of the balance between the hormones from copus allatum and from the thoracic glands. This may result in the production of an insect intermediate in form between a pupa and an adult or between a nymph and an adult.

Metathorax: The third (last) segment of the thorax, it bears the second pairs of wings in winged insects, and also a pair of legs.

Metecdysis: The period after a moult in insects or other erthropods, when the new cuticle is still soft.

Metazonite: The posterior portion of a millipede tergum when the tergum is divided by a transverse groove.

Metepimeron (Pl. Metepimera): The epimeron sclerite of the metathorax.

Metepisternum (Pl. Metepisterna): The episternum of the metathorax.

Methomyl: Lannate; Dunnate. A broad spectrum, systemic and contact insecticide and acaricide. It is also effective as a nematicide. It is a white crystalline solid, slightly soluble in water, soluble in acetone, ethanol, methanol.

Methyl bromide: It is a general poison with high insecticidal and some acaricidal properties. It is used for space fumigation and for the fumigation of plants and plant products in stores. It is soil fumigant also used for the control of nematodes, fungi and weeds. It is a colourless gas, scarcely soluble in water but soluble in most organic solvents. It is highly toxic to man.

Methyl parathion: Folidol; Metaphos; Dimethyl parathion; Metacid; Wafatxo. (0, 0-dimethyl-0-4-nitrophenylthiophosphate). It is an organophosphorus insecticide, similar to ethyl parathion except having a methyl group in place of ethyl.

Metinfraepisternum: A ventral subdivision of the metepisternum.

Metric system: A system of measurement used in the most parts of the world. The units of the metric system are meters (for length), grams (for weight), and litres (for volume).

Mevinphos: See phosdrin.

Mg: The abbreviation for milligram. 1/1000 of a gram.

Micraner: An abnormally small male ant.

Micrergate: A dwarf worker ant.

Micro: A combining form meaning small.

Microbial control: The control of pests, by the use of microorganisms, such as virus, bacteria, protozoa, fungi, rickettesia and nematodes or by their products.

Microbial insecticide: Bacteria or any tiny animals used to prevent damage or destroy insects.

Microfauna: Microscopic forms of animal life.

Microflora: Microscopic forms of plant life.

Microgaster: An alternative name for the small braconid fly, *Apantelas* sp. whose larvae are parasitic in caterpillars.

Microgeographic race: A local race, restricted to a very small area.

Microlepidoptera: The smallest moths including the primitive. Micropterygidae, only a few mm in length. Butterflies and moths were earlier classified according to the size. Macrolepidoptera and Microlepidoptera but this division is no longer used in classification. It has been replaced by the type of wing venation and other characters.

Micromathus: A genus of beetles whose immature larvae exhibit the phenomenon of *Paedogenesis.*

Micron: A unit to measure the diametere of spray droplets. Approximately 25,000 m (microns) equal an inch (0.001 mm).

Microorganism: A microscopic animal.

Micropterous: Insects having small and vestigial wings. The term is often used to various castes of insects of ants and termites. More extreme than Brachypterous.

Micropterygidae: A family of moths, seldom more than a few mm across and of world wide distribution. They belong to the Homoneura, a suborder in which the venation is almost identical in two pairs of wings. They feed chiefly on pollen by means of biting mouth parts not elongated into the normal proboscis. Some authors do not regard them as true moths and place them in a separate order, Zeugloptera.

Micropyle: A minute opening(s) or porecs) in the chorion of an insect egg through which sperm (spermatozoa) enters the egg.

Microscopic: Visible under microscope.

Microsiphum rosae: It attacks rose trees; *M. pisum* attacks peas and other leguminous plants. Both can carry the virus of mosaic and other diseases. The adults and the nymphs suck the cell sap by piercing—sucking mouth parts.

Microtermes: A genus of small termites that do considerable damage to the roots of young cotton plants in North Africa and other places.

Microthorax: A thin neck-like thorax of insects, e.g. dragon flies.

Microtrichia: Minute cuticular hairs on the wings of some insects which at a low magnification may produce an impression of cloudiness or colour.

Middle plate: The median strip of cells in germ band between the lateral plates.

Midge: A name given to many small gnat-like insects including biting or blood sucking midges (Ceratopogonidae), gall midges (Cecidomyiodae) and the smaller members of the family Chironomidae or non biting gnats whose aquatic larvae are known as blood worms.

Midgut: The mesentron; middle portion of the alimentary tract which is lined by epidermal tissue and in which most of the digestion and absorption of food normally takes place.

Migrating: Moving from place to place.

Migration: Many insects migrate in large numbers and travel very long distance in search of food or of more favourable conditions. Migration may take place on the ground, e.g. ant, or in air e.g. locust and numerous species of butterflies and moths. The desert locust *Schistocerca* migrates annually between the Middle East and Northern India; the milkweed butterfly *(Danaus plexippus)* has summer breeding grounds in the Northern United States and winter breeding grounds in Mexico and Central America; at times it flies the Atlantic and is found in Western Europe and Britain.

Migratory phase: A phase of locusts in which migration takes place in groups or swarms.

Millimeter (mm): 1/1000 m (0.001 meter), or 0.03937 inch; approximately 1/25 inch.

Millimicron: A unit of measurement which is equal to 0.001 micron.

Millipedes: Millipedes belong to class Diplopoda of phylum Arthropoda. A pair of stink glands in most of the body segments except some anterior most and some posterior most ones are present. The glands open on body surface and secrete foul smelling or poisonous chemicals to ward of their enemies. Some of the species of millipedes have been recorded as pests of crops.

Mimetic polymorphism: Polymorphism in Lepidoptera in which the various morphs resemble other species distasteful or poisonous to a predator; often restricted to females.

Mimicry: Resemblance of one species in colour or structure to other species that are distasteful or poisonous to a predator.

Minimize: To reduce to the smallest possible amount.

Minute: Very small; an insect a few mm in length or less would be considered minute.

Miotic poison: A chemical that disrupts cell division and resultant growth.

Mipafox: Pestox-15; Isopestox; N, N-diisopropyl phosphorodiamide fluoride. It is a strong systemic and contact insecticide of organophosphorus group. Now it is not used as it caused irreversible paralysis in man and animal.

Miscible: Able to be mixed.

Miscible liquids: Two or more liquids capable of being mixed and which will remain mixed under normal conditions.

Misdiagnose: To make a mistake in deciding what pest has caused the problem.

Mist blower: Sprayer that produces mist-like droplets.

Mite: Invertebrates belonging to the phylum Arthropoda and order Acarina except the tick.

Miticide: See Acaricide.

Molluscicide: A compound used to control slugs and snails which are intermediate hosts of parasites of medical importance to man, and some of them are crop pests.

Molya nematode: *Heterodera avenae* (Tylenchida: Heteroderidae). This nematode is widely distributed in Europe and Australia and has recently been recorded in Rajasthan, Haryana and the Punjab, in India. It infests wheat, barley, oats and rye, and the attacked plants remain stunted and give a shrivelled, unhealthy appearance.

Moniliform: Bead-like, with rounded segment.

Moniliform antenna

Monocrotophos: Nuvacron; monocil. It is a fast acting insecticide with systemic, stomach and contact action, used against a wide range of pests on a variety of crops. It is a crystalline solid, miscible with water, soluble in acetone and ethanol. It is dangerous to fish and livestock.

Monoculture: The extensive cultivation of a single species of plant.

Monoculture: Growing same crop

season after season.

Monoecious: Possessing both male and female organs; hermaphrodite.

Monocondylic joint: A joint with a single point of articulation between the adjacent leg segments.

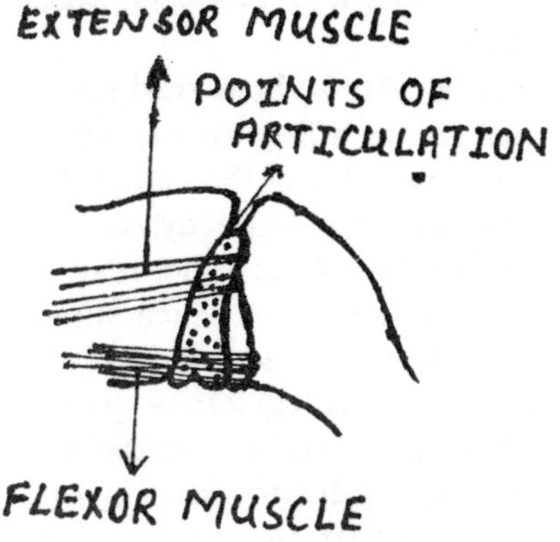

Monocondylic Joint

Monocular microscopes: These are suitable principally for a working a transparent mounts, by transmitting light, and they are normally fitted with a stage designed to take the standard glass slide, 3 x 1".

Monogenic: Determined by a single gene.

Monograph: In taxonomy, an exhaustive treatment of a higher taxon in terms of all available information pertinent to taxonomic interpretation; usually involving full systematic treatment of the comparative anatomy, biology, ecology, and detailed distributional analysis of all included taxa.

Monophagous Insects: Insects, which restrict their feeding to a single species of a food plant, such as the potato tuber moth.

Monophyly: The derivation of a taxon through one or more lineages from one immediately ancestral taxon of the same or lower rank.

Monothalamous galls: Galls with a single cavity, containing a single insect larva.

Monotrysia: An alternative name for Homoneura, i.e. moths having almost identical venation in two pairs of wings and having reduced mouth parts.

Monotypic: A taxon containing but one immediately subordinate taxon, as a genus containing but one species containing but one sub-species.

Morph: Any of the genetic forms (individual variants) that account for polymorphism.

Morphogenesis: The development of the adult and its organs within the pupa before hatching.

Morphology: The science which deals with the form of living organisms.

Morphospecies: A typological species recognized merely on the basis of morphological difference.

Mortality: Population decrease factor; death rate.

Morula: The mass of cells formed by holoblastic cleavage of the egg.

Mosaic evolution: Evolution involving unequal rates for different structures, organs, or other components of the phenotype.

Mosiac vision: See Aposed image and compound eye.

Mosquitoes: *Anopheles* spp., *Culex* sp., *Aedes* spp. (Diptera: Culicidae). Mosquitoes are found throughout the world and spread deadly diseases. Out of 45 species of *Anopheles*, only 9 are vectors of human malaria. The females carry the malarial parasite and inject them into the human blood stream. The *Culex* transmits filariasis to man. The *Aedes* spreads dengue yellow fever and other virus diseases. They are blood sucking insects of the group Nematocera and the family Culicidae. They have long legs, thin branched antennae which are larger and more bushy in the males, a long suctorial proboscis, within piercing stylets, small scales similar to those of butterflies and moths along the veins of the wings and in most cases also over the body. The larvae are aquatic, leg-less, breathing through siphons at the hind end.

Moths: All Lepidopterous insects other than butterflies. They generally fly at night, and have large body and have a mechanism for hooking the winge together when flying. They have antennae that are not clubbed but are usually hairy, branched or feather like.

Motor neuron: A neuron of which the axon terminates in an affector.

Moult: A process of shedding the exoskeleton; ecdysis; to shed the exoskeleton.

Moulting: The periodic process of loosening and discarding the cuticle, accompanied by the formation of a new cuticle, and often by structural changes in the body wall and other organs. This process accommodates the growth of the body. During the process the biochemical, physiological and morphological changes takes place.

Moulting glands: Epidermal glands that secrete a fluid which softens the lower layer of the cuticle prior to moulting. (see ecdysis).

Moulting hormone: A hormone or group of hormones which stimulates the production of moulting fluid from the epidermis glands. They are secreted by the thoracic glands or ecdysial glands or ring glands under the influence of neuro-secretions from the neurosecretory cells of the brain.

Mounts of wings of Lepidoptera: The following procedure should be followed. Take four dishes containing water, 90% alcohol, 10% Hydrochloric acid and finally a solution of bleaching powder (NaOCl) in water. Wet the wing in alcohol, then dip into the acid for few seconds, transfer to the bleach for a few minutes, until the wing veins are visible. Ringe the wing thoroughly in water, mount it. The wing may be mounted directly in Euparol, or dehydrate through a series of alcohols, to absolute alcohol, clove oil and canada balsam.

Mouth: The anterior opening of the stomodaeum, primarily located in the ventral wall of the head, but in most sucking insects retracted into the head with the transformation of the cibarium into a sucking pump, in which case the functional mouth is the enterance to the pump chamber; situated between the labrum and the hypopharynx.

Mouth parts: Insects have labrum, or upper lip, labium or lower lip and between these two pairs of side ways moving jaws, mandibles and maxillae. The labium and the maxillae bear sensory palps. These mouth parts are modified in bees, butterflies, bugs, flies, etc. according to their feeding habits.

Muellerian mimicry: Similarity (usually consisting of a similar warning colouration), of several species which are distasteful, poisonous or otherwise harmful; the two or more harmful or distasteful species resembling each other. Each benefits from the similarities because if predator tasting one, learn to avoid others also. Thus the loss from the predators is shared among the species.

Multi-dose poison: Chronic poisons, such as anticoagulants like warfarine.

Multiparasitism: When a single host is parasitized by two or more species of parasites.

Multivariate analysis: The simultaneous analysis of several variable Characters.

Muscidae: A family of Diptera order which includes housefly and other numerous similar type of flies belonging to the group Cyclorrhapha. The insects have short, aristate antennae and two halteres. The mouth parts are specialized containing a pair of labelle which have numerous tubes (pseudotracheae). The larvae are small, legless maggots and the pupae contained in a barrel shaped puparium. the adults are carriers of cholera, typhoid, dysentry, etc. The species injurious to crops are now seggregated into the related family Anthomiidae.

Mushroom body: A crowd of nerve cells grouped together into a mushroom shape, their axons forming a stalk. Large number of such bundles are frequently present in the fore-brains of insects.

Mushroom shaped gland: A cluster of glandular seminal vesicles at the junction of the two vasa deferentia with the ejaculatory duct of a male insect.

Mustard aphid: *Lipaphis erysimi* (Kalt.) (Hemiptera Aphididae). It is found world wide and is a serious pest of cruciferous crops in India. The nymphs and the adults cause damage by sucking the cell cap from the different parts of the plant.

Mustard sawfly: *Athalia lugens proxima* (Klug) (Hymenoptera: Tenthredinidae). It is found in Indonesia, Formosa, Burma, and in Indian subcontinent as a pest of cruciferous plants. The larvae are dark green, have 8 pairs of abdominal prolegs with wrinkled appearance. These larvae defoliate the leaves. The adults are orange yellow, small insects with black markings on the body and have smoky wings with blue veins.

Mutation: The genes are stable and self-replicating structures, but on rare occassions a change may occur simultaneously in them. This change is called mutation; in gentics a discontinuous change of a genetic factor, usually the replacement of several base pairs in the DNA.

Mutualism: The symbiotic relationship between two organisms in which both parties derive benefit.

Mycetocytes: Special cells in the bodies of insects containing symbiotic bacteria or protozoa necessary for the life of the host. These break down cellulose in termites and cockroaches. In some blood sucking insects these are the source of the necessary vitamins. Sometimes these cells are grouped together in a special organ called the mycetome.

Myiasis: Disease or injury caused by the invasion of dipterous larvae, most cases of human myiasis are accidental infection of the intestine through swallowing larvae in food. Other infections are in wounds and sores.

Myotome: A division of the body muscles corresponding to a metamere.

Myriapoda: A class of phylum Arthropoda comprising the millipedes and centipedes.

Myrmecophiles: Miscellaneous insects, mites, spiders etc. which inhabit the nests of ants either as true guests, fed and tended by ants or an scavengers.

Mystax: A patch of bristles or stiff setae above the mouth (In Diptera). Also called them oustache.

Myzus persicae: It is a serious pest of potato, sugarbeet and other plants in many parts of the world. It spreads the virus of mosaic, leaf roll and other diseases.

N

Naid: The term applies to the early stages of the mayflies, stoneflies and dragonflies; an aquatic gill breathing nymph.

Naiad

NALAD: (0, 0, -dimethyl-0-2, 2-dichloro-2, 2-dibromoethyl phosphate). It is a powerful organophosphorus insecticide obtained by the addition of chlorine or bromide at the double bond of dichlorovinyl group.

Naphtha: It is a petroleum product falling in the group of 5-10 carbon. These are seldom used as insecticides because of their low boiling point and hence, probable fire hazards.

Naphthalene: It is the most abundant single constituent of coal tar. It is also formed by high temperature cracking (breaking up of large molecules) of petroleum.

Nasutes: Soldier termites; specialized termites within the genus Nasutitermes, able to defend the colony by discharging an acid secretion from the glands situated at the end of a long snout or rostrum.

Nasutitermes: A genus of termites whose soldiers caste are nasutes, i.e. they have a long pointed snout from which they can discharge a poisonous fluid. It is a tropical genus that builds very large nests of woodypuld and sand, etc. above the ground.

Natality: Population increase factor; birth rate.

Natatorial leg: Legs specially adapted for swimming by having a fringe of long hairs, e.g. water beetles.

Natural control: The reduction of the pest population by the forces of nature, such as climatic factors, parasitoids, predators and diseases.

Natural enemies: The parasites, predators or other microorganisms in the environment which attack pest species.

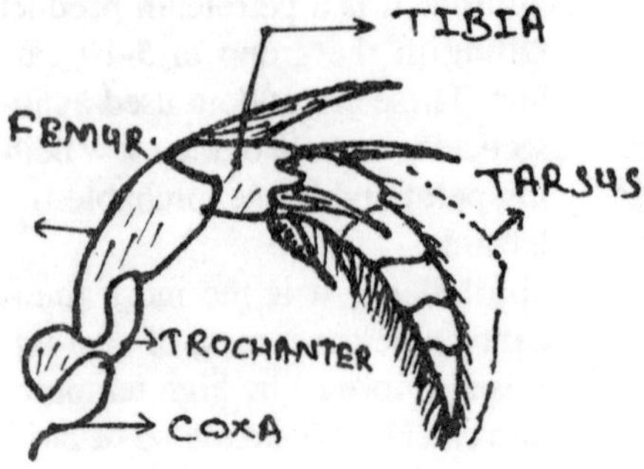

Natatorial leg

Natural selection: The unequal contribution of the genotypes to the gene pole of the next generation, through differential mortality and differences in reproductive success, caused by components of the environment.

Nausea: A feeding or inclination to vomit.

Naustus (Nausti): An individual of a termite caste in which head narrows anteriorly into a snout like projection.

Nearctic region: This geographical region includes the American continent down to the deserts of the United States/Mexico border.

Necrosis: Localized death of living tissue, i.e. death of a certain area of a leaf or of a certain area of an organ.

Necrotic: It refers to the dead parts of the living body.

Nematicide: A chemical that kills nematodes.

Nematocera: A suborder of Diptera. The insects have branched, many jointed and filamentous antennae. They are usually slim, longlegged. The larvae have well developed head and jaws but are without true legs though some (Chironomidae) have false legs on the abdomen. Many of these larvae are aquatic. Pupae are of the Orthorrhaphous type, i.e. they slit longitudinally along the middorsal line and are not enclosed in a puparium like that of the house fly, e.g. Craneflies, mosquitoes, midges.

Neotology: The science dealing with the life of recent organisms.

Neoprene: A synthetic rubber aften used to make gloves and boots which offer protection against most pesticides.

Neoptera: Insects capable of flexing or folding their wings.

Neotenin: A substance known as juvenile hormone secreted by the corpus allatum of larvae. It maintains the juvenile (larval) characters after each moult. At last larval stage, production of this hormone ceases and larvae transform to pupae.

Neoteny: Attainment of sexual maturity in an immature or larval stage.

Neotermes: A widely distributed tropical genus of termite. *Neotermes greeni* attacks tea plants in India.

Neotron: DCMP (Bis (p-choloro-phenoxy) methane). It is an acaricide, compatible with most of the common insecticides.

Neotype: A specimen selected as type subsequent to the original description in cases where the original types are known to be destroyed or were suppressed by the commission.

Nephotettix: A genus of leaf hoppers (Cicadellidae) that spread virus disease of rice in China, India and other countries. *Nephotettix binunctata* is a pest of rice in Malya and Ceylone.

Nephrocytes: Special cells scattered in groups in various parts of the body but especially abundant in the dorsal sinus, where they are known as pericardial cells, probably have some function in connection with excretion and have been termed as 'hepatic cells'.

Nepidae: Predatory aquatic insects of the group Heteroptera. Those insects have long breathing tubes resembling a sting at the hind-end and bent front legs like the pincers of a scorpion. They have wings but are unable to fly. Two commonest genera-*Nepa* (broad and stoutly built) and *Ranatra* (long and stick like) are world wide distributed; e.g. water scorpions.

Nerological: Pertaining to the nervous system and its disorders.

Nerve: Any one of the fibrous tracts of the peripheral nervous system, whether a single fiber or a group of fibers; a nerve trunk.

Nerve fiber: The axon or other branches of a neurocyte.

Nerve tract: A strand of nerve fibers; usually applied to tracts within a nerve centre.

Nerve trunk: A bundle of nerve fibers in the peripheral system; the usual nerves.

Nerve winged insects: See Neuroptera.

Nervous system: A double nerve cord runs ventrally along the body with a pair of ganglina in each segment in more primitive insects. Frequently the two ganglia of a pair are joined into one and in more highly specialized insects, many of these pairs may be fused together to form larger ganglia. Anteriorly a collar of nerve fibers sorrounding oesophagus and connecting the berral nerve cord with the brain. The brain is situated above the oesophagus and is formed from the fusion of three pair of ganglia. The first pair form the optic lobes and second pair the antennary lobes. In addition to these structures there is a primitive visceral or sympathetic nervous system consisting of small ganglia inrervating the various internal organs.

Nervure: One of the veins of an insect's wing. It has a chitinous tube usually containing a central trachea sorrounding by haemolymph.

Nettling hairs: Utricating hairs; Irritant hairs of some of caterpillars. These are hollow and have some poisonous glands often on their bases. The tip of the hairs is often barbed so that when it is broken the poison is released.

Net winged: having membranous wings with numerous cross veins, e.g. Odonata, Ephemeroptera, Neuroptera.

Neural groove: The median ventral groove of the embryo between the neural ridges.

Neural ridges: The two longitudinal ventral ridges of the embryo in which are formed the lateral cords of neuroblasts.

Neurilemma: The nucleated sheath of nerve tissue covering the ganglia, nerve trunk, and terminal branches.

Neurite: See axon.

Neuroblasts: The primitive nerve cells differentiated from the ectoderm.

Neurocyte (cyton): The cell body of a neurons usually called the "nerve cells".

Neurohormone: These originate in special cells of the central nervous system, the neurosecretory cells, and travel in the form of neurosecretory granules via the axons of these cells to special glandular organs (e.g. the corpora cardiaca). In these organs they are stored and subsequently passed into the blood.

Neurone: An entire nerve cell, including the neurocyte and axon and all their branches.

Neuropile: The medullary substance or mass of fibrous tissue within a ganglion.

Neuroptera: (Neuron, nerve; pteron, a wing), This order includes heterogenous group of alderflies, lace wings, ant lions, etc. which have two pairs of membranous wings, with many accessory branches and cross veins. Insects are small to large with soft bodies; larvae are campodeiform with biting or suctorial mouth parts. Insects are generally predaceous. The larvae may be aquatic or terrestrial.

Neutralisation number: It is an expression of the acidity in terms of mgms of KOH required to neutralize the acid in 1 gm of oil, under standard conditions.

Neutral term: A taxonomic term of convenience, such as form of group which may be employed without reference to the formal taxonomic hierarchy of categories and which has no nomenclature significance.

New name: A replacement name for a preoccupied name.

Nezara: See Soybean Green stink bug.

Niche (ecological): The precise constellation of environmental factors into which a species fits or which is required by species.

Nidus: A group of regenerative cells in the ventricular epithelium.

Night vision: The compound eyes of the nocuturnal insects have a large number of ommatidia. The crystalline cones in these are often short and their inner end are at some distance from the corresponding retinal cells. The black pigments sorrounds each crystalline cone like a sleeve, are either absent or can be retracted at night in nocturnal insects. The effect of this is that light entering through each cone can reach retinal cells of its neighbours as well as its own. Each retinal cell, therefore, receives multiple stimulation and the net result, instead of a mosaic image formed of dots, will be continuous, bright but rather blured image known as a *superposition* image. In many moths there is also a *tapetum* or mirror formed from a large number of tracheae sorrounding the retinal rods. The effect of this, besides giving a good oxygen supply to the eyes, is to reflect some of the light forwards from the back of the eye so that some of it reaches the retinal cells twice. Thus the nocturnal insects in general are extremely sensitive to small amounts of light at the expense of clarity of perception.

Nite: The egg of a louce, particularly of the human louce.

Nitrification: The conversion of the atmospheric nitrogen into the nitrates by thc bacteria.

Noctuidae: A family of order epidoptera, it includes large number of moths. The moths are medium sized, usually brownish, stoutly built, having a well developed proboscis. The larvae have biting and chewing type mouth parts and generally active at night; some of them called 'cutworms', e.g. *Heliothis aimicera, Spodoptera litura.*

Nocturnal: Active at night, i.e. during the period of complete darkness (see crepuscular).

Nodiform: In the form of a knob or knot.

Nodus: In Odonata, a notch marking the position of a stout prominent cross vein near the middle of the costal border (front margin) of the wing; the narrow waist between the thorax and the abdomen of an ant, wasp, ichneumon or similar member of Hymenoptera.

Nomenclator: A book containing the list of the scientific names assembled for nomenclatural, rather than taxonomic purposes.

Nomenclature: Latin: *nomen* (name), *calare* (to call) means literally to call by name; is the application of districtive names to each of the groups recognised in the zoological classification.

Nomen conservandum: A name preserved by action of the commission and placed on the appropriate official list.

Nomendubium: The name of a nominal species for which available evidence is insufficient to permit recognition of the zoological species to which it was applied.

Nomen oblitum: A name loosing its validity under the statute of limitations.

Nominalism: A school of philosophy, denying the existence of universals, and emphasizing the importance of man given names for the grouping of individuals.

Nomial taxon: A named taxon, objectively defined by its type.

Nominate: Of a subordinate taxon (subspecies, subgenus, etc.), which contains the type of the subdivided higher taxon and bears the same name.

Nonbiodegredable: Incapable of getting broken down in the biological system.

Nondimensional species: The species concept, represented by the noninterbreeding of species at a given place and time.

Non polar insecticides: The molecules of non-polar insecticides, such as DDT, BHC and synthetic pyrethroids, dó not have any separation of charges, because the electrons are uniformly distributed. Such insecticides are lipophilic, i.e. have affinity for fat or fat-like substances.

Non preference: In this case, the resistence of a plant to the attack of a pest is due to the lack of one or more of the preferred factors in the plant.

Non-selective: A chemical that is generally toxic to plants or animals without regard to species.

Non-toxic: Not poisonous.

Non-volatile: A pesticide chemical that does not evaporate.

Notapleural suture: A suture between the notum and the pleural sclerites.

Notatorial leg: These type of legs are flattened, hairy and have oar like structures and are found in giant water bug; legs modified for swimming.

Notaulix (Pl. Notaulices): A longitudinal suture of the mesonotum separating the median area from the lateral area (Hymenoptera). Incorrectly called "Notum" and sometimes mistaken for the parapsidal furrows.

Notonectal: Swimming on the back, e.g. water boatman.

Notopleuron (Pl. Notopleura): An area on the thoracic dorsum at the lateral and of the transverse suture (Diptera).

Notoptera: An alternative name for the order of primitive insects; also known as Grylloblattodea.

Notum: Pl. Nota; The cuticular plate of the segment (tergum), usually refers to the tergum of a thoracic segment. According to their position they are named pronotum, mesonotum and metanotum.

Nozzles: Devices which control drop size, rate, uniformity, thoroughness, and safety of a pesticide application.

Nuclear cytoplasm: The small mass of egg cytoplasm containing the egg nucleus.

Numbness: A combination of total loss of sensation, especially of sense of touch, and that of burning and pricking, anesthesia and paresthesia combined.

Numerical phenetics: The hypothesis that relationship of taxa can be determined by a calculation of an overall unweighted.

Nuptial flight: The mating flight of social insects, such as ants and bees; marriage flight.

Nurse cells: The trophocytes or nutritive cells of the ovary of testis.

Nurses: Young workers bees whose chief occupation is the feeding of the larvae.

Nymph: The early stage in the development stage of insects which has no larva stage. It is the stage between egg and adult during which growth occurs in such insects as cockroaches, grasshoppers, aphids, etc.

Nymphalidae: A family of the order Lepidoptera. It includes handsome butterflies, most of them are brightly coloured.

O

Oberea: *Oberea bimacuta;* Rasberry cone borer. A small beetle of the family Cerambycidae, whose larvae bore downward into rasberry cones (also see soybean girdle beetle).

Objective synonym: Each of two or more synonym based on the same type.

Obligate: Compelled to live under only one type of environment.

Obligate parasite: An organism that passes its entire life on a living plant/animal.

Oblique bein: A slenting cross vein, in Odonata where Rs crosses M_1x_2.

Obtect pupa: The type of pupa that has the antennae, wing pads, and legs glued to the body and is enclosed in tight fitting wrapper, e.g. Pupae of Lepidoptera, some Coleoptera and Primitive Diptera. A butterfly pupa of this type is called a chrysalis.

Occipital arch: The arch of the cranium between the occipital and postoccipital sutures; Its dorsal part is the occipital proper, and lateral parts the postgenae.

Occipital condyles: Processes on the margins of the postoccipital to which the lateral neck plates are articulated.

Occipital foramen: An opening in the back of an insect's, head through which the oesophagus, nerve cord, etc. enter the thorax.

Occipital ganglion: A single or paired postcerebral ganglion of the stomodaeal nervous system.

Occipital suture: A transverse suture in the posterior part of the head which separates the vartex from the occiput dorsally and the genae from the post-genae laterally.

Occiput (Occipital region): The dorsal posterior part of the head, between the occipital and postoccipital suture; In a restricted sense the area between the vertex and the neck.

Ocellar bristles: Bristles arising close to the ocelli (In Diptera).

Ocellar triangle: A slightly raised triangular area in which the ocelli are located (In Diptera).

Ocellus: Pl. Ocelli. A simple eye consisting of a single cuticular, bead like lens; even if several are placed together they are not combined into one organ. It is present in adults and in some larvae. The ocelli of larvae are lateral and those of adults are dorsal and have a different nerve supply. An eye spot or eye-like pattern on the wings of a butterfly or moth.

Ocular: Relating to the eye; visual; ophthalmic.

Ocular sclerite: A narrow bend of the cranial wall encircling the compound eye with in the ocular suture.

Ocular suture: The line of inflexion in the crenial wall around the compound eye, forming internally a circumocular ridge.

Odonata: *(Odonotos,* atooth),

Dragon flies, damselflies. These are insects with elongated and brilliantly coloured bodies. Eyes are large and mouth parts are strongly toothed. Wings are membranous with numerous cross veins. The aquatic nymphs or larvae are predatory and is characterised by an elongated prehensile labium forming the so called Mask. The order is divided into suborders viz., Zygoptera (Damsel flies) and Anisoptera (Dragon flies).

Odonoptera: A name occasionally used synonymously with Odonata.

Odontotermes: A genus of termites (Isoptera) in which the soldier caste have large heavily armoured heads and very powerful jaws, e.g. *Odonototermes obesus*, the Indian termite.

Oecophoridae: A family of small moths whose caterpillars live between two pieces of leaf which they fasten together with silk to make a portable protective shelter from which they feed.

Oedema: The pathological accumulation of fluids in tissues spaces.

Oenocytes: Large ectodermal cells occurring in the abdomen of most of insects sometimes within the epidermis, but generally free in lateral segmental groups or scattered through the fat body; rounded acidophil corpuscles of unknown function in the blood of insects.

Oesophageal ganglion: A pair of small ganlia behind the main parts of an insect's brain and joined to it by thin connecting nerves. They rest upon the dorsal surface of the oesophagus, and are also closely connected with the corpora allata and form part of the visceral or sympathetic system.

Oesophagus: A tubular part of the stomodaeum between the pharynx and the crop or sometimes extending to the stomach.

Oestrus: See hot and warble flies.

Official index: A list of names or works suppressed or declared invalid by the commission.

Oil bordeaux mixture: It is prepared by the addition of edible cotton seed oil to bordeaux mixture. It improves the retention of the copper on foliage.

Oil emulsion: It consists of lube oil (2 gallons), water (1 gallon), potash fish oil soap (2 pounds). It was in use in the beginning of 20th century for the control of pests.

Olethreutidae: A large family of small moths whose larvae are destructive pests of fruits trees, feeding on leaves, flowers and fruits. The Codling moth, *Cydia pomonella* or *Carpocapsa pomonella*, whose larvae attack apples and pears.

Olfactory organs: The organs of small.

Olfactory pits: Small pits on the antennae and sometimes also on the palps of insects, containing a number of hair-like receptors and nerve connections by which the insect has a sense of smell.

Olfactory sense: Many insects have an extremely efficient sense of smell by which by which they are attracted to the food plants or to a potential male. Some moths and butterflies fly very long distance apparently attracted by the scent glands of the opposite sex. Chemoreceptors of various kinds are generally distributed over the antennae but are sometimes also on the palps. The drone bee have as many as 30,000 receptors on antenna. The antenna of the house fly have definite olfactory organs, each consisting of a small pit with a number of hair-like cells connected to the nerve endings. The antennae may be held out in particular direction when the insect is flying so that these pits are exposed to the flow of air and the maximum sense of smell is achieved.

Oligogenic character: A character determined by only few genes.

Oligophagous insects: Insects such as spotted boll worm, which restrict their feeding to a small number of usually similar plants; having a very restricted range of food.

Oligopneustic: Having a reduced number of spiracles (see breathing).

Oligopod larva: This type of larva is active, and has a well developed head and thoracic legs but no abdominal processes. These are common in some Coleoptera and most of the Neuroptera. This type is subdivided into two, the Campodeiform, which moves actively and the Sacarabaeiform, which is C-shaped and soft bodied and has short legs and slow movement.

Ommatidium (Pl. Ommatidia): One of the component (optical) of a compound eye; each ommatium is cone-shaped structure consisting of a cutucular lens, a crystalline cone and a retinulae or sensitive cells at the base. The whole eye is roughly hemishperical with all the ommatidia converging inwards towards the optic nerve.

Ommochrome pigments: Red, brown or yellow pigments which give the characteristic eye colour of many insects but are sometimes also found in other parts of the body. These are derivatives of the amino acid trypotophane.

Omnivorous: Feeding generally on animal or vegetable food or on both.

Oncogenic: The property to produce tumors (not necessarily) cancerous in living tissues.

Oncopeltus: A genus of chinch bug. Plant sucking hemipteran bugs of the family Lygaeidae, e.g. *Oncopeltus fasciatus*.

Onion fly: *Delia antiqua* (Anthomyidae: Diptera). The larvae are the pest of onion and other similar plants. The adults lay eggs on the young leaves of the onion plants. The newly hatched larvae then move down the stem and enter the bulb on which they feed.

Onion thrips: *Thrips tabaci* Lindemann (Thysanoptera: Thripidae). It is a major pest of onion and garlic but also feeds on cotton, cabbage, potato, cuccumber, etc.

Damage is done by the adults and nymphs. It acts as vector of the steak virus disease of pea and yellow spot of pineapple. The adults are slender, yellowish brown; males are wingless, whereas, the females have long, narrow strap-like wings which are furnished with long hair along the hind margin.

Onisciform larva: A larva that is extremely flattened, as the larva of the Lycaenid butterfly (Also called Playform larva).

A flattened platyform larva, like wood-house in appearance.

Onomatophore: 'Name-bearer' = type.

Ontogeny: The developmental history of an individual organism from egg to adult.

Onychium: A term applied to a veriety of hook-like or pad-like structures associated with the feeth of insects.

Oocytes: The egg cells within an ovariole, oogonium, before maturation.

Oogonium: The first stage in the differentiation of an egg cell from a primary female germ cell.

Ootheca (Pl. Oothecae): The horny case or capsule or covering of an egg mass (ortheca) formed of secretion products of accessory genital glands (Collateral glands) of the oviducts.

Open blood system (Open blood circulation): The characteristic of insects and other Arthropods in which the blood is not contained in arteries, veins and capillaries but fills all the body cavities.

Open cell: A wing cell extending to the wing margin, not entirely sorrounded by veins.

Open coxal cavity: One bounded posteriorly by a sclerite of the next segment (front coxal cavities, Coleoptera) or one touched by one or more pleural sclerites (middle coxal cavities, Coleoptera).

Operculum (Pl. Opercula): A lid or the cover.

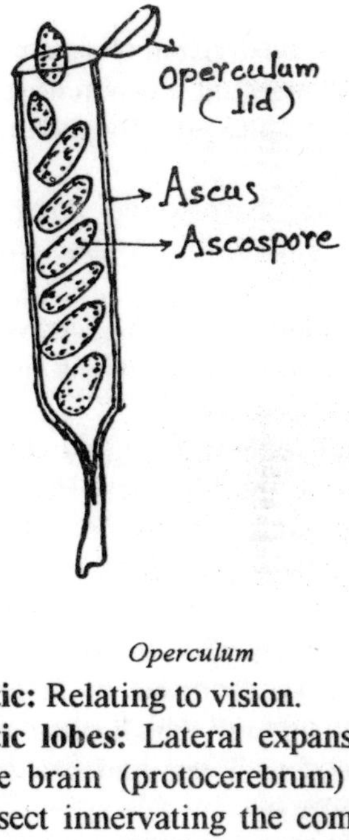

Operculum

Optic: Relating to vision.

Optic lobes: Lateral expansion of the brain (protocerebrum) of an insect innervating the compound eyes and ocelli; they form the largest part of the insect brain and are situated immediately above the oesophagus.

Opticon: The inner zone or internal medullary mass of the optic lobes.

Oral: Any structure connected with the mouth; pertaining to the mouth.

Oral toxicity: How poisonous a pesticide to man or animal when it is swallowed.

Oral vibrissae: A pair of stout bristles one on each side of the face near or just above the oral margin and larger than the other bristles on the vibrissal ridge (Diptera).

Orbital plate: A sclerite on the head next to the compound eye (Diptera); also called genovertical plate.

Order: A subdivision of class or subclass containing a group of related families and usually ending in-optera.

Organism: Any living thing, like plant, animal, fungus, bacteria, insects, etc.

Organophosphates: A group of synthetic organic insecticides containing phosphorus.

Oriental region: A region centered on India, the Malayan, Peninsula, Indo-China, and Indonesia, but with rather ill defined boundaries in all directions and overlapping considerably with the next.

Orifice: The opening or hole in a nozzle through which liquid material is forced out and broken up into a spray.

Orfice velocity: Velocity at which the spray leaves the nozzle orifice.

Original description: A statement of characters accompanying the proposal of a name for a new taxon in conformance with articles 12 and 13 of the code.

Orthogenesis: Evolution of phyletic lines following a predetermined rectilinear pathway, the direction not being determined by natural selections.

Orthokinensis: A simple reaction of insects by which a particular stimulus can cause them to move in a straight line but they come to rest as soon as the stimulus is removed.

Orthoptera: (Orthos, straight; petron, wing). It includes grasshoppers, locusts, crickets. Insects have biting and chewing mouth parts; forewings modified into tegmina; hind legs modified for jumping; stridulatory and auditory apparatuses well developed. The head is bent at right angles to the body. Cockroaches and mantides, which were formerly classed as Orthoptera are now placed in a separate order Dictyoptera.

Osmaterium (Pl. Osmeteria): A fleshy, tubular, eversible, usually Y-shaped scent glands at the anterior end of certain caterpillars (Papilionidae).

Osmoregulation: Regulation of the salt and water concentration in cells and organism.

Ostium (Pl. Ostia): A slit-like opening in the insect heart; one of the paired and usually lateral openings of the heart, by which the blood flows into the heart from the pericardial sinus in insects and other orthropods.

Ostium bursae: The opening of the bursa copulatrix in Lepidoptera, equivalent to the vulva of female insects having the genital opening on the eighth segment.

Ostiole: A small opening.

Othreis: *Othreis fullonia,* a noctuid moth whose fruit piercing larvae are pest of citrus plant.

Ovarial ligament: A ligmentous strand attaching the terminal filaments of an ovary to the dorsal diaphram or the body wall, sometimes united with that from the opposite side in a meidan ligament attached to the ventral wall of the dorsal blood vessel.

Ovariole: One of the secondary divisions (secondary egg tube) of the ovary, composed of a terminal filament, an egg tube, and a pedicel. These are of two main types: (i) Panoistic—in which there are no specialized nuse cells and (ii) Meroistic—which have nurse cellls or trophocytes. Meroistic ovariole again fall into two groups, i.e. (a) Teletrophic—with the tropho-cytes all gathered together at one end, and (b) Polytrophic—where each oocyte has its own nurse cells.

Ovary: The female reproductive organ containing the egg cells. A insect has two ovaries. Each ovary is composed of a number of separate branches or ovarioles in which eggs are produced. Cockroaches have 8 ovarioles in each ovary but the queen bees have more than 200. In each ovariole the oocytes are produced in the germarium near the distal tip of the tube and as they move down they develop into eggs.

Overall similarity: A value (usually numerical) of similarity calculated by the summation of similarities in numerous individual character.

Ovicide: A chemical which kills the eggs before they hatch.

Oviduct: The tube leading away from the ovary through which the eggs pass. Each of the two ovaries have its own duct into which eggs from the ovariole are discharged. These two ducts, join to form a median oviduct which in embryo and primitive insects has an opening (gonopore) in the 8th abdominal segments (7th in Dermaptera). In most present day insects this original aperture is closed and the oviduct becomes connected to a wide duct or vagina with an opening in the 9th abdominal segment. The colleterial glands or accessory glands open into the vagina and produce a glue-like substance which either forms a capsule to enclose the eggs or cement to attach the eggs to leaves, stems, etc.

Oviductus communis: The median duct of the female reproductive system, usually opening into the genital chamber or vagina.

Oviductus lateralis: One of the paired lateral ducts of the female system connected with the ovary, mesodermal in origin; but sometimes partly or entirely replaced by an ectodermal branch of the median duct.

Oviparity: Reproduction by laying eggs.

Ovoparous: Egg laying insects.

Oviporus: The posterior opening of the vagina in most Lepidoptera, serving only for the discharge of the eggs when there are two genital apertures.

Oviposit: To deposit or lay eggs.

Ovipositor: The egg laying organ formed of the gonopods of the eighth and ninth abdominal segments, or also in a functional sense, the egg laying tube of insects formed of the protractile terminal segments of the abdomen.

Oviposition: The deposition or laying of eggs. The act of depositing eggs.

Ovotron: 4-chlorophenyl-4-chlorobanzene sulphonate. It is a chlorinated hydrocarbon acaricide.

Ovoviviparity: Reproduction in which eggs hatch soon after they are laid.

Ovum: The mature (unfertilized) egg cell.

Outer fork: The fork of R_{4+5} (In Ephemeroptera).

Outer vertical bristles: The more laterally located of the large bristles on the vertex, between the ocelli and the compound eyes (Diptera).

OW type (Oil in water): It is a type of emulsion wherein water is the external or continuous phase and oil is the internal or dispersal phase. This is the one type generally adopted for insecticidal purposes.

Oxycarenus: A genus of plant bugs of the family Lygaeidae. It is a pest of cotton seed in India, Africa and other countries.

P

Paedogenesis: Precocious maturity; The production of eggs or youngs, usually parthenogenetically, by an immature or larval stage of an animal, e.g. gall midge (*Miastor* sp.) whose larvae may give birth to twenty or thirty larvae which eat their way out of the body of their parents. This process may be repeated for many generations and pupae and adults are rarely produced. It also occurs in some beetle larvae.

Page precedence: Occurrence of a name on an earlier page in the same publication than synonym or homonym of it.

Painted bug: *Bagrada cruciferarum* Kirkaldy (Hemiptera: Pentatomidae). It is a serious pest of cruciferous crops and widely distributed in Burma, Sri Lanka, India, Iraq, Arabia and East Africa. It also feeds on rice, sugarcane, coffee, etc. The nymphs and adults suck the cell sap from the leaves and pods and the plants gradually wilt and dry up. The adults are sub-obate, black and have number of orange or brownish spots.

Palaeartic region: This Zoogeographical region includes, Europe, Asia, and north Africa, down to desert belt that includes the Sahara, the Arabian and Persian deserts and those of central China.

Palaeoptera: A group of insects comprising the two orders Ephemeroptera (mayflies) and Odonata (dragonflies), both of which have primitive wing venation.

Paleontology: The science that deals with the life of past geological periods.

Palp: A sensitive tactile organ, usually near the mouth, on many invertebrates such as polychaetes, Molluscs and Arthropods. In insects they are on the labium and on the maxillae but never on the mandibles. They are often multiseg-mented and elongate, but may be short, inflated and equipped with obvious tactile and olfactory cells.

Paloifer: A lobe of the stipes of the maxilla which bears the maxillary palps.

Palpiger: A lobe of the stipital region of the labium or prementum, bearing the palpus.

Palpitation: The forcible palpitation of the heart, perceptible to the patient, usually with an increase in frequency with or without irregularity in rhythm.

Palpus (Palp): A segmented or unsegmented appendage process borne by the maxilla of labium; telopodite of a gnathal appendages.

Pampel's fluid: Glacial acetic acid,UCC; distilled water 30 cc; formaldehyde 40%, 6cc; 95% alcohal 15cc.

Pandemic: An epidemic that occurs over an extended geographical area.

Panoistic egg tube (ovaries): A

Panoistic egg tube

type of egg tube in which the vitellarium contains eggs only. It has no special feeding cells but secretes a nutritional fluid for the developing eggs. They are found in more primitive groups.

Pantropical: A species occurring widely throughout the tropical and sub-tropical parts of the world.

Papilionidae: A family of the order Lepidoptera consisting of butterflies characterized by having tails on the hind wings.

Papilionoidea: A super family of the order Lepidoptera containing all butterflies. Formerly classed as Rhopalocera, they are now included in the Heteroneura. They may be distinguished from the moths having clubbed antennae and by the absence of frenulum and jugum on the wings.

Papilla: A small nipple-like elevation.

Papule: A small solid elevation on the skin.

Parabiosis: This consists of joining two or more individuals in such a way that their blood system are connected. In insects this is usually achieved by cutting of the tip of the head of one animal and inserting the rest of the head in an incision in the cuticle of the other specimen.

Paraglosae: The two lateral (outer) lingular lobes of the labium each with a muscle arising in the prementum.

Paragnatha: A pair of lobes of the gnathal region of Crustacea situated between the mandibles and the first maxillae.

Parallelism: The independent acquisition of similar characters in related evolutionary lines.

Parameres: Lateral processes or lobes of phallobase (This term is usually used in Coleoptera, but it is also applied to the gonapothyses). A pair of inner processess, which together with the claspers and the penis comprise the external genitals of a male insect.

Paraneoptera: A group of insects consisting of the five orders Psocoptera (book-lices), Mallophaga and Anoplura (lice), Thysanoptera (thrips) and Hemiptera (bugs).

Paraneuroptera: An alternative

name for Odonata.

Paranotal lobes: Lateral lobes of the pronotum in certain fossil insects and theoretical lobes of the mesonotum and metanotum supposed to be the precursors of the wings.

Paranotum: (Pl. paranota). Notal flanges in fossil insects.

Parapatry: Of populations or species, in nonoverlapping geographical contact without interbreeding.

Paraprocts: A pair of ventral plates below the anus in some insects; the reduced sternal lobes of the eleventh abdominal segment.

Parapsidal furrows: Longitudinal grooves on each side of the mesoscutum of many Hymenoptera (Also called parapsidal suture or Notaulix).

Parapsides (Sing. Parapsis): Lateral area of the mesoscutum in some Hymenoptera laterad of the parapsidal furrows.

Parapsis: One of the two sclerites of the mesonotum immediately lateral of the parasidal suture (Hymenoptera). Also called scapula.

Paraptera: See epipleurites.

Parapteron: A small additional sclerite, sometimes present in the mesothorax and metathorax of an insect, just below the wings.

Paraptery: Of population of species, in non-overlapping geographical contact without interbreeding.

Parasite: A true parasite is a living organism which takes up its abode (home) temporarily or permanently on or within some other organisms and at whose expense it obtains its nourishment. Victim (host) usually lives long enough for the parasite to complete its development. Generally larval stages are parasitic and the adults are free living. The parasite is restricted to single host and is more specific in its choice. A parasite usually does not kills its host at one stroke. They are of following type:

(i) *Endoparasite:* Within the host body

(ii) *Facultative:* Which can live freely or parasitically

(iii) *Obligate:* Which can live only parasitically.

Parasitoid: The term coined by Reuter (1913) to differentiate from the typical parasite. This term is applied to those insects which kill their hosts in contrast to true parasite, which donot. The distinction is not always valid, hence Fiender (1973) recently referred to such insects as 'Carnivoroids'. An animal that feeds in or on another living organisms for a relatively longer time, consuming all or most of its tissues or blood and externally killing it.

Paratergite: See laterotergite.

Paratype: A specimen other than the holotype which was before the author at the time of preparation of the original description and was so designated or indicated by the original author.

Parietals: The lateral areas of the cranium between the frontal and occipital areas, separated above by the coronal suture.

Paris green: Emerald green, French green, Mitis green, Schweenfurter green. It is a complex compound of copper meta arsenite and copper acetate (3:1 to 2:1). It is aresenical (arsenite) insecticide.

Particle density: It is the actual density of solid material only as if there were no air space between them.

Parthenogenesis: Reproducing by eggs that develop without being fertilized. In bees, the males and droner are produced by parthenogenesis but the workers and queens come from fertilized eggs. In stick insects *(Dixippus)* parthenogenesis is the normal method of reproduction, male being very rare. The same is also true with scale insects and mealy bugs (Coccidae). Many other insects such as cockroaches and some moths, butterflies, and flies will occasionally develop by parthenogenesis. The aphid produces parthenogenetically in plains.

Parthenogenetic: Asexual.

Patagium: A short or winged-like outgrowth on each side of the prothorax in certain insects, e.g. in some noctuid moths. They were well developed in primitive fossil dragonflies of the order Palaeodictyoptera.

Patanga: *Patanga succincta*, a species of migratory locuss, common in China, Malayasia, Borneo and Philippines.

Patella: A segment between the meropodite (femur) and the carpodite (tibia) in the legs of Pycnogonidia, most arachnida and in the last legs of Xiphosura.

Pathogen: Any microorganism which can cause disease. Most pathogens are parasites but there are a few exceptions.

Patronymic: In nomenclature, a dedicatory name, a name based on that of a person or persons.

Paurometabolous (Paurometabola): Insects in which metamorphosis is gradual and wings develop externally with simple metamorphosis, the young and adults living in the same habitat, and the adults are winged; the young are called nymphs.

Peach aphid: *Myzus persicae*; a widely distributed, pink or green aphid and found on peach, potato, etc.

Peach fruit fly: *Dacus zonatus* (Saunders) (Diptera: Tephritidae). It is a pest of peach, fig, guava, ber, citrus, cuccurbits, tomato, brinjal and pear in the Indian subcontinent. The damage is caused by the maggots only which are found inside the fruits and are dirty white, headless and legless. The adult fly is a small insect about the size of a house fly, reddish brown with yellowish cross bands on the abdomen and with transparent wings, having a brown spot on the tip of each wing.

Peach leaf curl aphid: *Brachycaudus helichrysi* (Keltenback)-(Hemiptera: Aphididae). It is a world wide serious pest of peach, almond and other temperate fruits. The damage is caused by the nymphs and the females, which suck the cell sap from the shoots and leaves, leaving leaves curled. The aphid is very small and is generally yellow, with dark stripes on the head.

Pea leaf minor: *Phytomyza atricornis* Meigon (Diptera: Agromyzidae). The pest is widely distributed in northern India and feeds on a large number of cruciferous plants, pea, linseed, potato, etc. The larvae are only destructive which make prominent whitish tunnels in the leaves. The adults are two winged flies, which have a greyish black mesonotum and yellowish frons.

Pecten: Any comb like structure or organ, e.g. (a) a comb-like stridulating organ on some insects (b) a row of spines or hairs on an palp, or chelicera of various arthropods, on the respiratoy tube of a mosquito larva, on the antennae of some moths.

Pectinate: Comb-like with branches or processes like the teeth of a comb; pectinate antennae; pectinate tarsal claw.

Pectinate antenna

Pedicel: The second segment of the antenna; the stem of the abdomen, between the thorax and the gester (ants); one of the ovariole stalks, or short ducts from the egg tubes to the oviduct.

Pedipalps: The second pair of appendage of an arachnid.

Pelagic: Animals inhibiting the open sea; ocean dwelling.

Pellet: A dry formulation of pesticide mixed with other components in descreet particles, usually larger than 10 cubic millimeters.

Penetrant: Wetting agents that enhance the ability of a liquid to enter pores of substrate, to penetrate the surface, are termed penetrating agent or penetrants.

Penis: One of the paired intromittent or copulating organ of certain insects containing the gonopor from which the seminal fluid is discharged or the usual median phallic organ. It is formed from two elongated outgrowths of the 9th abdominal segment and is situated between the two claspers. In primitive insects (mayflies) the paired structures can be seen but in most insects these are fused together to from a single tubular structure.

Pentatomidae: Shield bugs, Large and sometimes brightly coloured shield shaped insects of the group Heteroptera; i.e. bugs in which the proximal half of the wing is horny and distal half membranous. They have long proboscis. Generally herbivorous but some are predatory also. These insects have distinct 5 segmented antennae and a triangular dorsal shield or scutellum formed by the tergum of the mesothorax extending back over abdomen.

Penultimate: Next to the last.

Pepipalps: The second appendages of adult chelicerata, corresponding to the mandibles of Mandibulata.

Perennial: Plants that continue to live for more than two years.

Pericardial cells: Strands of special cells along the sides of the heart, probably in most of cases nophrocytes. These are believed to be connected with excretion as they can absorb nitrogenous waste substances from the blood, but the whole process is not understood.

Pericardial sinus: A part of the haemocoel or body cavity sorrounding the heart in an insect or arthropods. Blood flows from this cavity directly into the dorsal tubular heart by way of number of ostia.

Periphallic organs: Peripheral genital processes of the ninth segments or also of other segments in the genital complex including the movable claspers or harpagones.

Peripheral nervous system: The outgoing parts of the nervous system in distinction to the central ganglia and connectives including the sensory neurocytes and their axons, and the axons of the motor neurons.

Peripneustic: With none or only a few spiracles closed in each lateral series; having lateral spiracles on all the segments of abdomen.

Periproct: The terminal piece of the body containing the anus, anterior to which segmentation takes place.

Peristome: The ventral margin of the head bordering the mouth.

Peristome: The ventral marginal part of the cranium formed by the clypeus or epistoma, the pleurostomata, and the hypostomata.

Peritreme: A sclerite of the body wall containing the spiracular opening.

Peritrophic membrane: A thin cuticular cylindrical membranous envelope sorrounding the food in the ventriculus and sometimes extending into the proctodaeum; generated from the ventricular epithelium either from all or a part of the length of the latter or from a ring of specialized cells at its anterior end. It protects the epithelial cells from damage by the passage of food, but is permeable to enzymes and digested foods. It is continuously produced by secretion, either at the junction with the fore gut or over the whole epithelium of the mid gut simultaneously.

Perivisceral sinus: The principle part of the body cavity between the dorsal diaphram and the ventral diaphram if the latter is present.

Permethrin: Ambus; Permasect;

Methyl (3-phenoxyphenyl±Cistrans 3-(2, 2, dichlorovinyl)-2, 2-dimethyl cyclopropane carboxylate. It is a synthetic pyrethroid insecticide being recommended against vide range of major pests of crops, stable in sunlight. It has low mammalian toxicity and is biodegradable.

Permissible level: A concentration expressed in ppm of the residue in or on a food when first offered for consumption. It is calculated from (i) Accepted daily intake (ADI), (ii) The food factor, and (iii) the average weight of the consumer. The ADI is the daily dose which during an entire life time will not cause any injury. It is set at a safe level as determined by data obtained in animal feeding studies. A safety factor of 100 is used in translating animal data to human beings. It is expressed in milligrams of the pesticide per Kilograms of the body weight per day (mg/km/day). The food factor refers to the total diet made up of a food or of various foods.

Persist: To stay; to remain.

Persistence: The property of a chemical to remain effective as a residue.

Persistent pesticide: A pesticide chemical (or the metabolites) that remains active in the environment more than one growing season. These compounds often accumulate and are stored in animal and plant tissues.

Pest: It is an organism whose activities cause it to be inimical to the welfare of human beings. It includes invertebrates, some insects, mites and ticks and some vertebrates such as birds, rodents.

Pest density: The population level at which a pest species causes economic damage.

Pesticide: Pest killer. It is generally used to refer to a chemical pesticide that is used for controlling any pest. In this sense, other terms that are used some times, such as physical pesticides (i.e. temperature, and irrigations), mechanical pesticides (e.g. fly flapper); biological pesticides (e.g. parasitioid and predators) and microbial pesticides (e.g. microorganisms or their products) are strictly not pesticides.

Pesticide tolerance: The amount of pesticide residue which may legally remain in or on a food crop.

Pest management: May be defined as the integration of all appropriate practices, procedures and techniques relating to crop protection into a single unified programme or system aimed at holding one or more pests at subeconomical levels with a minimal harmful impact of pest populations as long as they do not go above the economic threshhold levels.

Pest spectrum: The complete range of pests attacking a particular crop.

Petiolate: Attached by a narrow stalk or stem

Petiole: A stalk or stem; the narrow stalk or stem by which the abdomen is attached to the thorax.

pH: A measure of the acidity or alkalinity of a medium. A pH value of 7.0 indicates neutral-lower values indicate acidic and higher values indicate alkaline.

Phagocytes: Haemocytes having active ingestive and digestive properties.

Phallobase: The proximal part of the phallus highly variable in its development; sometimes a large structure supporting the aedeagus; often produced into thecal fold or sheath; about the aedeagus, sometimes represented only by basal phallic sclerites in the wall of the genital chamber.

Phallocrypt: A pocket of the phasllobase or the genital chamber wall containing the base of the aedeagus.

Phallomeres: Genital lobes formed at the side of the gonopore in the ontogeny of some insects, in most cases they unite to form the phallus, but in Blattidae and Mantidae, they develop separately into complex genital organ of the adult.

Phallotheca: A fold or tubular extension of the phallobase; about the aedeagus.

Phallotreme: The distal opening of the endophallus usually at the end of the aedeagus.

Phallus: The unpaired penis, or median, intromittent organ including the phallobase, the adeagus, the endophallus, and various processes of the phallobase and the aedeagus if present.

Pharate instar: In cases in which a new instar has been formed completely and the moulting fluid has been absorbed as in the puparium of higher Diptera. The decisive criterion is thus the presence of air between the old and the new cuticle.

Pharynx: The anterior part of the foregut, between the mouth and the oesophagus.

Phase contrast attachment: When a colourless, transparent object such as the cleared skin of an insect larva is examined as a slide mount, it is often difficult to study, because there is very little variation in tone or colour between different parts of the object. The phase contrast is a device which increases the contrast between different parts of the image by setting up optical interference between rays that have taken different paths between the lenses of the microscope system.

Phasmida (Phasma: an apparition): It includes stick insects and leaf insects, known for mimicry in body form resembling twigs or green leaves; Insects are large elongated and cylindrical often apterous, mouth parts biting type, living mostly in the tropics.

Phenetic ranking: Ranking into categories strictly based on degree of overall similarities.

Phenkapton: 8, 8-diethyl-S-(2, 5-dichloro-phenylthiomethyl) dithiophosphate. It is a selective organophosphorus acaricide of prolonged action and is similar to trithion.

Phenogram: A diagram indicating the degree of similarities among the taxa.

Phenon: A simple of phenotypically similar specimens; a phenotypically reasonably uniform sample.

Phenotype: Any measurable characteristic or distinctive trait possessed by an organism is called phenotype.

Pheromone: A substance secreted by an insect for giving information to other insects of the same species and thereby influencing their behaviour. Such a substance may be sexual attractants as in female moths; they may be warning substances serving to keep away other insects; others may be used as a means of social communication and the cementing together of a colony. Ants leave odour trails which other members of the colony will follow. The queen bee produces substance chemically identified as 9-oxydecenoic acid. The worker bee like this substance and pass it on to each other as a means of recognising the presence of the queen.

Phobotoxis: A reflex action by which as insect turns to avoid an adverse stimulus and by repeated trials and errors eventually moves away from it.

Phorate: Thimet. It is a persistant systemic insecticide used as granular formulation (10 G) for the protection of seedlings from sap feeding and soil insects. It has some fumigant action. It is a clear liquid, only slightly soluble in water, but miscible with carbon tetrachloride, dioxane, xylene, and vegetable oils. It is very poisonous and dangerous to fish and livestock.

Phoresy: It is the name given to the trick of using some other animals to carry you about, e.g. A fly *Dermatobia hominis* which uses mosquito as a transport animal. Some larvae of aquatic insects live on crabs. A small chalcid wasp is carried on the legs of ants, and the triungulin larvae of blister beetles are carried on the bodies of bees and wasps.

Phonoreceptors: A sense organ responsive to sound.

Phoridae: Humb backed flies; A family of small black flies with a humped thorax and distinctive wing venation. They rarely fly but run about on flowers and foliage or on wall, sand, windows.

Phosalone: 0, 0-diethyl-S-(6. chloro -benzolinyl-3-methyl) dithiophosphate. It is an organophosphorus insecticide and acaricide, recommended against wide range of pests.

Phosdrin: Mevinphos (0, 0-dimethyl-0-(1-methyl-2-carbomethyoxy yvenyl) phosphate). It is a derivative of phosphoric acid. It exists in Cis and Trans isomeric forms.

Phosphamidon: Dimecron. 0-0-dimethyl-0-(2-chloro-2-N, N-diethyl carbamoyl, methyl vinyl phosphate. It is a derivative of phosphoric acid and is highly soluble in water. It is systemic insecticide and acaricide and rapidly absereved by the plants, and is effective against sucking pests.

Phosphnion: Phospherion; 0, 0-diethyl-8-(2, 2-dichloro-1-chloroethoxy. vinyl phosphate. It is a derivative of phosphoric acid and is the closest analog of dichlorovos.

Photodecomposition: Destroyed by light.

Photogenic organs: See light producing organs.

Photoreceptors: Organ which respond to the light.

Photosynthesis: Process by which the energy of radiations (light) is used by the chlorophyll of green plants to manufacture food material from carbon and water.

Phototaxis: A directional movement in relation to source of light. Moths are positively phototactic, flying towards the light. The larvae of many insects show negative phototaxis, moving away from the light.

Phototonus: An action whereby the tonus of certain muscles is affected by the presence or absence of light so that an insect will take up a different attitude in the dark from that it takes up in the light. Some flies, raise their bodies above the ground in the light and collapse limply immobile in the dark or if their eyes are covered.

Phragma (Pl. Phragmata): A plate like apodeme or invagination of the dorsal wall of the thorax. They are hardened and serve for the attachment of muscles and other organs. The word Phragma and apodeme seem to be used interchangeably according to the context and taste of the author.

Phragmanotum: See postnotum.

Phyletic: Pertaining to a line of descent.

Phyletic weighting: Assessing the taxonomic importance of a character on the basis of its phyletic information content.

Phyllidae: A family of leaf insects (Phasmida) related to the stick insects but resembling a leaf or bunch of leaves. The body is flattened, hind wings are atrophied and the wings covers resemble a dead, brown leaf. The legs are expanded and flattened, so that they also resemble part of dead leaves.

Phyllocnistis: A genus of small moths whose larvae are shallow linear leaf miners. They tunnel between the upper and lower epidermis of leaves, scraping the cells and feeding upon the sap. The air trapped in the leaves of mines give the appearance of translucent or glistering white branching line with in the leaf, each with a dark central line caused by the deposition of frass or faeces, eg. Citrus leaf miner *(Phyllocnistis citrella)*.

Phylloptera: A name used in the system of classification proposed by Wardle (1937) for the order Phasmida comprising stick-insects (Phasmidae) and leaf insects (Phyllidae), both of which were formerly classed with Orthoptera.

Phylogenetic tree: A diagramatic presentation of inferred lines of descent, based on palaeontological, morphological as other evidences.

Phylogeny: The study of the history of the lines of evolution of higher taxa.

Phylogram: A tree-like diagram indicating degree of relationship among taxa.

Phylum: (Pl. Phyla) One of the dozen or so major divisions of the animal kingdom.

Physical control: Control of pests by physical means such as heat, cold, moisture, light, electricity and electromagnetic energy.

Physical pesticides: They refer to electromagnetic energy and heat sterilization for the control of pests.

Physical selectivity: Refers to the use of broad spectrum pesticides in such ways as it obtain selective action. This may be accomplished by using protective device, exploiting pest behaviour and feeding habits, and selecting pesticide formulations, dosages and application times.

Physiological manipulation: The use of natural or and synthetic growth regulating hormones and pheromones (hormonal signals) to suppress pest population, e.g. Insect growth regulators, attractants, repellents).

Physiological selectivity: Refers to pesticides that are inherently more toxic to some pests to the others.

Physogastric: Having the abdomen grossly swollen, and mainly or entirely membranous; the sclerites being small and widely separated, e.g. a queen termite.

Physopoda: An alternative name for the order Thysanoptera (thrips).

Phytocidal: Having an ability to injure or kill plants.

Phytomyza: A genus of leaf miner-flies of the family Agromyzidae of order Diptera whose larvae make tunnels between the upper and lower surfaces of leaves. (See Pea leaf miner).

Phytophagous: Feeding on plants or on materials of plant origin.

Phytosanitation: Measures requiring the removal or destruction of infested or infected plant material likely to form a source of reinfection or reinfestation.

Phytosaprophagous insects: A general name for insects that feed on dead or decaying plant material, e.g. Collombola and Thysanura, the larvae of many Diptera and numerous species of Coleoptera.

Phytotoxic: Injurious to the plants. A chemical liable to damage or kill plants or plant parts.

Pytotoxicity: Injury caused to plants notably by pesticides.

Pictured: With spots or bands.

Piercing-Sucking mouth parts: Found in bugs and mosquitoes. They consist of an outer tube formed by the labium which is merely a protective covering for the stylets inside. The mandibles, maxillae, labrum-epipharynx and hypotharynx are modified to form the stylets which are used for piercing and sucking.

Pieridae: A world wide family of butterflies of the order Lepidoptera which are characteristically white yellow or orange. They have six functional legs with a pair of forked claws on each foot. The caterpillars are usually green or yellowish and many of them are agricultural pests, particularly of Cruciferous plants, e.g. cabbage butterfly (*Pieris brassicae*).

Pigeonpea pod fly: Scientific name - *Melanogromyza obtusa* Mallas (Diptera: Agromyzidae). It is serious pod borer of pigeon pea in India but is more serious in northern India. The adult is a small metallic black fly whose tiny maggots bore into the pods and feed on seeds.

Pigments: The wide variety of colours in insects is due partly to chemical pigments and partly to the affect of optical interference caused by the physical structure of the surface. Following chemical pigments are present:

(i) *Melanins*: Black or brown derivatives of the amino acid tryosine

(ii) *Biliverdin*: A green substance similar to the bile pigment of humans, derived from cytochrome which is universally present and plays a part in cell respiration

(iii) *Ommochromes*: Red, yellow or brown pigments in the eyes of insects and sometimes also on other parts.

(iv) *Anthroquinones*: Bright red substances such as cochineal obtained from the scale insects.

(v) *Carotenoids*: Yellow and red pigments obtained from the plants and found in the wing covers of the lady bird beetles and the Colorado beetle.

(vi) *Pterines*: White, yellow, orange and red substances in the wing scales of butterflies.

Pile: Thick short hairs, giving the appearance of velvet.

Pilica basalis: The ventral fold of the wing or line of flection between the base of the mediocubital field and the axillary region forming a prominent convex fold in the flexed wings extending between the median plates from the articulation of radius with second axillary to the articulation of the vannal veins with the third axillary.

Pilica jugalis: The jugal fold of the wing in some of insects, or radial line of folding setting off the jugal region from the vannal region.

Pilica vannalis: The vannal fold of the wing, or redial line of folding usually between the cubital field and the first vannal vein, but somewhat variable in position.

Pilifer: One of the pair of lateral projections on the labrum (Lepidoptera).

Pilose: Covered with the hairs.

Pincers: Claspers or gonopods at the hind end of the abdomen of certain insects. They may be present in both sexes (eg. earwings), but more often they are on the males only. They may be used for attack or defence or for grasping the females at mating; pincers-like jawas in some insects.

Pinching: The bigger insects, notably butterflies and moths can be killed by pinching the thorax between thumb and finger in one neat, quick, decisive movement. This is a knack and is learned only with practice. This method kills the insects instantaneously, without giving it a chance to struggle and break its wings, or knock off the scales.

Pink boll worm: *Pectinophora gossypiella* (Saunders) (Lepidoptera: Gelechidae). It is most destructive pest of cotton in America, Africa, Australia and Asia. Damage is caused by the caterpillars only, which are pink and are found inside the flower buds, the panicles and the bolls of cotton or fruits of Okra. The adult moth is deep brown moth with blackish spots on the forewings and the margins of the hindwings are deeply fringed.

Piscicide: A pesticide used to kill fish.

Piston pump: It is a plunger pump that takes in definite volume of spray fluid from the inlet area and transfer it to the outlet.

Placoid receptors: Sensory organs, possibly olfactory, on the antennae of some insects. They are often called pore-plates and consist of a number of sensory cells beneath a very thin plate of cuticle.

Planidium larva: A type of first instar larva in certain Diptera and Hymenoptera which undergo Hypermetamorphosis; a larva that is legless and some what flattened.

Planta: A median ventral sclerite of the pretarsus distal to the unguitractor plate.

Plant bug: A general name for insects of the order Hemiptera. The mouth parts of these insects are piercing and sucking type with mandibles and maxillae in the form of long stylets lying in trough-like labium. Palps are reduced or absent. In most cases probascis can be folded back under the head and thorax, when not in use. Most insects are herbivorous and some are predators.

Plant lice: A general name for small sap sucking insects of the order Hemiptera and the suborder Homoptera. Hemimetabolous insects with piercing and sucking mouth parts. Many have 4 membranous wings and although wingless forms exist. Plants lice term generally applied to aphids but in broader sense may be extended to include white flies (Aleyrodi-dae), tree hoppers (Membracidae), leaf hoopers (Cicadellidae), frog hoppers (Cercopidae).

Plant resistance: It is a relative term not absolute. In practice it represents the ability of a plant to produce a high yield of good quality, when compared with a susceptible plant that produces a low yield of poor quality.

Plantulae: Adhesive pads on the tarsal sclerites of some insects.

Plasma: It constitutes insect blood. It consists mostly of water and also contains amino acids, proteins, fats, sugars and inorganic salts and it transports food materials, hormones and small quantity of oxygen and carbon dioxide.

Plastron: A film or bubble of air trapped in the curved hydrofuse hair on the body of aquatic insects. It is used for respiration. Oxygen withdrawn from the plastron by respiration is replaced by diffusion from the water.

Platyform larva: A larva that is extremely flattened as the larva of the Psepheridae. (Also called onisciform larva).

Plecoptera: (Plekein, to fold: pteron, a wing). The stone flies are softbodied insects and they have long thread-like antennae. Wings (four) are membranous and mouth parts are biting type. The adults of moth do not feed and live only for a short time.

Pleiotrophy: The capacity of a gene to affect several characters, that is several aspects of the phenotype.

Plenary powers: Special powers granted to the International Commission on Zoological Nomenclature (Commission).

Plesiomorph: Of characters, primitive or as found in ancestors.

Pleural: Pertaining to pleura, or latral sclerites of the body; lateral.

Pleural apophysis: The internal arm of the pleural ridge.

Pleural region: The podial region or ventrolateral parts of the body on which the limbs are implanted, metamerically divided into segmental pleural areas.

Pleural ridge: The endopleural ridge formed by the pleural suture, bracing the pleuron above the leg or between the coxal articulation and the wing support.

Pleural suture: A suture on a thoracic pleuron extending from the base of the wing to the base of the coxa, which separates the episternum and the epimeron.

Pleural wing process: The wing support of the pleuron at the upper end of the pleural ridge.

Pleurite : Any minor sclerite of the pleural area of a segment or one of the component sclerite of a pleuron; a lateral or pleural sclerite.

Pleuron (Pl. Pleura) : The sclerotization of the pleural area of a segment apparently derived from the proximal (subcoxal) part of the primitive limb basis and usually subdivided into pleurites.

Pleuropodia: Vestigial leg-like organs on the first abdominal segment of some insect embryos (e.g. those of grasshoppers and locusts), functioning as glandular organs secreting an enzyme that helps to dissolve the shell of the egg and to facilitate the hatching of the insect.

Pleurostoma: The subgenal margin of the cranium bordering the mandibles.

Pleurostomal suture: The part of the subgenl suture above the mandibles.

Pleurotergite: A sclerite containing both tergal and pleural elements.

Pleuroventral line: The line of separation between the pleural region and the ventre; lying mesad of the limb bases but obscured when the latter are fused with the sterna.

Plodia: *Plodia interpunctella,* the Indian meal moth; a world wide pest of flour, cereals, etc. (See Indian meal moth).

Plume moth: Scientific name- *Exelastis atomosa* Walshingham (Lepidoptera: Pterophoridae). It is a specific pest of pegionpea in many parts of India, particularly in Andhra Pradesh, Assam, Madhya Pradesh, Tamil Nadu, Maharashtra, Karnataka. It has also been recorded in Natal, and new Guinea. The caterpillars cause damage by boring into the pods and eating the grains. The hind wings are deeply cleft and fringed making them look like a bunch of feathers.

Plumose: Feather-like; Plumose antennae.

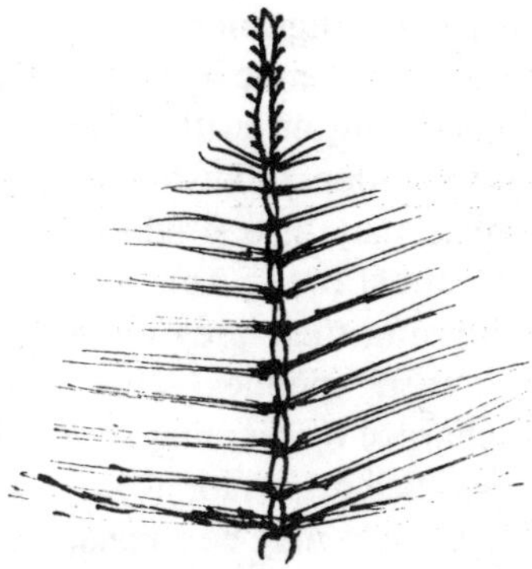

Plumose antenna

Plutella: *Plutella maculipennis* (The diomond back moth). It is a pest of cabbage and other cruciferous crops in several countries.

Pneumatic pump: See pneumatic sprayer.

Pneumatic sprayer: The sprayer which is fitted with an air pump is known as compression sprayer or pneumatic sprayer. The air pressure is employed in the sprayer for forcing a liquid through the nozzle for forming a spray.

Podeum (Podeon): A narrow stalk or petiole connecting the thorax with the abdomen in some insects.

Podex: The pygidium or hind part of the abdomen of an insect consisting of the last two or three segments sometimes fused together or modified for specific functions.

Podical plates: Latero-ventral plates formed by the modification of the both abdominal segment of certain insects. They are on each side of the anus and may bear anal cerci.

Podite (Podomere): A limb segment of an insect or other orthropods but only those which have independent musculature.

Point: A small triangle of stiff paper, on which the small insects are mounted.

Poison: Assimilation of these substances in small quantity through inhalation, ingestion or absorption results in systemic injury, acute toxic conditions or death within a reasonable time following exposure. Insects may also produce poisonous substances for killing or paralysing their prey or for repelling their enemies. This may be discharged by piercing sting (often a modified ovipositor as in bees and wasps) or it may be squirted (as in ants). It may present in the saliva (bugs, fleas, mosquitoes) or may enter the wound by a bite from the large mandibles (as in certain beetles). Some arctiid caterpillars have stinging hairs, others discharge poison from special glands in their skin. The majority of insect poisons contain a mixture of soluble proteins and enzymes, amino acids, histamins, various acids, alkalies, esters, blistering oils, nerve depressants, alkaloids and numerous other organic substances.

Poison bait: It consists of relatively small quantity of a poison combined with a food material of the pest.

Pollen basket: A smooth area on the outer surface of the hind tibia, bordered on each side with a fringe of long curved hairs, which serve as a pollen basket (bees) (Also called corbicula. pl, corbiculae). In leaf eating bees and some other solitary species the pollen basket is situated underneath the abdomen.

Pollen brush: Also called a scopa. A bursh formed from rings of stiffened setae on the first tarsal segment of the hind leg of a honey bee or similar species. It is used for brushing pollen off the body and into the pollen basket of the corresponding leg on the opposite side.

Pollen collecting leg: In these legs, the metatibia is a concave basket into which pollen grains are filled by a brush-like tarsus or pollen comb of the opposite leg, e.g. honey bees.

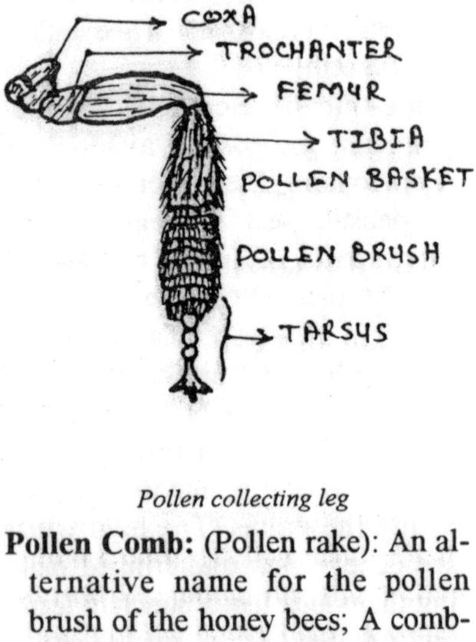

Pollen collecting leg

Pollen Comb: (Pollen rake): An alternative name for the pollen brush of the honey bees; A comb-like structure in the joint between the tibia and tarsus of the first leg of a honey bee used for cleaning pollen off the head and antennae.

Pollen Compressor: Sometimes called the auricle on the hind leg of a bee. It is a flattened plate forming the base of a concavity at the upper end of the first tarsal segment. Pollen is brushed into here and compressed by flexing the leg before being pushed into the pollen basket on the tibia.

Pollination: Transfer of male gametophyte (pollen) to female flower or male part of flower.

Pollutant: A harmful chemical or waste material.

Pollute: To render unsafe or impure by carelessness or misuse. To add an un-wanted material (often a pesticide) which may do harm or damage.

Polyandry: The condition of one female having neumerous male mates.

Polyembryony: The production of two or more off springs (embryo) from a single egg, by division of the embryo at very early stage, e.g. parasitic Hymenoptera.

Polygamy: The condition of single male having many female mates.

Polygenic: Of a character controlled by several or numerous genes.

Polymorphism: The simultaneous occurrence of several discontinuous phenotype or genes in a population with the frequency even of the rarest type higher than can be maintained by recurrent mutation. These may be simple sex differences or age differences but there are also other genetic and environmental factors which may bring about polymorphism. Bees, Ants, Wasps, and Termites have different 'caste', e.g. workers, drones, queens, soilders. Few insects have multiple larval forms, e.g. blister beetle. The larval instar (triungulin) is very active, later instars may be relatively inactive. Aphids have winged sexual and wingless parthenogenetic forms.

Polynominal nomenclature: A system of nomenclature consisting of a scientific designation of a species through more than two words; the antecedent of the Linnaean "binomial system".

Polyphaga: A suborder of the order Coleoptera whose larvae are either legless or have legs with only one claw. The antennae are of various forms and there is generally an absence of cross veins in the hind wings.

Polyphagous insects: Insects, such as grasshoppers and locusts which feed on variety of plants or parasitizing a number of different hosts, however, they may still be selective.

Polyphyletic: A term applied to a composite taxon derived from two or more ancestral sources; not of a single immediate line of descent.

Polyploidy: A condition in which the nuclear compliment of chromosome is an integral multiple (greater than two) of the haploid number.

Polypod larva: This type of larva is free, active and has six legs on the thorax and a number of abdominal processes (prolegs) which may be used for locomotion for, e.g. common in Lepidoptera. Sometimes also called Eruciform larva.

Polypod larva

Polythetic: Of taxa, in which each member has a majority of a set of characters.

Polytopic: Occurring in different places as for instance, a subspecies composed of widely seperated populations.

Polytrophic egg tube (Ovariole): A type of egg tube consting of a successive series of chambers, each of which contains trophocytes for nourishing the developing egg cell.

Polytypic: A taxon containing two or more taxa in the immediately subordinate category, as a genus with several species or a species with several subspecies.

Polyvinyl lactophenol: Polyvenyl alcohol (2.5 g); distilled water (12-20 cc); lactophenol solution (25-32 cc). The poyvenyl alcohol is obtainable in varying viscosities and the quantity of other two constituents are increased to the viscosity used; this gives a medium of decreasing refractive index. Also used as a clearing agent.

Polytrophic egg tube

Population: A group of individuals of the same species living in limited defined area or ecosystem.

Porcupine: *Hystrix indica*. It is found throughout India, and prefers to inhabit rocky and nearby regions. It is noctural and lives in long tunnels. It inflicts damage to field crops, orchards, etc.

Pore canal: The channel of the cuticle beneath the seta or other external part of many sense organs.

Porrect: Extending forward horizontally; porrect antennae.

Porthetria: An alternative name for Lymantria.

Postabdomen: The modified posterior segments of the abdomen which are usually more slender than the anterior segments (Crustacea).

Postalar bridge (Postalare): A lateral extension of the postnotum of a wing-bearing segment behind the wing base, generally united with the epimeron.

Posalar callus: A rounded swelling on each side of the mesonotum between the base of the wing and the scutellum (Diptera).

Postantennal appendages: The appendages of the tritocerebral somite; the chelicerae of Chelicerata, the second antennae of Crustacea, embryonic rudiments in some Hexapoda, absent in Myriapoda.

Postcoxal bridge (Postcoxale): The postcoxal part of the pleuron often united with the sternum, behind the coxa.

Postcubitus: The usual sixth vein of the wing, represented by an independent trachae in most nymphal wings, associated basally with the cubitus in the adult.

Posterior: Hind or rear.

Posterior cell: One of the cells extending to the hind margin of the wing, between the third and sixth longitudinal veins (Diptera).

Posterior cross vein: A cross vein at the apex of the discal cell (Diptera).

Posterior intestine (Rectum): The terminal section of the proctodaeum commonly termed the "rectum" but usually divided into an anterior rectal sac, and a posterior rectum proper.

Posterior mesentron rudiment: The posterior group of cells of the ventral endoderm remnant that regenerates the mesenteron.

Posterior notal wing process: A posterior lobe of the lateral margin of the alinotum supporting the third axillary sclerite of the wing base.

Posterior pharynx: A pharyngeal chamber of the stomodaeum behind the brain present in Orthoptera, Coleoptera and some other insects.

Post frontal suture: Facial sutures present in some insects diverging from the coronal suture laterad of the antennal base.

Postgena (Pl. Postgenae): A sclerite on the posterior lateral surface of the head, posterior to the gena.

Postgene: The lateral and ventral parts of the occipital arch or the areas of the cranium posterior to the genae.

Posthumeral bristle: A bristle on the anterolateral surface of the mesonotum near the margin of the humeral callus (Diptera).

Postmarginal vein: The vein along the anterior margin of the front wing beyond the point where the stigmal vein arises (Chalcidoidea).

Postmentum: The postlabium, or basal part of labium proximal to the stipital region, or prementum; when sclerotized containing either a single postmental plate, or a distal mental plate and a proximal sub or a distal mental plate and a proximal submental plate.

Postnodal cross veins: A series of cross veins just behind the costal margin of the wing, distad of nodus, and extending from the costal margin of the wing to R_1 (Odonata).

Postnotum (Phragmanotum): The postscuteller, phregma-bearing plate, often present in the dorsum of an alate segment derived from the anterior part of the following tergum. Also called post-scutellum.

Postoccipital suture: The posterior submarginal groove of the cranium having the posterior tentorial pits in its lower end, internally it forms the post occipital ridge on which are attached dorsal prothoracic and neck muscles of the head.

Postocciput: The extreme posterior rim of the head, between the postoccipital suture and the foramen magnum.

Post scutellum: A small transverse piece of a thoracic notum immediately behind the scutellum; in Diptera, an area immediately behind or below the mesoscutellum.

Poststergite: The narrow post costal tip of a definitive sternal plate that includes the intersegmental sclerotization following.

Posttergite: The narrow post-costal lip of a postnostal thoracic plate.

Postural hairs: Rows of specilized hairs or setae with sensory receptore (proprioceptors) at their bases so that a change of position of the hair gives the insect information about its posture. They are chiefly situated at the joint specially between the head and thorax and between thorax and abdomen.

Postverticle bristle: A pair of bristles behind the ocelli, usually situated on the posterior surface of the head (Diptera).

Potable: Water suitable for drinking.

Potato tuber moth: *Gnorimoschima operculella* (Zeller) (Lepidoptora: Gelechiidae). It is a pest of potato and also attacks tobacco, tomato, egg plant and solanaceous weeds. The larvae which are pinkish white or greenish in colour with dark brown head cause damage to tubers. The adults is small narrow winged, nocturnal moth with greyish brown molted with dark brown wings. Another species of tuber moth, *G. heliopta* (Lower) has also been reported on potato tuber.

Potency: The strength of something. Example, how deadly a poison is.

Potentiation: The joint action of two pesticides to bring about an effect greater than the sum of their individual effects.

Potential pests: The pests which have never caused important damage because their population have always been kept low by density dependent mechanism.

Potsan: (Zinophos; 0, 0-diethyl-0-(4-methyl counarinyl-7) thiophosphate). It is a selective organophosphorus insecticide with stomach and contact action.

Poultry louse: *Menopon gallinae* Linn. (Mallophaga: Menoponidae). It is also called the 'shaft louse' of fowl because it is seen running towards the body along the shaft of a feather when the bird part its wings. The body is dorso-ventrally flattened and wingless and biting moth parts. It does not suck blood but feeds on epidermal scales, bite of feather, etc. Along with louse are also found the *Eumenacanthus stramineus* Hitzch, *Goniodes gigas* Tasch. and *Lipeurus caponis* Linn.

ppm: Parts per million (1:1000,000). A way of expressing amounts of chemicals in foods, plants, animals, etc.

Preantennae: Theoretically a pair of primitve procephalic appendages anterior to the antennae; possibly represented in *Sclopendra* and *Dixippus* by a pair of embryonic preantennal lobes absent in all adult arthropods.

Preapical: Situated just before the apex.

Prebaiting: Tempting rats to feed on unpoisoned baits for a day or two before the use of poison bait.

Prebsilas: A narrow transverse sclerite, just posterior (basal) to the mentum, in the gnathochilarium of some millipedes.

Precautions: Warnings; safeguards.

Precipitate: To settle out.

Precipitation: The amount of rain, snow, sleet or hail that falls.

Precipitin reaction: The formation of a visible precipitate at the interface when an antigen and the corresponding antiserum are brought together.

Precosta: A small first vein of the wing in certain fossil insects.

Precoxial bridge (Precoxale): The precoxal part of the pleuron anterior to the trochantin, usually continuous with the episternum, frequently united with the sternum sometimes as distal sclerite.

Predaceous: Feeding as predators.

Predacide: A pesticide used to control vertebrate predator pests.

Predator: An animal that attacks and feeds on other animals (its prey), usually animals smaller or less powerful than itself. The pray is usually killed quickly, and mostly or entirely eaten; many prey invididuals are eaten by each predator.

Predator bridge (Prealare): A lateral extension of the presctal area of the abnotum before the wing base, sometimes connected with the episternum.

Predictive value: The capacity of a classification to make predictions on newly employed characters or newly discovered taxa.

Predisposition: Making plant more susceptible to a pest or disease, usually as a result of genetic, cultural or environmental defects.

Preference: The factor by which certain plants are more or less attractive to insects by virtue of their texture, colour, aroma, or taste.

Pregenital: Anterior to the genital segments of the abdomen.

Prelabium: The distal part of the labium, comprising the prementum, the ligula and the papi.

Pre-Linnaean name: A name published to Jan. 1, 1878, the starting point of zoological nomenclature.

Prementum: The free distal parts of the labium of an insect, bearing the labial palps and the ligula.

Preoral: Anterior to or infront of the mouth.

Preoviposition period: The period of time between the emergence of adult female insect and the start of its egg laying.

Prepectus: An anterior or marginal sclerite of the sterno-pleural area of a segment, set off by a transverse suture continuous through the sternum and episterna.

Pre-plant: Application of pesticide prior to planting crop.

Prepupa: A quiescent stage between the larval period and the pupal period; the third instar of thrips. It is sometimes considered to be a pharate pupa (pharate instar).

Prescutum: The anterior of three sclerites which typically form the notum or dorsal cuticular covering over each of the three thoracic segments of an insect.

Prestomal teeth: Small chitinous teeth round the prestomum or outer mouth of a blow fly or similar insect, distinct from pseudotracheal teeth, which are formed from the chitnous rings of the pseudotracheae.

Prestomum: A mouth to which the food channels or pseudotracheae converge at the bases of the two labella or oral lobes on a blow fly, house fly. It is not a true mouth since this is further back at the entrance to the pharynx.

Presutural bristles: Bristles on the mesonotum immediately anterior to the transverse suture and adjacent to the notopleuron (Diptera).

Prescutal suture: A transverse groove of the mesonotum or metanotum behind the antecostal suture, setting off a prescutum from the scutum, and forming internally a prescutal ridge.

Prescutum: The anterior area of the mesonotum or metanotum between the antecostal suture and the prescutal suture, when the latter is present.

Presternum: A narrow anterior area of the sternum sometimes set off from the basisternum by a submarginal suture of the eusternum (not the acrosternite).

Pretarsus (Pl. Pretarsi): The terminal part (segment) of the leg distal to the tarsus, including the median remnants of thc dactylopodite, and the lateral claws, or ungues; in most larvae a simple claw-like segment.

Prevailing: The predominant or general occurrence or use.

Preventative: A measure applied in anticipation of pest attack.

Primary homonym: Each of two or more identical species-group names which, at the time of original publication, were proposed in combination with the same generic name.

Primary intergradation: A zone of intermediacy between two phenotypically different populations, having developed *in situ* as a result of selection.

Primary segmentation: A segmental division of the body corresponding to the embryonic metamerism.

Primary zoological literature: Literature dealing with animals or zoological phenomena, not merely a listing of names.

Primates: The Rhesus monkey (*Macaca mulatta* Zimmermann), Bonnet monkey (*Macaca radiata* Geoffroy) and the Langur, (*Presbytis entellus* Dufresne) are the chief primates, which cause damage to vegetation and crops.

Primitive insects: Insects have evolved from a wingless arthopod which is intermediate between insects and millipedes. The insects, such as springtails (Collembola), silverfish (Thysaneura), proturans (Protura), and campodeans and japygids (Diplura) are nearest to the hypothetical ancestor. Thesc insects which are not regarded as insects by some authorities, are called primitive insects.

Priority: The principle that of two competing names for the same taxon (below the rank of an infraorder) ordinarily that is valid which was published first.

Proboscis: The lengthened mouth parts of an insect specially adopted for piercing, sucking or other specialized modes of feeding. The proboscis may be formed by a lengthening of any or all of general mouth parts (eg. labium, maxillae, mandibles, hopopharynx and labium). In butterflies, the tubular tongue is fomed by the two lengthened, interlocking maxillae, the mandibles having disappeared. In bees, the labium is lengthened for sucking nector and there is a sensory pad (flabellum) at the end of it. In plant and blood sucking bugs, the labium forms a lengthened trough-like lower lip with the mandibles and maxillie in the forms of fine piercing stylets resting in it. In mosquitoes, the labrum, maxillae, mandibles and hypopharynx are all in the form of piercing stylets; the hypopharynx forming a chamber down which saliva is discharged. In the house fly, the proboscis is highly specialized for sucking and filteration of food. The main part is the labium developed into a large pair of oral lobes (labella) in the form of pads containing suction tubes (Pseudotracheae) which can be pressed against the surface on which the insect is feeding.

Procaryotes: Those microorganisms (virus, bacteria, etc.) that lack a well developed and defined nuclei and meiosis.

Procephalon: The region of the definitive head anterior to that of the gnathal segments, representing the primitive protocephalon, formed in the embryo from the cephalic lobes and the tritocerabral segment; bearing the labrum, the mouth, the eyes; the antennae, and the second antennae when the last are present.

Procerebrum: The protocerebrum (the forebrain of an insect) plus parts derived from embryonic presegmental ganglion. Sometimes also used synonymously with protocerebrum.

Proclinate: Inclined forward or downward.

Proctodaeum: The hindgut, or the hindmost of the three major divisions of the alimentary tract, from the malpighian tubules to the anus.

Procuticle: The inner portion of the cuticle, consisting of endocuticle and exocuticle.

Produced: Extended; prolonged; projecting; born.

Product: A term used to describe a pesticide as it is sold. It usually contains the pesticide chemical plus a solvent and additives.

Prospimeron: The epimeron of the prothorax.

Proepisternum: The episternum of the prothorax.

Profile: The outline of as seen from the side or in a lateral view.

Prognathous: Having the head horizontal and the mouth projecting forward.

Prolegs: Short unjoined legs on some of the abdominal segments. Except for the loopers (Geometridae), which have only two pairs, most caterpillars of moths and butterflies have five pairs of prolegs. The micro-pterygidae have eight, and sawflies usually have 6 or 7 pairs. Prolegs are used for grasping and this is helped by hooks called crochets, which are absent in sawflies.

Prolonged exposure: Contact with a pesticide chemical or its residue for a long time.

Prominent: A raised, produced, or projecting portion.

Pronotal comb: A row of strong spines borne on the posterior margin of the pronotum (In Siphonaptera).

Pronotum: The dorsal sclerite of the prothorax.

Pronymph: A newly hatched first instar nymph of a dragonfly or other hemimetabolous insects.

Propellent: An agent that propels i.e. drives away; a driving agent.

Prophylactic control: The control of pests by adopting preventing measures.

Propleural bristles: Bristles located on the propleuron (In Diptera).

Propleuron (Pl. Proleura): The lateral cuticular plate, or pleuron, of the prothorax.

Propneustic: The insects with only the first pair of spiracles (those on the prothorax) are functional, e.g. pupae of certain Diptera.

Propodeum: The narrow posterior portion of the thorax, which is actually the first abdominal segment united with the thorax (Hymenoptera, Suborder Apocrita).

Propodite (Tarsus): The penultimate segment of a generalized leg.

Proprietory chemical: A chemical made and marketed by a person having the exclusive right to manufacture and sell it.

Proprietory name: Distinguishing name given by the manufacturer to a particular formulated product.

Propupa: The instar preceding the pupa, in which the wing pads are present and the legs are short and thick (Thysanoptera) (sometimes called the prepupa).

Prosoma: A term referring to the anterior part of the body, usually applied to the cephalothorax; the anterior part of the head or cephalothorax.

Prosternum: The sternum or ventral sclerite of the prothorax.

Prostomium: The anterior preoral unsegmented part of the trunk of a segmented animal.

Protaptera: A suggested name for a hypothetical common ancestor of insects, millipedes and centipedes.

Protarsus: The tarsus of the first leg of an insect.

Protectant: A chemical applied to the plant or animal surface to prevent infection or injury by the pest.

Protection: The prevention of a pest or pathogen from producing economic injury to a crop by placing a chemical or physical barrier between the pest or the inoculum of the pathogen and the host or suscept.

Protective colouration: Any pattern or arrangement of colours in an animal which enables it to escape detection by resembling or blending into its sorroundings.

Prothetely: Precocious appearance of an adult characters as a result of insufficient juvenile hormones.

Prothorax: The first segment of the thorax.

Prothoracic glands: These are paired glandular structures located in the ventolateral areas of the prothorax (extending into the mesothorax in some species), and are sometimes partially fused in midline. These secret moulting (ecdysone) hormone which helps in moulting.

Prothorax: The first segment of the thorax; the anterior of the three thoracic segments. It bears the first pair of legs.

Protocephalon: A general early stage in the evolution of the arthropod head, corresponding to the cephalic lobes of the embryo, comprising the prostomium and usually the first postoral somite, forming the procephalic region of the definitive insect head.

Protocerebrum: The first part of the arthropod brain, containing the occular and other association centres lying anterior or dorsal to the antennal (deutocerebral) centres.

Protonymph: The second instar of a mite.

Protopamchloride: An antidote for organophosphate poisoning, but not for carbamate poisoning.

Protopodite: The basal stalk of some crustacean limbs composed of the united coxopodite and basipodite.

Protopod larva: The most primitive type of insect larva with unsegmented abdomen and rudimentary appendages. Such larvae are usually endoparasites and can only survive where food is abundant making locomotion unnecessary, e.g. minute parasitic Hymenoptera.

Protopad larva

Protura: Minute wingless insects, 1 to 2 mm long, without eyes or antennae and with piercing mouth parts, contained in pockets from which they can be extruded at feeding. They live under bark and stones or in rotting vegetation.

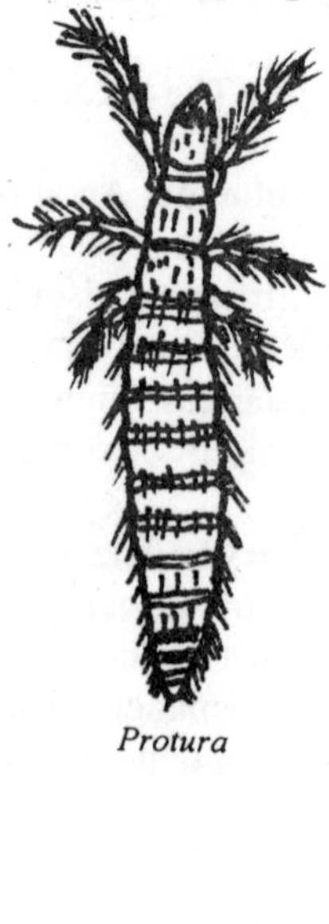

Protura

Proventriculus: The posterior portion of the foregut; the gizzard; a specialized part of the stomodeum immediately anterior to the ventriculus.

Proximal: Near to the body or to the base of an appendage. Towards the end.

Prozonite: The anterior portion of an millipede tergum, where the tergum is divided by a transverse groove.

Pruinose: Covered with a whilte waxy powder.

Przibram's rule: An empirical rule observed in certain insects according to which the weight of an insect doubles during the each instar and the linear dimensions are increased at each moult by the ratio 1:26. This can be explained on the assumption that during each stage every cell divides once and grows to its original size.

Pseudaposematic characters: False warning colourations (See Batesian mimicry).

Pseudarolium (Pl. Pseudarolia): A pad at the apex of the tarsus resembling an arolium.

Pseudocone eyes: Compound eyes such as those of the house fly in which the crystalline cones are perfectly formed and consist of a mass of liquid. In the eyes of this type the cuticular lenses form the main part of the optical system.

Pseudocubitus: A vein appearing as the cubitus, but actually formed by the fusion of the branches of M and Cu_1 (In Neuroptera).

Pseudogamy (Gynogenesis): Parthenogenetic development of the egg cell after the egg membrane has been penetrated by a male gamete.

Pseudohalteres: Reduced or vestigial forewings of Strepsiptera in the form of small club shaped processes. They resemble but are not homologous with the halters of Diptera, the latter being reduced hind wings.

Pseudomedia: A vein appearing as the media but actually formed by the fusion of the branches of Rs (In Neuroptera).

Pseudoplacental viviparity: A phenomenon found in certain insects in which the egg has very little yolk and the embryo is nourished through placenta-like structure in contact with material tissues. It occurs in *Hemimerus,* the parasitic earwing, in some aphids, and Psocoptera and in the Polyctenidae (Hemiptera).

Pseudopod (Pl. Pseudopodia): A fleshy, abdominal leg, not a true segmented appendage (also called proleg).

Pseudopupa: A coarctate larva; a larve in a quiescent pupa like condition, one or two instars before the true pupal stage (In Coleoptera, Meloidae).

Pseudotracheae: Food channels resembling tracheae in the membrane covering the labellar lobes of house flies, blow flies, etc. Each lobe has a large number of these tubes which are supported and kept open by incomplete sclerotized rings. They converge into collecting ducts which led to the prestomium and hence to the mouth.

Pseudotracheal teeth: Crushing organs formed from the chitinous rings of pseudotrecheae. Contrast with prestomal teeth.

Psocoptera: This order includes book lice and their allies. Insects are small, soft bodied, winged or apterous. Mouth parts are of biting type and have no cerci.

Pteralia: The articular small sclerites of the wing base, including the humoral plate and the axillary plate or axillary sclerites which make up the articulation.

Pterergate: An exceptional form of worker ant having the rudiments of wings.

Pteridine: The basic chemical constituent of pteroic acid of the pterin pigments in the wings of insect.

Pterin pigments: White, yellow or red pigments in the bodies and wings of insects, chemically related to uric acid and possibly waste products of metabolism.

Pterioc acid: An essential substance for growth; a constituent of the vitamin folic acid also important for the growth of the feathers of birds and the formation of pterin pigments (Pterines) in insects.

Pteropleural bristles: Bristles on the pteropleuron (In Diptera).

Pteropleuron (Pl. Pteropleura): A sclerite on the side of the thorax, just below the base of the wing, and consisting of the upper part of the mesepimeron (Diptera).

Pterostigma: A wing spot; a thickened, darkly pigmented opaque spot along the costal (fore) margin of the wing, near the wing tip of the vein R_1 (also called the stigma in Odonata).

Pterostigma

Pterotheca: The part of a pupal skin that covers the developing wings of an insect.

Pterothorax: The two wing bearing segments of the thorax, often closely united or connected with each other, i.e. second and third thoracic segments.

Pterygota: A subclass of Class Insects. In this insects are primarily winged, but sometimes secondarily wingless.

Pterygote: Winged insects, a member of the subclass Pterygota.

Ptilinal suture: The crescentric groove cuting across the frons above the antennal bases in Diptera where the ptilinum has been withdrawn.

Ptilinum: An eversible sac in the head of some Diptera used in ruptering the puparium. In the adults it is withdrawn through the frontal (ptilinal) suture leaving a U shaped suture.

Pubescent: Covered with short fine hairs; Hairy.

Pulsatile organs: Accessory pumping organs that are sometimes present at the bases of the antennae and the wings for pumping blood into them.

Pulsating membrane: Small muscular membranes found in the thorax, head and appendages of various insects, rhythmic contractions of which probably contribute to the circulation of the blood.

Pulse beetle: *Callosobruchus chinensis* Linnaeus (Coleoptera: Bruchidae). This insect is found in the U.S.A., Mauritius, Formosa, China, Africa, the Philippines, Japan, Indonesia, Sri Lanka, Burma and India. It is a pest of gram, mung, urid, tur, peas, cowpea, lentil, cotton seed, sorghum and maize. The larvae cause the damage by feeding inside the grain. The adult beetle is oval, chocolate or reddish brown and has long serrated antennae. Another species, *C. analis* has also been reported on various pulses.

Pulvilliform: Lobe-like or pad-like; shaped like a pulvillus; pulvilliform empodium.

Pulvillus (Pl. Pulvilli): A pod or lobe beneath each tarsal class (In Diptera); lateral lobes of the pretaraus arising beneath the bases of the claws; ventral lobes of the tarsal subsegments (euplan-tulae) are sometimes called tarsal pulvilli.

Puctate: Pitted or beset with punctures.

Puncture: A tiny pit or depression.

Pupa: Pl. Pupae. The stage between the larva and the adult and usually an inactive stage. The pupa or chrysalis is formed after last larval moult. Numerous internal changes takes place in it. These involve the breaking down of larval organs and tissues and the growth of the adult organs. Three principal types of pupae are present in insects. They are (i) Exarate—those with free limbs and mouth parts and also are capable of certain amount of movement, e.g. bees. (ii) In some cases an exarate pupa is enclosed in a puparium formed from the dried skin of the previous larval instar, e.g. house fly. This is called coarctate pupa. (iii) Obtect—The pupa of moths and butterflies are called chrysalis. It is less mobile because the lcgs, wings and mouth parts are firmly glued to the body. In many cases it is enclosed in a silken cocoon.

Pupa coarctate: See coarctate pupa.

Pupa libera: A free pupa, i.e. one with soft integument and freely moving appendages; an exarate pupa.

Pupal moulting: Moulting to pupal form. If a second pupa is produced as a result of experimental JH treatment, we call this 'second pupal moulting'.

Puparium (Pl. Puparia): A case formed by the hardening of the next to the last larval skin, in which the pupa is formed; a hard barrel shaped case enclosing the pupa of certain Diptera.

Pupate: Transformed to pupa.

Pupation: The act of changing of larvae into the pupae.

Pupation hormones: The pupation appears to be controlled by the balance of several hormones. The corpus allatum secretes a juvenile hormon (neotenin) and as long as this is being produced the insect will remain in larval stage after each moult. After generally a definite number of instars the production of this hormon slows down, the pupation occurs, however the ecdysone produced by the ocdysial glands (prothoracic glands) is produced at this stage.

Pupiparous: Giving birth to the larvae that are full grown and ready to pupate.

Pygidial glands: Glands opening in the pygidium or posterior tip of the abdomen of certain insects. They may secrete poisonous or obnoxious substances as a protection for the insects, eg. Ground beetle (*Carabus)* secrete butyric acid.

Pygidium: The last dorsal segmente of the abdomen, or the hind tip of the abdomen.

Pyloric caeca: Enteric caeca. Sac like branches or diverticula from the midgut just behind the gizzard in certain insect (e.g. cockroaches). They provide additional surface area for secretion and absorption.

Pyloric valve: A valvular fold usually situated in the pyloric region of the proctodaeum but some times formed by the posterior end of the ventriculus.

Pylorus: An anterior part of the proctodaeum usually containing the pyloric valve, sometimes distinctly differentiated from the true intestinal region and forming anatomically the rear end of the stomach.

Pyralidae: A family of the order Lepidoptera having small moths, generally classed with Microlepidoptera although some species are larger. They have narrow forewings and broad, fringed hind wings. Many of them have well developed palps resembling those of the snout moths. They include meal moth (*Ephestia, Plodia*), sugarcane moth (*chilo*).

Pyrethrins: Pyrethrum. The pyrethrins is a general term usually including the cinerins as well as pyrethrins extracted from flowers of *Pyrethrum cinerariaefolium*. They are powerful contact insecticide, non-systemic, causing a rapid paralysis or knockdown.

Pyrethrum: It is a natural organic insecticide found in the flower heads of certain species of chrysanthemum genus. The active principles are Pyrethrin I and II, Cinerine I and II, and Jasomolin II.

Pyriform: Pear-shaped.

Pyrilla: *Pyrilla perusilla.* The sugarcane leafhopper found in India. The nymphs and adults suck the juice of the sugarcane and deteriorate juice quality.

Pyrolan: (1-phenyl-3-methylphyazoly-5-dimethylcarbamate). It is a systemic insecticide.

Pyrophoric: Tendency for a material to ignite spontaneously.

Pyrophyllite: It is the best diluent for most of the dust preparations. It has a lamellar structure and adheres well to plant foliage.

Q

Quadrangle: A cell immediately beyond the arculus (Odonata, Zygoptera).

Quadrate: Four sided.

Quarantine laws: Laws that enable enforcement of restriction on the introduction of pests through transportation from foreign countries (foreign quarantine) and transportation of pests within the country from one place to another (domestic quarantine).

Queen: A fertile female in social insects. These are the only individuals capable of laying eggs in colony of bees, wasps, ants and termites. In a colony of hive bees the queen develops in a special queen cell and is produced by feeding a female larva on partly digested food called 'royal jelly'.

Queen ant: The fertile female and founder of a colony of ants. It is a winged insect similar in appearance to the winged males but usually larger than the wingless workers or incomplete females. The full grown females (queens) during favourable weather conditions fly out on her nuptial flight and these are followed by the males. This is the only time when she flies, and after being fertilized, the wings shed off and she will either return to the same nest or starts a new one.

Queen bee: The fertile female and founder of a colony of bees. In *Apis mellifera*, the queen is slightly longer than the worker but with relatively shorter wings. There is normally only one queen in a hive. After being fertilized during the 'nuptial flight' she returns to the hive and remains there unless driven out by a young queen. She lays thousands of eggs during the life span of 4 to 5 years. In the humble bees (Bombidae), the queen hibernates and is normally the only member of a colony to survive the winter, a new colony being started each year.

Queen substance: A chemical, identified as 9-oxydecenoic acid is produced by the mandibular glands of queen bees. During the 'nuptial flight' the scent of this substance probably attracts the male bees. In the hive, the worker bees lick it and pass it on to each other as a means of recognising the presence of the queen.

Queen termite: The queen is only fertile female. After fertilization and having lost her wings, she stays underground for the rest of her life. She hardly moves, and acts as a egg laying machine. Her abdomen is full of eggs and thus grows many times to its normal size.

Queen wasp: The fertile female or queen is larger than workers and males. In temperate climates, the queen hibernates and is the only member of the colony to survive the winter. In the spring, she lays few eggs and start building a new nest.

Quiescence: A state of inactivity, brought about by unfavourable conditions of temperature, humidity or other climatic conditions, by lack of some essential vitamins or other food substances, by failure to secrete particular hormone or by disturbance of the balance between opposing hormones. In this condition insects hardly move and may go without food for long period. This may occur at any growth stage and even in adult stage of insects.

Quinones: A group of chemical substances (cyclic diketones) in the formation of some of the yellow and red pigment of insects and also in poisonous or repellent secretions such as those emitted by cockroaches, earwigs, etc.

R

Race: Sub species.

Radial cell: A cell bordered anteriorly by a branch of the radius; the marginal cell (Hymenoptera).

Radial cross vein: A cross vein connecting R_1 and the branch of the radius immediately behind it.

Radial sector: The posterior of the two main branches of the radius.

Radius: The longitudinal vein between the subcosta and the media; the third vein of the wing, its base flexibly attached to the second axillary.

Rami valvularum: The proximal, often slender, parts of the first and second valvulae by which the latter are attached to the valvifers.

Ranatra: A genus of elongated water scorpions; predating bugs of the family Nepidae and the order Hemiptera having the front legs modified for seizing their prey (tadpoles and small fish). It lives among pond weeds and breaths through a long respiratory tube present at the end. The eggs are laid in the stems of various aquatic plants.

Ranking: The placement of a taxon in the appropriate category in the hierarchy of categories.

Raptorial: Fitted for grasping pray; raptorial front legs.

Raptorial leg (Preying type): The legs are adopted for catching and holding the prey in between the strong femur and the thin tibia, e.g. preying mantids. These are generally anterior pair but the other are also used occassionally for this purpose.

Rasping-sucking mouth parts: These are intermediate instructures between the piercing—sucking and chewing types and are found in thrips. In this type, the left mandible, maxillae and hypopharynx are elongated which move in and out through and opening in the apex of the cone shaped head. The insects lacerate the plant epidermis and suck the sap as it exudes.

Rassenkreid: Synonym for polytypic species.

Rate: Rate refers to the amount of active ingredient of a pesticide applied per unit area. Rate is preferred to the occassionally used terms, dosage and application.

Ratflea: *Xenopsylla cheopis* (Rothschild) (Siphonoptera: Pulicidae). Also called 'Oriental flea', and is distributed world over. It is a direct transmitter of the bubonic plague, caused by the bacterium, *Pasteurella pestes* to human beings. The adult rat-flea is a hard skinned, very spiny insect with piercing-sucking mouth parts, concealed antennae, no wings and with very large legs for jumping.

Ratooning: The raising of a crop from new shoots that develop stumps (stubbles) of the harvested crop.

Rat-tailed maggot: The legless larvae of various Syrphidae. It has very long tetescopic 'tail' or respiratory siphon that reaches upto the surface of the water. It has two tracheal tubes with their spiracles (openings) at the tip sorrounded by fringe of the floating hairs, e.g. larvae of Drone fly (*Eristalis tenax*).

Recapitulation: The theory that ontogeny recapitulates phylogeny.

Receptaculum seminis: Spermatheca; A sec in a female reproductive system for storing sperm cells required for fertilizing the eggs; the semen recepticle of the female. This facilitates females to mate only once in her life span and retain enough sperm cells in her body to fertilize all subsequently produced eggs.

Recent: In taxonomy, of taxa which still exist; antonym of Fossil; new.

Receptive apparatus: The part of a sense organ primarily responsible to the stimulus transmitted by or through the peripheral parts, formed of the sense cell or cells.

Receptors: So-called sense organs or sensory cells specialized structures of the integument responsive to external stimuli. They include Photoreceptors (e.g. retinulae of simple or compound eyes).

Recessive gene: A gene, not affecting the phenotype of the heterozygote.

Reclinate: Inclined backward or upward.

Recruitment: The addition of new individuals to a population, usually by either birth or immigration.

Rectal gills: The gills within the rectum of nymphs in anisopteran dragon flies. These are also called 'Tracheal gills' and have thin walled projections well supplied with tracheae, arranged in longitudinal rows.

Rectal glands: Thickened pad-like or papililform structures formed from masses of glandular tissues projecting into the wall of the rectal sac or hindgut of many insects. They probably absorb water and chloride or some other ions from the faeces to some extent, counterbalance the excretion tubules and preventing undue loss of water from the insect.

Rectal papillae: See rectal glands.

Rectal sac: The enlarged anterior part of the rectum, sometimes produced into a large rectal caecum.

Rectal valve: A circular or lobate fold of the proctodaeal wall between the anterior intestine and the rectum.

Rectum: The posterior intestine, including the rectal sac, and the posterior narrow rectum proper opening at the anus.

Recurrent nerve: The median stomodeal nerve extending posteriorly from the frontal ganglion.

Recurrent vein: One of the transverse veins immediately posterior to the cubital vein (Hymenoptera); a vein at the base of the wing between the Costa and Subcosta, extending obliquely from the subcosta to the costa (Neuroptera).

Recurved: Curved upward or backward.

Red cotton bug: *Dysdercus koenigi*

(Fabricious) (Hemiptera: Pyrrhocoridae). It is a pest of cotton, wheat, maize, pearl millet, clovers, etc. in India. The adults are elongated, slender, crimson red with white bands across the abdomen. Both adults and nymphs feed on cell sap.

Red flour beetle: *Triboleum castaneum* (Tenebrionidae: Coleoptera). A small reddish brown beetle, about 5 mm long and which is a common pest of cereals and flour all over the world.

Red hairy caterpillars: *Amesacta moorei*. Butler (Lepidoptera: Acctiidae). It is a polyphagus pest, damage is caused by caterpillars whose colour varied from reddish amber to olive green and the body is covered with hairs arising from fleshy tubercles. The moths are stout having white wings with black spots. The outer margins of the fore wings, the anterior margins of the thorax and the entire abdomen are scarlet red. Abdomen had black bands and dots.

Red pumpkin beetles:*Raphidopalpa foveicollis (Lucas)* (Coleoptera: Chrysomelidae): Synonym *Aulacophora foveicollis* Lucas. The two species, red pumpkin beetle, *R. foveicollis* and the blue pumpkin beetle, *R. atripenis* are common in north western India, the former being more important. It is distributed in Asia, Australia, Southern Europe and Africa and is a serious pest of cuccurbitaceous vegetables. Damage is done by grubs and beetles; the grubs bore the roots and stems; the beetles feed on leaves.

Red tea mites: Red coffee mite. *Oligonychus coffeae* (Nietner). It is a most serious pest of tea in north eastern India. The mites are minute and roudish. They form a thin web and live in colonies under it. The damaged area turn brown and the attacked leaves become bronze and fall off in severe cases.

Reduced front legs: Insects normally have six legs (one pair on each of the three thoracic segments). However, in two families of the butterflies (Satyridae and Nymphalidae), the front pair legs are reduced, brush-like and useless for locomotion, thus insects appear to have only four legs. Water bugs (Pondskaters) of the family Gerridae also have reduced forelegs but these are highly efficient for catching and holding prey.

Reductionism: The erroneous belief that complex phenonema can be entirely expalined by reducing them to the smallest possible component parts and by explaining these.

Redundant characters: Characters so closely correlated with other; already used characters that they do not contribute new information to the analysis.

Reduviidae: A family of Hemiptera order, group Heteroptera. It includes nocturnal, predatory, and blood sucking insects. Most of them prey on other insects which they catch by the powerful front legs and pierce the sharp pointed proboscic and suck the blood completely. The nymphs also suck the blood. *Rhinocoris fuscipes* is a predator of several lepidopterous and Coleopterous insect pests in India.

Red Vegetable mite: Tetrahychus telarius Linnaeus (Acarina: Tetranychidae): It is also called two spotted spider mite and is distributed throughout the world. It is a polyphagous pest and is known to feed on 183 species of plants including cuccurbits, brinjal, Okra, on which it is sometimes, very serious. The nymphs and the adults both suck the cell sap of leaves from ventral side by forming a web over the leaves, which interferes in photosynthesis. Full grown nymph is light brown and has two eye spots, four pairs of legs and is quite active. The adult mite is oval, pyriform and variable in colours, which may be ferruginous, red, greenish amber or rusty green.

Reflexes: A reaction; an involuntary movement or exercises of function in a part excited in response to a stimulus applied to the periphery and transmitted to the nevous centres in the brain or in spinal cord.

Regenerative cells: The cells that generate the replacement cells of the ventricularc (midgut) epithelium.

Regenerative crypt: A pocket of the ventricular epithelium containing a group (nidus) of regenerative cells.

Regressive characters: A character which is being reduced or lost in the course of phylogeny, sometimes independently in several related lines.

Regulatory control: See legal control.

Reinfestation: The return of insects or other pests after the left or were destroyed.

Relapsing fever: An acute infectious disease caused by bacteria. The disease is marked by alternating periods of fever and recovery from illness.

Relationship: Linkage of pest and host.

Relative Humidity: The amount of moisture in the air compared to the total amount that the air would hold at that temperature.

Remigial region (Remigium): The wing area anterior to the vannal fold, containing the costal, subcostal, radial, medial, cuboital, and postcubiotal veins.

Remote: Very slight or faint chance; far apart or distant.

Reniform: Kidney-shaped.

Repellent: A chemical that keeps insects away because of its offensive odour.

Repletes: A name sometimes given to the distended and honey filled members of a colony of Honeypot ants (*Myrmecocystus*). They gorge themselves so that the abdomen is enormously enlarged and then hang from the roof of the next as living stores of food for the rest of the colony.

Reproductive capacity: The insect generally produce much more offspring, however, only a small portion of these survive to maturity and remaining engulfed by the parasites, predators and other biotic and abiotic factors. A queen bee may lay 3000 eggs a day and termite may more than this. A wingless female of aphids reproduce parthenogenetically many millions descendents during a season. Chalcid wasps, lay eggs in the eggs of some other insects can increase their number by polyembryony (each produce many embryos). However, some insects produce very little offsprings. But, the population is generally kept to at low level due to biotic and abiotic factors.

Reproductive control: See autocidal control.

Reproductive isolation: A condition in which interbreeding between two or more population is prevented by intrinsic factors.

Reproductive potential: The ability of a population of organisms to increase in density under normal conditions (see reproductive capacity).

Reproductives: The casts of kings and queens in termites.

Residual toxicity: A characteristics of a pesticide to persist after application in amount sufficient to kill the pests for usually several days to several weeks or even longer.

Residue: It is that part of the pesticides which remains in or on a plant or animal after application. It is expressed as ppm and is based on fresh weight of the sample. Apart from the residues which result from the use of pesticides on food crops, there are also unintentional residues (residues which occur in food as a results of circumstances not designed to protect the food againt the attack of pests). Such residues occur: (i) in crops grown in previously treated soil and (ii) in animal products (milks and its products, meat and eggs) as a result of feeding of animals on fodder crops which have earlier received pesticidal treatment.

Resilin: An elastic protein forming an important part of the cuticle of an insect.

Resistance: Of insects to insecticides the ability in a population of insects to tolerate a dosage of pesticide which kills majority of individuals in normal population of the same insect species; Or plants to insects attack, the ability of certain verieties to produce high yields of good quality as compared with those of ordinary varieties at the same level of insect infestation.

Respiration: The entire series of physical and chemical processes accomplishing oxidation and removal of carbon dioxide. Insect normally breathe through spiracles leading into tracheae and tracheoles. Some times, however, some insects have special organs for respiration, tracheoles. The tracheoles are minute tubules which panetrate deeply into the every part of the body, so that oxygen can be defused directly into the cell.

Respiratory system: The anatomical structures (adaptations) of the animal that facilitate respiration (See respiration).

Reticulate: Like a network; covered with a network of veins or ridges.

Reticulate evolution: Evolution 'dependent on repeated intercrossing between a number of lines, and thus both convergent and divergent at once'.

Retina: The receptive apparatus of an eye.

Retina cells: The cells composing the retina.

Retinaculate: A locking mechanism beneath the base of the fore wing in some Lepidoptera; part of the wing coupling apparatus.

Retinaculum: A hook-like structure of which the fore wing of a moth engages with a bristle (frenulum) on the hind wing, thus coupling the wings together. A hook beneath the abdomen of Collembolans (Springtails) by which the forcula of the tail is held down beneath the insect. When released the tail propels the insect forwards into the air. Other attachment structures, e.g. on the ovipositor of some Hymenoptera and the mandibles of some beetles.

Retinal pigment cell: Pigment cells in the retinal region of the eye.

Retinal rod: Rhabdom; It is a transparent rod sorrounded by and secreted by a ring of seven or eight light sensitive cells (retinulae) beneath each crystalline cone in the compound eye of an insect. In each ommitidum (optical unit) light enters through the cuticular lens (cornea) is fused by the crystalline cone beneath this and impinges on the end of the retinal rods passing down through it and stimulates the sorrounding light sensitive cells. A similar arrangment but in a much simple form may be present in some ocelli also.

Retinule (Pl. Retinulae): A group of two or three or more light sensitive cells in a single ommatidium of a compound eye which sorround and secrete a longitudinal light receptor unit (optical rod or rhabdom).

Retractile: Capable of being pushed out and drawn back in.

Revision: In taxonomy, the presentation of new material or new interpretations integrated with previous knowledge through summary and revaluation.

Rhabdom: See retinal rod.

Rhabdomere: The receptive area of a retina cell that is one of the component parts of a rhabdom; one of the 7 or 8 segments making up a retinal rod (rhabdom).

Rhabdura: Diplura.

Rheotactic receptors: Sensory cells, hairs or other structures that detect changes in the flow of the water over the surface of a body present in the body of aquatic insects.

Rheotaxis: A response where an aquatic insect moves in a definite direction in relation to the flow of the water.

Rheotropism: Orientation (distinct from locomotion) in relation to the direction of flow of water, as in the case of some aquatic larvae.

Rhinocoris: *Rhinocoris fuscipes*. A predatory bug of Reduviidae family—Feeds on several Lepidopterous larvae and Coleopterous adults and larvae in India. The nymphs and adults both suck the haemolymph of the insects.

Rhynchophora: A large suborder of beetles (Coleoptera order) whose head is extended to form a rostrum or snout in front of eyes. They include weevils and some beetles.

Rhynchota: A name formerly used for Hemiptera order or bugs.

Ribaga's organ (Berlese's organ): A sac-like organ for receiving spermatozoa on the ventral surface of the abdomen of the female bed bug, *Cimex* sp.

Rice weevil: *Sitophilus oryzae* Linnaeus or *Calendra oryzae* (Coleoptera: Curculionidae). It is distributed world wide and throughout the India. It is distinct from another allied species the grain weevil; *Sitophilus gramarium* (L.). Both the species are similar in size and appearance and are found together feeding upon rice, wheat, maize and other grains. Both the grubs and the adults cause damage.

Rickettsie: An obligate intracellular parasite of arthropods, many types of which are pathogenic to human beings, and other animals. They are thought to be intermediate between bacteria and viruses, because they have features in common with both.

Ringer solution: It contains NaCl (0.65 g); KCl (0.025 g); $CaCl_2$ (0.03 g), made upto 100 ml with distilled water.

Ring gland: (Weismann's gland): A special endocrine organ which combines the source of all three metamorphosis hormones (occur in higher Diptera).

Riker mount: A thin, glass-topped exhibition case filled with cotton.

Rodent: All animals of the order Rodentia, such as rats, mice, squirrels, gophers, woodchucks.

Rodenticide: A chemical that kills rodents.

Rodolia: *Rodolia cardinais*. A coccinelid beetle used to control the cottony cushion scle (*Icerya purchasi*) introduced from Australia into other countries.

Rogor: See Dimethoale.

Roguing: The removal of unhealthy or unwanted plants from a crop.

Ronnel: Trolens, Etrolen, Koralan, Nankar; 0, 0-dimethyl-0, 2, 4, 5, -trichlorophenyl thiophosphate. It is a general purpose organophosphorus insecticide with low mammalian toxicity.

Roosting: Resting and sleeping at night.

Root borers: There are many insects that bore into the roots of

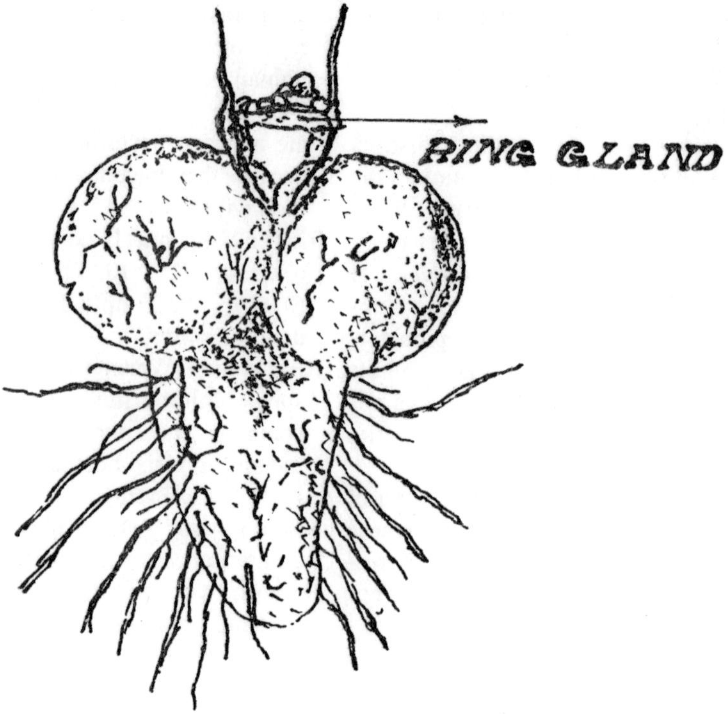

Ring gland

plants. They include the larvae of flea beetles, other coleopterous larvae and some larvae of Diptera and many larvae of moths (Lepideptera).

Rose aphid: *Macrosiphum rosaeformis* Das (Hemiptera: Aphididae). It is found attacking roses in north India from November to April. It has been recorded in Punjab, Delhi, Mysore, Andhra Pradesh. Besides this species, cotton aphid, *Aphis gossypii*, also infests rose plants from September to December. Both nymphs and adults suck the cell sap from leaves and tender shoots. Some other genara include *Siphocoryna Anuraphis, Capitophorus.*

Rostrum: A long, beak or snout of proboscis; the mouth parts as a composite organ.

Rotenone: This is the name given to the main insecticidal compound of certain *Derris* Spp. and *Loncho-carpus* Spp. It is a selective, non systematic insecticide with some acaricidal properties. It is insoluble in water but soluble in polar organic solvents. It is effective against aphids, caterpillars, thrips, some beetles, and red spider mite.

Rotund: Round, spherical.

Royal chamber: The space reserved for the accomodation of King and queen termites, or the space reserved for the queen of the entire colony.

Royal jelly: Brood food. A predigested food substance on which the larvae of bees must be fed in order to develop into queen bees. It is produced from the pharyngeal glands of the workers and contains a high proportion of proteins.

Rudimentary: Reduced in size; vestigial; atrophied; underdeveloped; poorly developed.

Rugose: Wringled.

Runaway pest: Any pest organism which enter a new territory where they have no natural enemies, and therefore, reproduce with little interference, resulting in a large population which can over run an area.

Run-off: The process of spray shedding from a plant surface during and immediately after application, when droplets coalesce to from a continuous film and surplus liquid drops from the surface.

Ruptor ovi: See Egg burster.

Russetting: Rough, brownish markings on leaves, fruits or tubers. This can be caused by some insects, fungi, pesticide chemical, or weather.

Rust red flour beetle: *Tribolium castaneum* Herbst. (Coleoptera: Tenebrionidae). It is a common pest of wheat flour and found world over. It also feeds upon dry fruits, pulses. Both the grubs and adults cause damage. The young larva is yellowish white and it turns red yellow at full grown up. The adult is small, reddish brown beetle. Its antennae are bent and bear distinct club formed by three enlarged terminal joints. Another allied species, *T. confusum* is also often seen in the wheat flour.

S

Sabadilla: *Schoenocaular officinale.* It has toxic insecticidal properties against certain pests.

Safener: A material added to a pesticide to eliminate or reduce phytotoxic effects.

Safety period: See waiting period.

Sagittae: Proximal lateral processes of the aedeagus in Hymenoptera.

Salivarium: The pocket between the base of hypopharynx and the labium into which opens the salivary duct; in higher insects converted into a salivary pump or a spinning apparatus.

Salivary glands: Glands that open into the mouth of most of the insects and secrete a fluid with digestive, irritant or anticoagulatory properties.

Salivary chromosomes: See Giant chromose.

Saltation: Discontinuous variation produced in a single step by major mutation.

Saltatoria: A name formerly used to denote those Orthoptera whose hind-legs are long and modified for jumping, i.e. locusts, grasshoppers and crickets, in contradistinction to Cursoria or running Orthoptera, which then included the cockroaches (now put in a separate order Dictyoptera).

Saltatorial leg (Jumping type): The femur is highly developed and posses strong muscles with which the elongated tibia is struck against the ground and the insects make a big hop. In grass-hoppers and the flea-beetles the femora is enlarged whereas in fleas the coxae also.

Sample: That portion of a true population which is actually available to the taxonomist.

Sand-flies: *Phlebotomus argentipes* Brunetti (Diptera: Psychodidae). Several species of sand-flies are known in the world. They bite man and act as a carriers of Kala-azar, three day fever, tropical ulcers, etc. In India, *P. argentipes* is most wide spread, and acts as a vector of Kala-azar virus. Only female flies suck blood and other warm blooded animals.

Sanitation: Cleaning up; keeping gardens, fields, animals, or buildings clean in order to reduce the number of insects and other pest and disease problems. This is one way to cut down on the use of pesticides.

San jose scale: Scientific name—*Aspidiotus pernicious* (Comstock) (Hemiptera: Coccidae). It is world wide and is most serious pest of nearly 200 different species of fruits, shrubs and ornamental plants belonging to 26 families. The damage is caused by numphs and females which suck the sap from twigs, branches and fruits. Body is covered with black or brown scale, underneath, a lemon-yellow insect is present.

Saprophagous: Feeding on dead or decaying animals or vegetable matter.

Saprophyte: An organism that obtains its food from dead plants or animals.

Saprophytic: Living on dead or decaying organic matter.

Sarcophagous: Feeding on flesh.

Sarothrum: The pollen brush on the basitartus of the hind-leg of a bee.

Sawflies: A group of Hymenoptera consisting of several families all characterized by having no 'wasp-waist' and having an ovipositor in the form of two small 'saws' between two lobes that act as sheaths. The saws are used for cutting into leaves or wood to make a hole in which the eggs are laid. The larvae of some of the sawflies are the serious crop pests. The larvae resemble the lepidopterous caterpillars but differ from them in having more abdominal legs.

Scabies mite or Mangemite: *Sarcoptes scabiei, Psorptes communis* (Acarina: Sarcoptidae). The first mentioned species of mite *(S. scabiei)* lives in subcutaneous gallaries made in the skin of man and warm blooded animals. It has remarkable range of hosts and in the course of its evolution various varieties have appeared which are confined to the specific hosts. S.S. Var. *hominis* found on man, S. S. Var. *bubulus* infests cattle; S.S. Var. *equi* infests horses, mules and donkeys; S.S. Var. *dromendarii* on camels; S.S. Var. *canis* on dogs. *P. communis* has also various varieties which attack buffalo, sheep, goat and horse.

Scale: A flat, unicellular, external process of the body wall, probably a modified setae; out growth of the body wall as in lepidopterous insects.

Scale insects: Members of the family Coccidae, the females of which often lose their limbs and become covered by a scale formed of cast-off skins glued together. In this state, it is difficult to recognise them as insects. Many are harmful to agricultural crops and some yield useful products such as shellac and cochineal.

Scales: The scales of insects are modified hairs or setae, flattened and of various shapes so that they overlap one another forming a protective covering for the wings and body. Macrotrichia are formed from a hollow outgrowth of cuticle containing an enlarged extensioin of an epidermal cell. Such scales are found in butterflies, moths, but they are also present on the wings and bodies of mosquitoes, some beetles. The scales of most insects are pigmented to varying degree with melanin, so that they are coloured differing shades of brown and black. The butterflies scales have colours due to *pterine* and other pigments, some butterflies have special scent-scales *(androconia)* which secrete a substance that attracts insects of the opposite sex.

Scalpellum: A piercing, needle-like stylet formed from the mouthparts, (e.g. mosquitoes and bugs).

Scansorial legs: Legs with a single large claw which can fold againt the end of the tibia and act as a climbing or grasping organ, e.g. in the lice (order Anoplura).

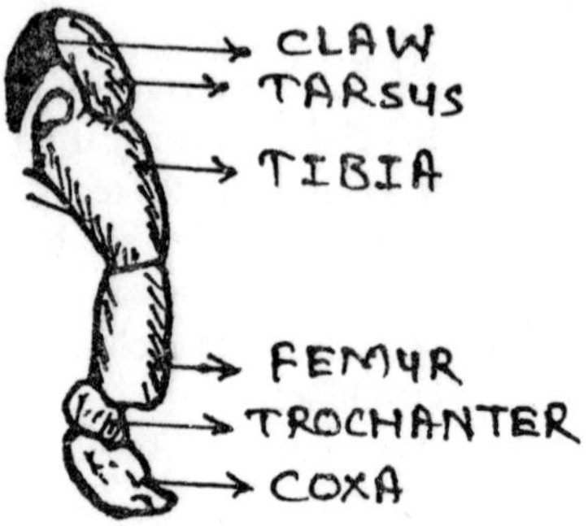

Scansorial legs

Scape: The basal segment or stalk of the insect antenna.

Scapula: Pl.=Scapulae. One of two sclerites of the mesonotum immediately lateral of the parapsidal suture (Hymenoptera) (Also called *parapsis).*

Scarabaeidae: Large beetles belonging to the group Lamellicornia whose antennae have clubs made of separate movable plates. The larvae live for several years and are very destructive to plants. They have large and horny head and the abdomen is inflated and soft with conspicuous spiracles. The family includes scarab bettles, dung beetles.

Scarabaeiform larva: A grub-like larva, i.e. one with body thickened, curved and cylindrical, with a well developed head and thoracic legs, without prolegs, and usually sluggish, e.g. larvae of Scarabaeidae family (Coleoleoptera order).

Scatter diagram: A bivariate or multi-variate graphic method of population analysis.

Scavenger: An animal that feeds on dead plants or animals, decaying organic matter, or animal wastes.

Scent gland: A gland producing an odorous substance. These are of various forms and are present on the bodies, wings or legs of insects. The secretions are of two kinds viz., (a) Repellent or unpleasant scents excluded in self-defence usually from glands on the abdomen, e.g. odours produced by the stink-bugs, bed bugs, earwigs, some beetles and ants. (b) Attractive substance-chiefly secreted before mating and often recognizable in extremely minute amounts and at long distances by insects of the opposite sex, e.g. queen bees substance, male butterflies scent scales (*androconia*) and secretions from the abdominal glands of some female butterflies.

Schistocerca: See Desert Locust and Locustidae.

Schizophora: A large group of cyclorrhaphous Diptera; all insects have a bladder or *ptilinum* on the head, by which the fly can push open the barrel shaped puparium in order to emerge. This group include House-flies, Blow-flies, Flesh-flies, Warble flies.

Schradan: (Octamethyl, OMPA,

Pestox III, Systam, Octamethyl pyrophosphamide, Octamethyl tetramide of pyrophosphoric acid). It is a systemic organophosphorus insecticide. It is non-phyto-toxic to the plants but toxic to animals.

Scholechiasis: The vomiting or evacuation of caterpillers.

Scientific name: A Latinized name, internationally recognised of a species and sub-species. The scientific name of a species consists of generic and specific name and the name of the describer of the species, and that of a sub-species Scientific names (excluding author's names) are always printed in italics.

Sclerite: A hardened body wall plate founded by sutures or membranous area in insects.

Scleroma: The sclerotic annulus of a body segment in distinction to the membranous conjuctiva.

Sclerotin: A hard nitrogenous substance impermeable to water, found together with chitin in the cuticles of insects, crustaceans, etc. It is composed of tanned proteins.

Sclerotization: Hardening and darkening of the cuticle. The hardening is owing to the deposition of sclerotizing substances in the exocuticle.

Sclerotized: Hardened.

Scolops (Scolopale): A sense rod or minute rod-like capsule enveloping the distal end of the sense cell in the certain sense organs.

Scolytoid larva: A fleshy larva resembling the larva of a scolytid beetle.

Scopa (Pl. Scope): A pollen transporting apparatus either a corbicula or a brush of hairs on the ventral side of the abdomen (bees).

Scopula (Pl. Scopulae): A small dense tuft of hairs.

Scorpionida: A order of the class Arachnida. It includes the true scorpions in which the abdomen is well segmented and has a tail like prolongation at the end, terminating in a sting, or the telson.

Scraper: The sharpened anal angle of the front wing (tegmen) of a cricket or long horned grass-hopper, a part of the stridulating mechanism.

Scrobe: A groove or furrow (antennal scrobe).

Scutellum: A sclerite of the thoracic notum; the mesoccutellum appearing as a more or less triangular sclerite behind the pronotum (Hemiptera, Homoptera, Coleoptera).

Scutoscutellar suture: The external suture of the V-shaped notal ridge of the alinotum, the arms divergent posteriorly, dividing the notum into scutum and scutellum.

Scutum: The middle and largest of the three sclerites which typically form the dorsal cuticular covering or *notum* of each thoracic segment of an insect.

The term is loosely used for any shield shaped place. The latin words *sectum* (= a shield) and *Scutellum* (a little shield) are often interchangeable and used to signify the dorsal covering of an insect which may be formed by fusion of a number of sclerites differing in origin in different insects.

Seasonal Dimorphism: Many insects produce two or more generations a year and there is often a difference in size, shape or colour between the spring broods and the autumn broods or between those hatched in the dry season and those of the wet season, e.g. (1) European nymphalid butterfly. *(Araschnia levana)*, that emerge in the spring have a typical fritillary pattern of orange-brown with black spots, those that hatch in mid-summer are dark brown or black with a broad white band across each wing. (2) The African Nymphalid *(Precis octavia)* in dry season have blue with black red, red and white markings, however, in the wet season a totally different pinkish-orange form emerges.

Sebaceous glad: A gland secreting fatty or oily material.

Second antennae: The appendages of the tritocerebral somite of Crustacea.

Secondary homonym: Each of two or more identical specific names which, at the time of original publication, were proposed in comfination with different generic names but which, through subsequent transference, reclassification or combination of genera have come to bear the same (or an identical) combination of a generic and specific name.

Secondary intergradation: A zone of hybridization or strong steepening of character gradients where two separately differentiated populations have reestablished contact.

Secondary pest: Species whose numbers are usually controlled by biotic and abiotic factors which becomes due to increase in numbers. A pest that preferably attacks such a host which has been damaged before by other (primary) pests.

Secondary segment: Any apparent segmentation produced by further sub-division of the primary sclertes.

Secondary segmentation: Any form of body segmentation that does not strictly confirm with the embryonic metamerism; the usual segmentation of orthropods having a well developed exoskeleton in which the membranous intersegmental rings are the posterior parts of the primary segments.

Secondary sexual characters: Characters which distinguish the two sexes of the same species but which do not (like gonads or accessory sexual characters) function directly in re-production.

Second axillary: The pivotal plate of the wing base resting on the pleural wing process, connected with the base of the radical vein.

Second maxillae: The third pair of gnathal appendages, in insects united the labium.

Second Median plate: A viable sclerotization at the base of the

Mediocubital field, folding convexly on the outer edge of the first median plate along the plica baselis; often absent, or represented by the united bases of the medial and cubital veins.

Second spiracle: The metathoracic spiracle, located near the anterior margin of the metapleuron, between the mesopleuron and the metapleuron, or in the posterior margin of the mesopleuron.

Second Trochanter: The second segment of the tolopodite, often not distinct from the base of the femur; in insects usually fused with the first trochanter.

Sectorial cross vein: A cross vein connecting two branches of the radial sector.

Sedentary: Remaining in one place.

Seed borers: Many insects bore into seeds either to feed on them or to lay their eggs in them. Some attack fresh growing seeds, others dried seeds such as stored grains. Codling moths *(Laspeyresia pomonella)* lay their eggs on leaves and flowers of apple trees and the newly hatched caterpillars bore into the developing seeds. Grain weevils and Bean Weevils use their snouts for boring into dried seeds. Pulse beetles *(Callosobru-chus chinensis)* lay their eggs on pulse seeds and the larvae bore into the seeds.

Seed Corn maggot: *Hylemia* (Delia) *cilicrura Hond* (Anthomyi-dae: Diptera). This is serious pest of germinating seed and seedlings of soybean, pea, cruciferous vegetables, cereals, potatoes, cucuerbits, maize, tobacco, onions, pepper and lucerne in Canada, Italy, Germany, the U.S.A., England and Japan. The maggot is the damaging stage which is pale white with pointed head and black curbed mandibles and the posterior end thicker and rounded.

Seed protectant: A chemical applied to seed before planting to preventing disease and or insect attack of seeds and new seedlings.

Seed treatment: The application of a pesticide chemical to seeds before planting in order to protect them from injury or destruction by insect, fungi, and other soil pests.

Seed Tick: The larvae of the ticks.

Segment: A sub-division of the body or of an appendage between are of flexibility associated with muscle attachment body segment is a *Somite*, a segment of an appendage is a *Podite*; and an embryonic body segment is a *metamere*.

Segmental appendages: The paired ventrolateral segmental outgrowths of the body wall serving primarily for locomotion.

Selective pesticide: A pesticide that kills the target pests with minimal adverse effects on the eco-system.

Selective pesticides: One that, while killing the pest individuals spares much or most of the other fauna and flora, including beneficial species, either through differential toxic action (*Physiological selec-*

tivity) or through the manner in which the pesticide is used (physical selectivity).

Semelparous: A type of life history in which the animal produces only once during its life time.

Semiaquatic: Living in wet places, or partially in water.

Semilooper: The larva possessing three pairs of abdominal legs. They are located on the fifth, sixth and tenth abdominal segments. The larva crawls by forming small loop of the body. Such larvae are found in certain moths of the order Lepidoptera.

Seminal vesicle: A structure, usually saclike, in which seminal fluid of the male is stored before being discharged, usually an enlargement of the vas deferens.

Semispecies: The component species of superspecies; also, populations that have acquired some, but not yet all, attributes of species rank; borderline cases between species and subspecies.

Senior homonym: The earliest published of two or more identified names for the same or different taxa.

Sense cell: The neurocyte of a sensory neurone. A sense cell of Type I has an unbranched distal process going to a specific sense organ; a sense cell of type II has one or more branched processes with a diffuse distribution.

Sense cone or peg: A minute cone or peg, sensory in function.

Sense organs: Any specialised structure of the body associated with nerves to receive external stimuli. Most of the sense organs in insects are innervated setae, by which an insect can see, smell or hear.

Sensillum: A simple organ, or one of the structural units of a compound sense organ.

Sensillum basiconicum: A sense organ of which the external part has the form of a minute projecting cone or peg.

Sensillum campaniformium: A sense organ without an external process, in which the cuticular part has typically the form of a bell or hollow cone receiving the distal process of the sense cell.

Sensillum chaeticum: A sense organ of which the external part is spine shaped.

Sensillum cocloconicum: A sense organ in which the external process is sunken in a cavity of the body wall.

Sensillum opticum: A photoreceptor of the eye type of structure, or one of the ommatidia of a compound eye.

Sensillum placodeum: A sense organ having a flat, plate-like external part.

Sensillum scolopophorum: A sense organ in which the sense cell contain "Sense rods" of the scolops type.

Sensillum squamiformium: A sense organ of which the external part is scale-like in shape.

Sensillum trichodeum: A sense organ of which the external part is a seta.

Sensitivity: The labile property of protoplasm that makes it responsive to stimuli, highly developed in nerve tissue.

Sensory hairs: Many of hairs or setae of insects have nerve endings to their bases, are extremely sensitive to touch, vibration, sound, etc. e.g. hairs on the antennae of mosquitoes.

Sequelae: A diseased condition following as a consequence of another disease.

Series: In taxonomy, the sample which the collector takes in the field or the sample available for taxonomic study.

Serology: The study of the nature and interactions of antigens and antibodies.

Serosa: A membrane covering the surface of the yolk in the egg's and later enclosing thc amniotic cavity with its germ band.

Serrate: Toothed along the ridge like a saw; serrate antenna.

Serricornia: A term formerly used for beetles with saw-like antennae including Elateridae, Buprestidae and Lampyridae.

Sesame gall fly: *Asphondylia sesami* Felt. (Diptera: Cecidomyi-dae). It is also known as sesamum gall fly and is distributed in Tamil Nadu, Andhra Pradesh, Kerala, Madhya Pradesh, Gujarat, Rajas-than and Maharashtra state of India. The adult is mosquito-like. The young maggots feed floral parts and cause malformation of buds which fail to develop into seeds.

Serrate antenna

Sesamum gall midge: See sesamum gall fly.

Sesame leaf and pod caterpillar: *Antigastra catalaunalis* Duponchel (Lepidoptera: Pyralididae). A serious and regular pest of til, distributed in Europe, Africa, Cyprus, Malta, Indonesia and South East Asia. The caterpillars damage by webbing the apical shoots and bore young pods. The moth is small insect, having dark brown markings on the tips of the wings.

Sesamia: A genus of noctuid moths (Lepidoptera) whose moths bore into the stems of cereals and of sugarcane in North Africa, India and South East Asia, e.g. *Sesamia inferens*.

Sessile: Attached or fastened; incapable of moving from place to place; attached directly, without a stem or petiole.

Seta (Pl. Setae): A hair-like unicellular external process of the body wall or of any derivative of the latter; a hair; a bristle.

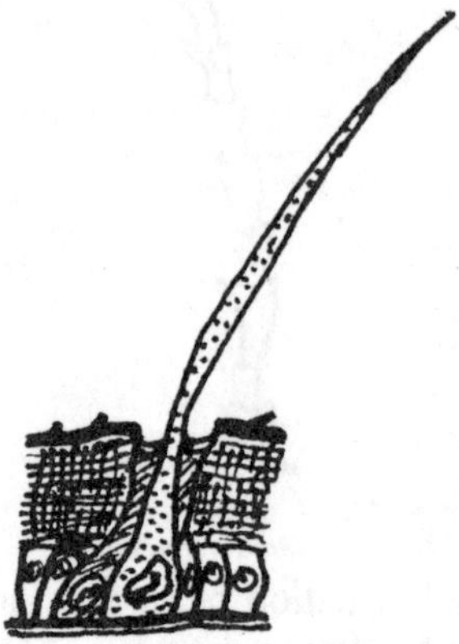

Seta

Setaceous: A bristle-like; Setaceous antenna.

Setal Membrane: The membranous floor of the hair socket, a alveolus, supporting the seta.

Setate: Provided with bristles.

Setulose: Bearing short, blunt bristles.

Setiform: Hair-like or bristle-like.

Seven-spot ladybird beetle: *Coccinella septempunctata.* A very common, fairly large ladybird beetle having a black thorax and head, yellowish markings on the sides of the thorax and bright orange-red elytra with seven black spots; i.e. three and a half on each elytron. It is a predatory on many Aphids.

Sevin: See Carbaryl.

Sex chromosome: A special chromosome, not occurring in identical number or structure in the two sexes and usually concerned with sex determination.

Sex-limited character: A character occurring in one sex only.

Sex linked characters: A character controlled by a gene located in sex chromosome.

Sex Pheromone: A substance secreted by one sex of an insect for attracting the opposite sex of its own species to facilitate mating.

Sexuales: A term used with references to plant lice and similar insects to denote the generation which contains male and female individuals capable of normal sexual reproduction.

Sexual dimorphism: Differences in size, form, colour, between the males and females of the same species. The phenotypic differences between the two sexes of a species.

Sexual reproduction: Reproduction resulting in diploid Zygote with a maternal and a paternal chromosome set.

Sexuparae: A term used with reference to plant-lice and similar insects for the generation of parthenogenetically reproduced females whose off spring can develop into normal males and females.

Shelf life: It is period of time during which the pesticide may be stored without becoming unfit for use; storage life.

Shieldbug: A common name for number of bugs, mostly plant feeders, belonging to the Pentatomidae, Scutelleridae and a few other related families in the order Heteroptera. All are flattened, roughly shield-shaped with a dorsal thoracic shield or *scutellum* covering most of the body. Some are brightly coloured. The wings are half membranous and half thickened, mouth parts are piercing type.

Shoot-borers: A name for a great variety of insects that bore into shoots of plants. They include weevils, beetles and the larvae of moths.

Short term pesticide: A pesticide which breaks down almost immediately, after application, into non-toxic by-products.

Sibling species: Pairs or groups of closely related species which are reproductively isolated but morphologically identical or nearly so.

Side-dressing: To put fertilizer, or a pesticide in granular form on or in the ground near plants after they have started to grow.

Sigmoid: Shape like the letter 'S'.

Signal words: Words which must appear on pesticide labels to denote the relative toxicity of the product. The signal words are "DANGER-POISON" (for highly toxic) "WARNING" (for moderately toxic) and "CAUTION" (for low order toxicity). The symbol of the skull and crossbones must appear on the labels of high toxic pesticides alongwith the words "DANGER-POISON".

Silica aerogel: A sorbtive dust that is insecticidal killing insects by desiccation. It is prepared by treating sodium silicate with sulphuric acid.

Silique: A long pod divided by membranous partition and that splits at both seams, characteristic of fruit of the mustards and related plants.

Silk: A tough material in the form of extremely fine threads secreted by various glands of insects and spiders. Chemically it consists mostly of a strong elastic protein *fibroin* covered by a layer of a more gelatinous protein *sericin*. In the silkworm larvae it is produced by modified salivary glands, in Embioptera and some Diptera the silk glands are on the tarsi of the front legs, in some Coleoptera the Malpighian tubules are modified to produce silk.

Silk Moths: The larvae of many moths produce silken cocoons but these are all not suitable for commercial silk production. The Chinese Silkworm, *Bomyx mori*, is now cultivated on a large scale in many parts of the world. The caterpillars feed on mulberry leaves and can each produce a continuous thread of silk upto 300 meters long. One Kg. silk is produced by about 50,000 cocoons. Oak Silkworm, *Antherea pernvi (China)* and *A. yamanai* (Japan) are also known for silk production. Cynthia moth (*Philosamia cynthia*) has long been bred for silk production in China, Japan and South East Asia. It was introduced into Europe and North America about 1860.

Silverfish: *Lepisma saccharina* Linn. and *Thermobia domestica* Packard (Thysanura: Lepismidae). It is a minor pest of house-hold but distributed world over. They are usually found in hot and moist parts of a house. *L. sanccharina* is carrot shaped, soft bodied, glistening wingless insect and covered with silvery scales. The other species called firebrat, *T. domestica* resembles silverfish except for mottled black and white patches on the upper surface of the body. The insect with chewing mouth parts may eat book bindings, paper, wall paper, fabrics and starched clothes, etc.

Simple eye: An ocellus. A very simple photosensitive organ consisting of a single cutiucular lens, a transparent corneagen layer below it and few sensitive cells (retinulae) below this. Sometimes there are also few rhabdoms and pigment cells which are rather indefinitely arranged. Most larvae have paired ocelli, sometimes a single pair and sometimes in clusters. Adult insects, in addition to their compound eyes, usually have three ocelli arranged in a triangle on the top of the head. They inform the insect of changes in the intensity of the light, but are incapable of forming any kind of image.

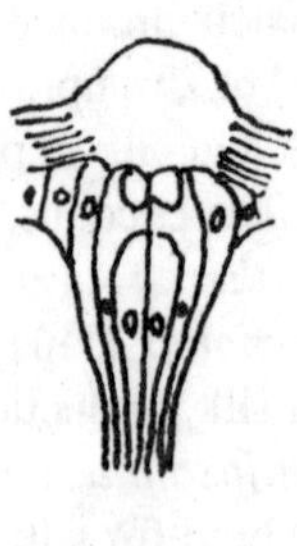

Simple eye

Simple metamorphosis: Metamorphosis in which the wings (when present) develop externally during the immature stage. There are three stages, namely egg, nymph and adult; gradual metamorphosis, incomplete metamorphosis, direct, metamorphosis. The insect involved in this type of metamorphosis are called Hemimetabola.

Singhara beetle: *Galerucella birmanica* (Jacoby) (Coleoptera: Chrysomelidae). It is widely distributed in Pakistan, Sri Lanka, Burma and India and is a serious pest of water nuts (*Trapa bispinosa* Rox (b). Damaged is caused by the grubs and the beetles by feeding on the leaves. The upper surface of body of grub is black and lower surface yellow. The adult beetles are brown to dark brown, with black eyes and a large hump in the middle of the body.

Single-dose poison: Acute poisons such as Zink Phosphide.

Siphon: A breathing tube, usually with two spiracles and two tracheae, projecting from the hind-end of an aquatic insect, e.g. mosquito larvae. The siphon attains extreme size in the aquatic larva

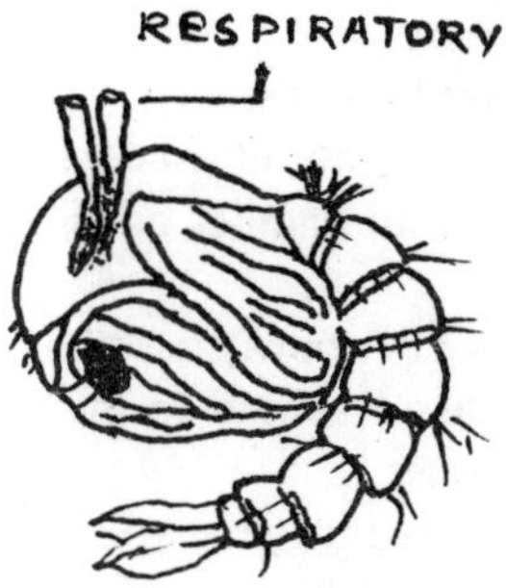

Siphon

of the Drone-fly *(Eristalis).*

Siphonaptera: Siphon=a tube; apterous=wingless. This order includes fleas which have piercing and sucking mouth parts with which they suck blood from the body of warm-blooded animals. The adults are ectoparasites with the body laterally compressed. The rat flea, *xenopsylla cheopis* (Pulicidae) frequently migrates to man and transmits the bacillus of the bubonic plague.

Siphunculata: Siphunculus. A little tube. This order includes sucking lice, which are important ectoparasites of man and domestic animals. They are wingless insects and have piercing and sucking type of mouth parts, with which they suck the blood of their host. Their thoracic segments are fused and they have I-segmented tarsi, and single claw with the help of which they cling to hair of their host.

Siphunculi: The paired protruding organs near the terminal end of the abdomen of Aphidoidea, through which a waxy secretion is secreted; also called cornicles.

Sitotroga: See Angonmois grain moth.

Skeckled: With spots.

Sleeping sickness flies: Tsetse flies of the genus *Glossina*. These are blood sucking flies that transmit the parasite *Trypanosoma* and cause African sleeping sickness.

Slurry: A thick suspention of pesticide made from wettable powder and water; paste-like liquid used as a seed dresser.

Smoke: Particle of pesticide chemical between 0.001 and 0.1m (micron) in diameter. The particles are released into the air by burning.

Snails and Slugs: These are soft bodied creatures, and belong to phylum Mollusca. Most of these are marine forms and a few are fresh water as well as land forms. They are generally oviparous but in some aquatic forms a trichophore or veliger larva in found.

Snout moths: A group of moths having long palps projecting in front of the head like a snout. These are some noctuids and some pyralids.

Social: Living in more or less organised communities of individuals

Socii: Lateral appendicular processes of the tenth segment in Trichoptera and Lepidoptera, possibly homologous with the cercus like appendages of the tenth segment in lower Hymenoptera.

Sogata: A genus of leaf-hoppers of the family Delphacidae. *Sogata furcifera* is a pest of rice and *S. distincta* of sugarcane in India.

Soldier Termites: A sterile caste of termites specially equipped for defending the colony. Usually they have large, thick sclerotized heads with powerful jaws. In some species there are *Nasute* forms in which the head is prolonged into sharp snout from which poison can be squirted.

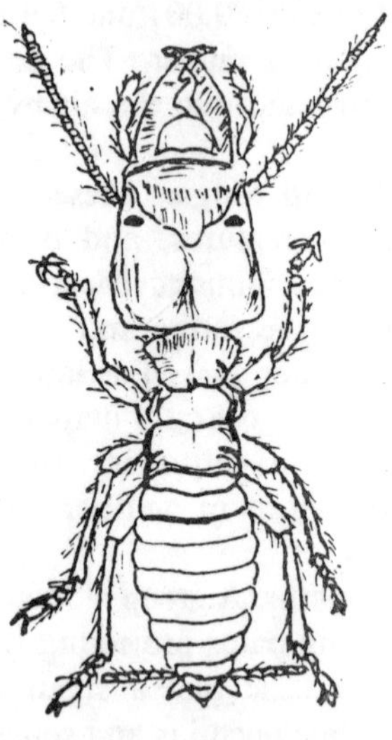

Soldier terminte

Soil application: Application of a chemical to the soil rather than to vegetation.

Soil drench: To soak or wet the surface of the ground with a pesticide chemical.

Soil incorporation: Mechanical mixing of the pesticide with the soil.

Solitaria: Locusts in solitary, non-migratory phase.

Solitary phase: Living singly.

Solubility: The ability of a chemical to dissolve in another chemical or solvent.

Soluble powder: A powder formulation that dissolves and forms a solution in water.

Solution: A preparation made by dissolving a solid, liquid, or gaseous substance into another substance (usually a liquid) with a chemical change taking place.

Solvent: A liquid which will dissolve a substance forming a true solution (liquid in molecular dispersion).

Soma: The body of an animal as distinguished from the germ cells.

Somatic cells: The body cells as distinguished from the germ cells.

Somatic layer: The external layer of the mensoderm applied against the body wall.

Somite: A primary body segment of an insect (usually restricted to the segments of adult insects) or metamere. A primitive or primary, body segment corresponding to an embryonic metamere or myotonic.

Sonagrm: A graphic representation of the vocalization of an animal.

Sooty moulds: A sooty coating of foliage and fruit formed by fungi that live in the honey dew secreted by insects.

Sorghum earhead bug: Scientific name-*Calocoris angustatus* Leth. (Hemiptera: Miridae). It is one of the most destructive pests of sorghum in southern India. Both the nymphs and the adults feed in the green earhead. It has been reported feeding on various cereals, millets and grasses, but its breeding is restricted to Sorghum, on which it assumes the status of a pest.

Sorghum midge: *Contarinia sorghicola,* a common midge in North America, West Indies, Sudan, Australia and India causing damage to sorghum.

Sorghum shootfly: Scientific name-*Atherigona varia soccata* Rondani (Diptera: Anthomyidae). It is widely distributed in Europe and Asia. In India it is more serious in southern parts. It also infests maize, wheat, broom corn, small millets and grasses. It causes damage to the seedlings as well as to the early stages of the crop. The maggots bore into the stem, cut the main shoots. The high yielding hybrids are more susceptible.

Sound of insects: Many insects produce sounds audible to humans (30-30,000 cycles per second) and many more produce sounds which are of frequencies beyond the range of human perception but perceptible to other insects. The insects produce sounds by following ways:

(1) *Stridulation*: This sound is made by rubbing the hind femora against the edges of the wings or by rubbing the two tegmina together, e.g. grasshoppers and crickets.

(2) *Wing beating sounds*: These are produced by the rapid vibrations of the wings; the faster the beat, the higher the note, e.g. Bee 200-400, mosquitoes 600, small midges 1,000 or more.

(3) *Drumming:* The cicada, has two resonating drums with air spaces beneath them situated on the base of the abdomen. It is known to produce loudest sound.

(4) *Spiracular sounds:* A few insects can produce sounds by rapidly expelling air from the spiracles and in some cases there are leaf-like folds within the tracheae. The vibrations of these help to produce the sound, e.g. some Diptera, Coleoptera and larvae of some moths.

(5) *Mechanical sounds:* The clicking of the click beetles and drumming of Death-watch beetles as it taps its head against the burrow are sounds of this type.

Soybean black spiny caterpillar: *Syntomis fortunei* Orja (Syntomidae: Lepidoptera). This insect was found feeding on Soybean in Madhya Pradesh, India. The adult moth's body bear three transverse orange coloured bands on the back of the head and abdominal segments. The wings are transparent and covered with several white specs. The larvae consume leaves which bear black, erect hairs over the body.

Soybean coreid bug: *Riptortus linearis* Fabr. (Coreidae: Hemiptera). It has been reported as a pod sucking pest of soybean, rice, and also on wide range of plants of Graminae, Legumineceae, Solanaceae and Convolvulacae in India, Philippines, the Netherlands and Malaya. The nymph and the adult suck the cell sap from the pods. The adult bug is brown, elongate, with flattened hind femora bearing tooth-like projections on the inside. The young nymphs are dark brown and almost ant-like in appearance.

Soybean Brown striped semilooper: *Mocis undata* Fabr. (Noctuidae: Lepidoptera). It is a pest of soybean and french beans in India and Philippines. The larva is slender, elongate and brownish with beautiful longitudinal strips of pink-brown and red which gave almost velvety appearance. It causes damage to soybean in August-September months. The adult moths are large sized, stout and brown in colour.

Soybean girdle beetle: *Oberea brevis* swed, (Lamiidae: Coleoptera). It is a serious pest of soybean in Madhya Pradesh, India infesting 20 to 80% soybean plants during rainy season. The female adult forms parallel girdles on the stem, petiole and petilate and places an egg near lower girdle inside the stem. The grub bores the stem completely and thus reduces about 70% grain yield. The larvae are in diapause from October-November to June-July. The grub is yellow with protuberances on the lateral sides.

Soybean green stink bug: *Nezara viridula* Linn. This is pest of soybean in India and the U.S.A. The adults and the nymphs feed on soybean pods by sacking the cell sap from the grain thus adversely affecting the grain size and yield. The adults are green and stout. It was also found feeding on cabbage, okra, tomato, potato and cotton in India.

Soybean leaf folder: *Hedylepta indicata* Fabr. (Pyralidae: Coleoptera). This pest is noted feeding on soybean, lucerne, green gram, black gram, horse gram and on cow in India, on soybean in Malaya, Rhodesia, Puerto Rico, on and limabeans in Barbados. The larva has the habit of folding the soybean leaves either from the tip onward or along the midribs and securing the folds with the help of a white resinous gum exuded by the larva. A closely resembling species, *H. diemenalis* was also found in association with the above species. The adult moth is delicate, slender, creamy yellow to light brown insect with long legs.

Soybean leaf miner: *Aproaerema modicella* Zeller (Gelechiidae: Lepidoptera). It is a serious pest of soybean and groundnut in India. The larvae of the pest mine the leaves under the epidermis, and such infested leaves show scorched, crinkled surface. In China and Indonesia it was also found damaging soybean. The adult moth is small, dark brown with wings bearing fringes of hairs on the inner and outer margins and the forewings bearing two white specks near the costal margin.

Soybean looper larva: *Scopula remotata* Guen. (Geometridae: Lepidoptera). It is a pest of soybean in India. The larvae of the pest make sieve-like network of holes in the leaves in the second week of September. Because of their curious form and mimicry in respect of colour and appearance, they are difficult to detect. The adult is medium sized, dirty white moth with numerous small black dots scattered on the wings.

Soybean stemfly (Stem miner): *Ophiomyia phaseoli* (Tryon); *O. Soae* Zeh. (Agromyzidae: Diptera). Both species of stemfly are the serious pests of soybean, green gram, black gram, french bean, cowpea in India, Sri Lanka, Egypt, Taiwan and Germany. In India it infests 20 to 96% soybean plants and cause 13 to 20% reduction in yield. The maggots bore the stem by making zig-zag tunnels which are redbrown and filled with excneting matter.

Space sprays: Sprays that are used for killing the flying insects.

Spatha: A dorsal lobe of the aedegus in Hymenoptera.

Spatulate: Spoon-shaped; broad apically and narrowed basally and flattened.

Speciation: The splitting of phyletic line; the process of multiplication of species; the origin of discontinuties between populations caused by development of reproductive isolative mechanisms.

Species: A group of individuals or populations that are similar in structure and physiology and are capable of inter-breeding and producing fertile offsprings, and which are different in structure and/or physiology from other such groups and normally do not interbreed with them.

Specific name: The second component of the binominial name of a species.

Speed of insects: The speed at which insects fly is variable in different insects. The horseflies are fast flying insects and can fly at the speed of about 40-48 Kms/hour and the bees can fly at about 8-15 kms per hour.

Spermatid: An immature spermatozoon.

Sperm cyst: One of the cellular capsules in the testis containing the spermatocytes.

Spermatheca (Receptaculum seminis): A sac or sperm receptacle of the female in which sperm cells are received from the male and stored until required for fertilizing the eggs.

Spermatheca (Pl. spermathecae): The sac-like structure in the female in which sperms from the male are received and often stored.

Spermatocyte: The sperm cell differentiated from a spermatogonia, before maturation.

Spermatogonium: The first stage in the differentiation of a sperm cell from a primitive male germ cell.

Spermatophore: A capsule containing sperm, produced by the males of some insects.

Sperm tube: One of the secondary divisions of the testis.

Spermatozoon: A mature germ cell.

Sphingidae: A family of Lepidoptera which includes Hawk-moths that are large stoutly-built moths whose fore-wings are much larger than the hind ones. The proboscis is long and the antennae end in hooks. The larvae have ten prolegs and usually bear a horn or spine at the hind and, e.g. til Hawk moth.

Sphinx moth: See til Hawk moth.

Spiders: Small animals closely related to insects. They have 8 jointed legs. 2 body regions, no antennae and no wings.

Spillage: The leaking, running over, or dripping of any pesticide chemical.

Spina: The median apodemal process of a spinasternum.

Spinasternum: One of the spina bearing intersegmental sclerites of the thoracic ventre, associated or united with the sternum may become a part of the definitive prosternum or mesosternum, but not of the metasternum.

Spindle shaped: Elongate and cylindrical, thickened in the middle and tapering at the ends.

Spine: A multicellular external process of the body wall.

Spinneret: A small tubular appendage from which silk threads are exuded by the larvae of many insects and spiders. In spiders, it is one at the hind end of the abdomen but quite different in structure in Lepidoptera larvae where it opens beneath the labiaum (lower lip). It is connected from a pair of modified salivary glands which produce silk in the form of a fluid which hardens in contact with the air. The insect uses silk for the sewing leaves together, construction of a sac or cocoon, or formation of a rope to which larva or pupa is hanged.

Spinose: Beset with spines (a spinose costa in Diptera).

Spinulae: Hamuli (A series of minute hooks on the anterior margin of the hind wings which catch into the uprolled hind-margin of the fore-wings (In Hymenoptera).

Spiracle: It is one of the several small openings in the segments of insects connected to a system of internal tubes through which oxygen is inhaled and carbon-dioxide is exhaled (also see respiration).

Spiracular bristle: A bristle very close to a spriacle (Diptera).

Spiracular plate: A plate-like spiracle next to or surrounding the spiracle.

Spiracular Sound Organ: See sounds of insects.

Spiracular trachea: The short, usually unbranched trachea arising directly from the spiracle.

Splanchnic layer: The inner layer of mesoderm applied to walls of the alimentary canal.

Splitter: In taxonomy, one who divides taxa very finely, to express every shade of difference and relationship, through the formal recognition of separate taxa and their elaborate categorical ranking.

Sponging mouth parts: Found in flies, on the lower side of the head. This insect has a fleshy, elbowed proboscis which can be folded and withdrawn. It is the modified labium and has at its base the rostrum which is modified clypeus and the distal lasutallum at the end of which is the sponge-like structure labellum.

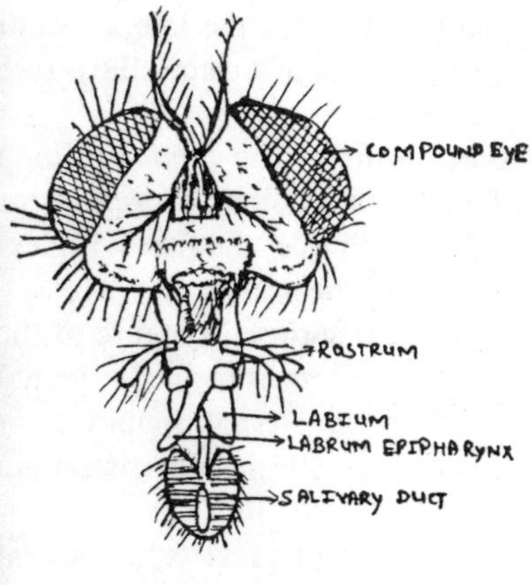

Sponging mouth parts

Sporadic grass-hopper: *Oxya nitidula* Walker (Orthoptera: Acrididae). It is widely distributed in Orient and is a pest of sporadic occurrence. The damage is done by the nymphs and adults which feed on various kharif crops. Nymphs are shining green and have a broad black band which extends from the base of the eyes to the wing pads.

Spotted boll worms: *Erias insulana* (Boisduval) and *E. vitella* (Fabricious) (Lepidoptera: Noctuidae). These two species of the bollworms are distributed in North Africa, India, Pakistan and are serious pest of cotton and Okra. The larvae of the pest bore into the growing shoots, flower buds, flowers and fruits of cotton and okra, either killing the plants or causing heavy shedding of the fruit bodies. The moths are yellow green, whereas that of *E. vitella* are of the same size but have a narrow light longitudinal green band in the middle of the forewings.

Spray: A mixture of pesticide with water or other liquid applied in tiny droplets.

Spray drift: The movement of airborne spray particles from the intended area of application.

Spreader: It improves the coverage of the treated surface.

Spring tails: See Collembola.

Spur: A thick, cuticular, movable spine; or hair on the legs of insects.

Spurious claw: A false claw; a stout bristle that look like a claw (Spiders).

Spurious vein: A vein-like thickening of the wing membrane between two true veins; an adventitious longitudinal vein between the radius and media, crossing the r.m. cross vein (Deptera).

Squama (Pl. squamae): A scale-like structure; a calypter; the palpiger (Odonata); a doral lateral lobe of the phallobase in Hymenoptera; one of the two small wing lobes close to the alula in housefly, alar squama and thoracic squama.

Stability: The ability of a pesticide formulation to resist chemical degradation over a period of time.

Stable-fly: *Stomoxys calcitrans* Linnaeus (Diptera: Muscidae). This fly and many other species of the sub-family Stomoxydinae are true blood suckers and are commonly known as biting flies. They are distinguished from other muscids by their having an elongated apical portion of the proboscis, which is longer than the length of the head, swellen at the base and is tappering at the apex.

Stabilizing-agents: They prevent the decomposition of pesticides during storage.

Stadium (Pl. Stadia): The period (interval) between two moults in a developing insect.

Stages: Distinct and sharply different periods in the development of an insect, e.g. egg stage, larval stage, pupal stage and adult stage.

Staging: This term is usually restricted to the type of double mount, where the insect is pinned on to the support, or stage.

Stalked: With a stalk or stem; with a narrow stemlike base; fused together to form a single vein.

Standard deviation (S.D.): The square root of the sum of the squared deviations d from the means, divided by N:

$$SD = \sqrt{\frac{d^2}{\sqrt{N}}}$$

Standard error (of mean): Standard deviation divided by the square root of the sample size, N.

$$SE = \frac{SD}{\sqrt{N}}$$

Statute of limitation: A provision in the Zoological Code (Art. 23 b) to protect universally adopted junior names against the revival of forgotten Serior Synonyms.

Stemma (Pl. Stemmata): An alternative name for the lateral ocelli (Simple eyes of a caterpillar). (see simple eye)

Stem mother: The *fundatrix* or a founder of a colony of asphids or other plant lice; the first wingless female of the season.

Steracostal suture: A suture of the thoracic sternum; the external mark of the sternal apophysis or furca, separating the basisternum from the sternellum.

Sternal apophyses: The lateral apodermal arms of the eusternum, in higher insects united on a medium base, the whole structure forming a furca.

Sternauli (Pl. Sternaulices): A longitudinal furrow on the lower part of the Mesopleuron (Hymenoptera).

Sternellum: The part of the eusternum posterior to the sternocostal suture.

Sternite: A sub-division of a sternal plate, or any one of the clerotic components of a difinitive sternum.

Sternacosta: The transverse internal ridge of the sternal suture through the basisternum from the sternellum.

Sternacostal suture: The external suture of the sternacosta, separating the basisternum from the sternellum.

Sternellum: The area of the eusternum posterior to the bases of the sternal apoplyses or the sternacostal suture.

Sternopleural bristles: Bristles on the sternopleuron (Diptera).

Sternopleurite: The infracoxal sclerotization of a generalized thoracic pleuron, generally united with the primary sternum in the definitive eusternal plate.

Sternopleuron (Pl. sternopleura): A sclerite in the lateral wall of the thorax, just above the base of the middle leg (Deptora).

Sternum: Either the primary ventral plate of a body segment or a composite definitive sternum.

Sticker: It improves the retention of spray or dust deposits.

Stigma (Pl. Stigmata): A dark thickening of the wing membrane along the costal border of the wing near the apex in many insects. An alternative name for a spirale or breathing pore of an insect.

Stigmal vein: A short vein extending posteriorly from the costal margin of the wing, usually a little beyond the middle of the wing (ohalcidoidea).

Stimulus: Any change in the environment or in the internal conditions of the animal that produces activity in tissues.

Sting: See poison.

Stink bugs: Petatomidae or shieldbugs, usually brightly coloured and possessing two scent glands on the underside of thorax. The nymphs have dorsal abdominal stink glands. When disturbed these insects discharge a foul or diagreeable smelling liquid from these glands and are thereby protected from being eaten by birds or other predators.

Stink glands: Many insects secrete substances which are foul smelling. In some insects these substances come from hypodermal cells concentrated in the pockets under the thorax or abdomen, in others the salivary glands or the accessory sex glands may be modified for the purpose. These substances contain formic acid, butyric acid, hydrochloric acid, selicylic acid, salicylaldehyde, oxides of nitrogen, oils, esters, etc. The insects which secrets these substances are: bedbugs, stinkbugs, oil beetles, some ground beetles and some caterpillars.

Stipes (Pl. Stiptes): The second segment or division of the maxilla which bears the palp, the galea, and the lacinia; lateral lobes of the millipede gnathochilarium.

Stomach: The ventricular or Mesenteron.

Stomach Mouth: A term sometimes given to the proventricular apparatus forming a mouth-like entrance to the ventriculus.

Stomach poison: Applies to the poisons which act through the stomach.

Stomodaeal nervous sytem: The nervous system centering in the garglia of the stomodaeum.

Stomodaeal valve: The cylindrical or funnel-shaped invagination of the posterior and of the stomodaeum into the cardiac part of the ventriculus.

Stomodaeum: (Foregat, fore-intestine). The anterior ectodermal part of the alimentary canal.

Stomixys: See stable fly.

Storage excretion: It is the temporary or permanent accumulation of excess of waster products, not dealt by the usual excretory organs, in special cells called the urate cells in the fat body.

Storage life: See shelf life.

Strepsiptera (Stylopida): The *stylip SP.* is the most important parasite of the nymphs of the sugarcane pyrilla. Its larvae are endoparasitic and the male adults are free living. In the adults, the forewings are modified into small clubs and are known as pseudohalteres. The hind wings are large and fan shaped. The females live in the host inside puparia. It is a small order of about 300 species.

Stria: Pl. Striae. A groove or depressed line, striote with grooves or depressed lines.

Striated border: The inner cytoplasmic layer of the ventricular epithelial cells with fine lines perpendicular to the surface.

Strickland code: A code of nomenclature prepared by a committee of the Britain Association for the Advancement of science under the secretaryship of H.E. Strickland and first published in 1842.

Stridulate: Stridulation. To make a noise (chirp or creak) by rubbing two structures or surface together.

Strigilis (Strigil): In Hymenoptera a curbed comb-like movable spur on the distal end of the front tibia opposed by a hair-fringed cavity at the base of the first tarsal segment, used for cleaning the antennae.

A similar structure on the forc-tibia of some Lepidoptera.

A process on the dorsal surface of the abdomen in Corixidae, doubtfully associated with sound production.

Stripe: A longitudinal colour marking.

Striped squirrel: *Funambulus pennanti* and *F. palmarum.* The former species occurs in north and north-eastern India, whereas the latter one inhabits southern region of the sub continent. Squirrels are diurnal and orboreal in habit. They cause damage in gardens and orchards.

Strobane: It is a chlorinated hydrocarbon similar to the toxaphene and prepared by the chlorination of a mixture of camphene and pinore.

Structural colours: Brilliant iridescent colours produced, not by pigments, but by reflection, refraction, diffraction or interference of light of differing wave lengths. The optical effect is brought about by the presence of numerous extremely thin, transparent plates on cuticle and scales.

Structural pests: Pests that attack and destroy buildings and other structures, clothing and manufactured and processed goods.

Stubbles: The stumps of plants left standing after harvest.

Stylate: With a style; stylelike; stylate antenna.

Style: A short antennal appendage of some Dipterous flies. Compore *arista*; a short more or less cylindrical appendage in insects.

Stylet: A hollow tube-like organ specially in aphids, through which the sugary contents of the phloem vessels are sucked by the insect.

A medium dorsal element in the shaft of the ovipositor formed of the united second valvulae.

Stylus (Pl. Styli): A short, slender, finger-like, process (Thysaneura); also see style and stylet.

Sutantennal suture: A suture or the face extending ventrally from the base of the antenna.

Subapical: Located just proximad of the apex.

Subalare: The epimeral epipleurite giving insertion to the posterior pleural muscle of the wing.

Subbasal: Located just distal of the face.

Sub-class: A major sub-division of a class containing a group of related orders.

Subcosta: The usual second vein of the wing associated basally with the anterior end of the first axillary sclerite and forms the anterior edge.

Subcoxa: A secondary proximal sub-division of the coxopodite differentiated from the coxa, present in some arthropods, forming a support in the pleural wall of the body segment for the rest of the appendage. In majority of insects, this segment is fused with the body wall and the second segment (coxa) forms the functional limb base.

Subcuticle: The innermost part of the endocuticle, representing newly secreted cuticle (see endocuticle also).

Subdiscal (or subdiscoidal) vein: The vein forming the posterior margin of the third discoidal cell (Hymenoptera).

Subequal: Approximately, or almost, equal in size or length.

Subfamily: A major sub-division of a family, containing a group related tribes or genera. Subfamily names end in *enae*.

Subgalea: A lobe or sub-division of the maxillary stipes bearing the galea.

Subgenal sutures: The lateral submarginal grooves of the cranium just above the bases of the gnathal appendages, forming internally a subgenal ridge on each side, continuous anteriorly with the epistomal suture when the latter is present.

Subgeneric name: The name of an optimal category between the genus and the species, enclosed in parentheses when cited in connection with a binominal or trinominal combination and, therefore, excluded from consideration when determining the number or words of which a specific or subspecific name is composed.

Subgenital plate: Usually the ninth abodminal sternum of the male extended beneath the genital chamber, but sometimes the eighth or the seventh sternum; a plate to be sternite that underlines genitalia.

Subgenus (Pl. subgenera): A major division of a genus, containing a group of related species. In scientific names, subgeneric names are capitalized and placed in perentheses following the genus name.

Subimago: The first of the two winged instars of a mayfly after it emerges from the water.

Subjective synonym: Each of two or more synonyms based on different types, but regarded as referring to the same taxon by those zoologists who hold them to be synonyms.

Submarginal vein: A vein immediately behind and parallelling the costal margin of the wing (Chalcidoidea).

Submedian cell: The cell behind the median cell in the basal posterior portion of the wing (Hymenoptera).

Submarginal cell: One or more cells lying immediately behind the marginal cell (Hymenoptera).

Submentum: A proximal (basal) plate of the post-labium; when continuous with a gular plate the submentum lies distal to the posterior tentorial pits.

Subocular suture: A suture extending ventrally from the compound eye.

Suboesophageal commissure: The commissure of the tritocerebral ganglia, which goes below the stomodaeum.

Suboesophageal ganglion: The composite ventral nerve mass of the head situated beneath the oesophagus in insects, formed of the united primitive ganglia of the gnathal somites. It connects the ventral nerve cord with the nerve collar.

Suborder: A major sub-division of an order, containing a group of related families.

Subphylum (Pl. Subphyla): A major sub-division of a Phylum, containing a group of related classes.

Subquadrange: A cell immediately behind the quadrange (Odonata, Zygoptera).

Subspecies: A sub-division of a species usually geographic race. The different sub-species of a species are ordinarily not sharply differentiated and integrade with one another and are capable of interbreeding.

Substitute name: A name proposed to replace a preoccupied name and automatically taking the same type and type-locality.

Subterranean: Living in the grounds.

Subtriangle: A cell or group of cells proximal to triangle (Odonata, Anisoptera).

Successions: Group of species that successively occupy a given habitat as the conditions of that habitat change.

Sucking insects: The insects that suck blood, nectar, plant sap or other liquid food in various ways. Insects contain *labrum, mandibles, maxillae, labium* and *hypopharynx* as mouth parts. Any of these may be lengthened and placed in juxtaposition or inter-locked to form a tube by which liquid food is sucked, other may form a second tube for the downward flow of saliva, while others again form piercing stylets. In butterflies and moths, the long, coiled proboscis is formed from the *galeae* of the maxillae, in bees the lengthened *glossae* are fused to form a spoon like flabellum. In bugs, the maxillary stylets form both food canal and salivary canal, fleas use *lacineae* or maxillae. All Diptera have *labella* or sponging lobes at the tip of labium, but these are highly developed in Cyclorrhapha with pseudotracheae.

Suctorial: Mouth parts adopted to sucking liquids or powdery substances.

Sudan III: It is an indicator which is used for distinguishing the ow and wo type emulsion. Sudan III is soluble in oil and potassium permanganate in water phase.

Sugarcane root borer: *Emmelocera depressella* Swinhoe (Lepidoptera: Pyralididae). This is a serious pest of sugarcane in Eastern India, but in Pakistan, Haryana and the Punjab it is of minor importance. The caterpillars cause damage by feeding on the underground portion of plants resulting in the drying up of the central whole of leaves. The caterpillar is creamy white with yellowish brown head. The moths are pale yellow brown and have white hind wings.

Sugarcane shoot borer: *Chilo infuscatellus* Snellen (Lepidoptera: Crambidae). It is a serious pest of sugarcane in India and Pakistan and also reported from Afghanistan, Burma, Indonesia, Formosa and the Phillippines. The larvae cause damage resutling into formation of "dead heart". The larvae are dirty white and have five light violet longitudinal stripes on the body. Moths have straw coloured forewings and whitish hind wings with apical light buff areas.

Sugarcane Stalk borer: *Chilo auricilius* Dudgeon (Lepidoptera: rambidae). Commonly known as 'Tarai borer' is most destructive pest of sugarcane in northern India. It also feeds on paddy, wheat, oats and also has been reported from Bihar, Uttar Pradesh, Haryana and Punjab in India. The damage is caused by the larval stage by feeding inside stem. The larva is blueish-pink with dark brown head and five longitudinal violet stripes on its back. The moth is straw coloured with forewings having golden spot and the hind wings are white like sibier.

Sugarcane top shoot borer: *Scirpophaga nivella* Fabricius (Lepidoptera: Pyralidae). It is a most serious pest of sugarcane in India and Pakistan but also distributed in Burma, Sri Lanka, China, Formosa, Japan, the Philippines and Thiland. Damage is caused by the larvae by boring the top nodes of cane resulting into a "bunchy top".

Sulcate: With groove or furrows.

Sulcus (Pl. Sulci): A suture formed by an infolding of the bodywall.

Sulphonation: Indicates the degree of refinement of oil in terms of the percentage of oil which does not react with sulphuric acid. This has been used as a measure of safeness for application to plant and animal life.

Sulfotepp: (Pirofos, Bladafum, Tetraethyl dithiopyrophosphate). It is closely related to TEPP and similar to parathion in insecticidal action.

Sumatrol: It is similar to tephrosin but differs in being phenolic. It has the same toxicity as toxicarol.

Sumithion: (Folithion, Metathion, Fenitrothion, Methylnitrophes; 0, 0-dimethyl-0, 4-nitro-3-methylphenyl thiophophate). It is similar to methyl parathion in chemical properties and is an active synergist for many organophosphorus insecticides like parathion, methyl parathion etc.

Sunnhemp Capsid: *Ragmus importunita* D. (Hemiptera: Miridae). This is a small active green bug which sucks plant sap. The pest is found on sunnhep in Southern India.

Sunnhemp hairy caterpillar: *Uteth-esia* pulchella Linnaeus (Lepidoptera: Arctiidae). This is a most important pest of sunnhemp in Tamil Nadu, India. The caterpillar feed on leaves and also bore into the capsules. The adult moth is pale whitish with red black spots on the upper wings and black margin blotches on the lower wings.

Superfamily: A taxonomic category immediately above the family and below the order. A group of closely related families, superfamily names end with in-*oidea.*

Superior appendage: One of the two upper appendages at the end of the abdomen; a cercus (Odonata).

Superlinguae: A pair of ventral lobes of the insect and similar in some respects to the paragnatha of Crustacea, developed from the mandibular somite and united with the lingua in the hypopharynx of adult insects.

Superposition image: Erect visual image produced by the overlapping points of light. This phenomenon is common in nocturnal insects in which the pigmented sheaths surrounding optical units (Ommatidia) can be retracted so that the points of light which normally form a mosaic image will merge together and overlap giving a less distinct but brighter image.

Superposition eye: An eye which permits the passage of light through the non-pigmented wall of one visual unit (Ommatidium) to the neighbouring visual unit.

Superspecies: A monophyletic group of entirely or largely allopatric species.

Supervised control: The control of pests as soon as it becomes necessary so that the pest does not increase its density above the *economic threshold.* Such a control necessitates keeping a regular record of the pest population.

Suphenone: (4-Chlorophenylsulphone: para-Chlorophenyl-phenyl sulph-one). It is chlorinated hydrocarbon occurring in two different forms. It is toxic to mites.

Supplement: An adventitious vein formed by a number of cross veins being lined up to form a continuous vein, located behind and more or less parallel to one of the main longitudinal veins (Odonata).

Supplementary reproductives: In termites the caste of males and females with light pigmentation, and small compound eyes. The females lay eggs in the colony supplementing the work of queen.

Suppress: To keep a pest from building up in numbers. To stop pest from causing injury or destruction.

Supraolar Bristles: A longitudinal row of bristles on the lateral position of the mensonotum immediately above the wing base (Diptera).

Supraoesophageal ganglion: Suprapharyngeal, hyperpharyngeal or cerebral gonglion. A pair of nerve centres situated on the dorsal side of the pharynx or oesophagus in insects and connected to the main ventral nerve cord by a collar formed of two commissures. Sometimes these ganglia are fused with others such as the optic and antennary and may be sufficiently complex to be regarded as 'brain'.

Supraspecific: A term applied to a category or evolutionary phenomenon above the species level.

Surface grasshopper: *Chrotogonus trachypterus* Blanchard (Orthoptera: Acrididae). It is distributed in Orient and Africa and is polyphagous in nature. The damage is done by the nymphs as well as by adults by feeding on leaves and also by cutting germinating plants. *C. chypterous* is common in the north and *C. oxypterous* occurs in southern regions.

Surface tension: Due to molecular forces at the surface, a drop of liquid forms an apparent membrane that causes it to ball up rather than to spread as film.

Surfactant: A material which reduces surface tension between two unlike materials such as oil and water. A spreader or wetting agent is used to increase coverage of the surface being sprayed.

Susceptibility: The degree to which an organism is affected by a pesticide at a particular level or exposure.

Susceptible: Capable of being easily infested or infected, not resistant.

Suspended: The particles that are dispersed or held in a liquid. A pesticide use that is no longer legal and the remaining stocks cannot be used.

Suspension: The dispersion of a sold in a liquid.

Suspensory muscles: See dilator muscles.

Sutur (Suere=to sew): A seam or impressed line indicating the division of parts of the body wall or appendages; or a narrow membranous area between sclerites; the line of juncture of the elytra (Celeoptera).

Swallow tail: A butterfly *(Papilio* Spp) with prolongations of the hind-wings.

Swarming: The movement of a large number of insects together. It may be for mating (mosquitoes), to relieve over crowding (bees), finding new feeding and breeding grounds (locusts) and settling together when hibernating (some species of tiny flies; coccinelid predators).

Swath: Width of target area sprayed at one pass.

Swimmeret: An abdominal appendage that functions as a swimming organ Crustacea.

Symbiont: An organism living in symbiosis with another organism.

Symbiosis: The living together, in a more or less intimate association, of two species, which benefits both.

Symmetry: A definite pattern of body organisation; bilateral symmetry, a type of body organisation in which the various parts are arranged more or less symetrically on either side of a median vertical plane, i.e. where the left and right sides of the body are essentially similar.

Symmetrical: Evenly developed on both sides.

Sympatric: Having the same or overlapping, areas of geographical distribution.

Sympatric speciation: Speciation without geographic isolation; the acquisition of isolating mechanisms within a deme.

Sympatry: The occurrence of two or more populations in the same area; more precisely, the existence of a population in breeding condition within the range of individuals of another population.

Symphyta: It is a sub-order of Hymenoptera and includes wood-boring wasps whose females have, ovipositors adapted for sawing and boring, as in the metallic blue, *sirex noctilio* (Siricidae), an important pest of timber in many parts of the world.

Sympatric hybridization: The occasional production of hybrid individuals between two otherwise well defined sympatric species.

Symplesiomorphy: The sharing of ancentral characters by different species.

Synapomorphy: The sharing of derived characters by several species.

Synapse: The central mechanism of inter-communication between the terminal fibres of two or more neurons.

Syncephalon: A secondary composite head, formed of the prostomium and one or more succeeding somites.

Synchronic species: Species which occur at the same time level.

Synergism: Increased pesticide activity of a mixture of pesticides above that at the sum of the values of the individual components.

Synergist: It is a material that has a little or no toxicity, but is capable of increasing the toxicity of a pesticide.

Synonym: In nomenclature, each of two or more different names that have been applied for the same taxon.

Synonyms: If more than one name has been given to the same genus or species.

Synonymy: A chronological list of the scientific names which have been applied to a given taxon, including the dates of publication and the authors of the names.

Synopsis: In taxonomy, a brief summary of current knowledge of a group or topic.

Syntype: Every specimen in a type-series in which no holotype was designated.

Systematics: Taxonomy; A branch of entomology; a practice of naming and classifying anything; the science dealing with the diversity of organism.

Systemic: A compound that is capable of absorption into plant sap or animal blood. The systemic pesticides in this way render the sap or blood toxic to those pests that feed in or on the treated host.

T

Tabanidae: A family of the order Diptera: It includes two winged insects; most females suck the blood by piercing type mouth parts. Most adults are parasitic on cattles, e.g. house flies, gad flies and clegs.

Tachinidae: A family of two winged insects of the order Diptera of which all have parasitic larvae. The adult flies are stout bristly insects with grey or striped bodies. The larvae (maggots) are parasite on other insects, spiders, wool-lice or centipedes.

Tachygenesis: A shortened development involving the ommission of several larval instars.

Tactile organs: Tactile receptors. The organs of touch. These consist of simple hairs, bristles or spines with sensory cells and nerve endings at their bases. These receptors are situated all over the body but are most abundant on the antennae and also on cerci. They may also be sensitive to low pitched sounds which cause them to vibrate.

Tactile receptors: See tactile organs.

Tactile spines: See tactile organs.

Taenidium (Pl. Taenidia): A circular or spiral thickening of chitin in the inner walls of a trachea strengthening the walls of air tubes (trachea).

Taeniothrips: A genus of thrips (Thysanoptera). It is a pest of several crops world over. Several species exist.

Tagma (Pl. Tagmata): A group of successing segments forming a distinct unit in the body of orthropod. In insects there are three stagmata, viz. the head, thorax and abdomen, but in spiders there are only two.

Tagoreceptor: An organ of touch having a single sense cell.

Taint: Unwanted flavour in fresh or processed food from a pesticide used on the growing crop.

Tandem: One behind the other; two connected or attached together.

Tangoreceptor: A sense organ responsive to touch.

Tanytarsus: A genus of the tiny, non biting mosquito-like insects of the family Chironomidae whose larvae live in running streams.

Papetum: A mirror-like reflecting surface (layer) formed either of pigment or of densely massed tracheae surrounding the retinal rods in the eyes of night-flying insects. These reflect the light back into the retinal cells, so that each receives a double stimulus and the general effect is greater sensitivity in dim light. Because of presence of tapetium the eyes of notunal insects shine.

Tarsal claw: A claw at the apex of the tarsus.

Tarsal formula: The number of tarsal segments on the front, middle and hind tarsi, respectively.

Tarsal taste organs: Chemosensory organs on the legs of some butterflies and on some other insects also. On painted lady these organs are 200 times more sensitive than human tongue.

Tarsomere: Tarsite. One of the subsegments of the tarsus, usually numbering two to five.

Tarsonemid mites: These mites are very small (100 to 300 m). The body and posterior legs are rather sparsely beset with setae. The anterior pairs of legs, especially their terminl segments are more densely clothed with setae.

Tarsus: The penultimate segment (last or fifth) of the leg. It is divided into two or more segments or joints or tarsomeres, none of which is individually provided with muscles except the first. This bears apical structures like claws, pulvilli, arolium or empodium.

Taxis (Pl. Taxes): A direct response involving the movement of an animal towards or away from a stimulus of an animal towards or away from a stimulus, e.g. *Phototaxis* (light), *Chemotaxis* (Chemicals), *Thigmotaxis* (touch), etc.

Taxon: Pl. Taxa. A taxonomic group that is sufficiently distinct to be worthy of being distinguished by name and to be ranked in a definite category.

Taxonomic category; Designates rank or level in a hierarchic classification. It is class, the members of which are all taxa assigned a given rank.

Taxonomic character: Any attribute of a member of a taxon by which it differs or may differ from a member of a different taxon.

Taxonomy: The science of classification into categories of varying rank, based on similarities and differences, and the describing and naming of these categories.

Tea Mosquito bug: *Helopeltis theivora* water house and *H. antonii* Signoret (Hemiptera: Miridae). This is a pest of tea in India, Sri Lanka, Vietnam and Indonesia. In Africa, another allied species , *H. berrogothia* is prevalent. The female bug is orange accross the shoulder and the male is almost black. Both the nymphs and adults suck the cell sap from young leaves, buds and tender stems. *H. theivora* is commoner in eastern India and *H. antonii* in southern India.

Technical material: The pesticide as it is manufactured by a chemical company. It is then formulated with other materials to make it usuable in wettable power, dust, liquid or other form.

Tegmen: Pl. Tegmina. The thickened, leathery front wings of an orthopteran insects, e.g. cockroaches and grasshoppers.

Tegula: Pl. Tegulae. A small scale-like structure overlying the bases of the front wing, e.g. in Lepidoptera and Hymeneptora.

Tegumen: A dorsal hood-like structure covering the male genitalia of some insects, an enlargement of the ninth abdominal segment.

Telopodite: The primary shaft of the limb distal to the coxopodite, the basal segment of which is the first trochanter.

Telotarsus: The distal of the two principal segments of the tarsus in Arachnida and Chilopoda.

Telotaxis: A reaction in which one insect or other organism orientates itself and moves in a definite direction towards a source of stimulu at a distance, e.g. point of light. Orientation which gives the impression of a purpose.

Telson: A twelfth abdominal segment in the embryos of some insects and in the adults of the primitive order, Protura. The insect has more than eleven and the last two or three are usually modified, fused together or reduced so that there are usually nine or ten visible segments.

Temik: (Aldicarb, 2-methyl-2-methylthiopropional-oxime-0-N-methyl carbamate). It is a systemic carbamate nematicide and insecticide.

Tenacity: Tendency of a pesticide deposit to resist removal by weathering.

Tenebrionidae: A family of Coleoptera order. It includes dark brown, elongated, noctural beetles which are usually unable to fly owing to the fact that the wings are vestigial or absent and the wing cases firmly fused together, e.g. mealworm beetles.

Tenent hairs: Clusters of minute tubular hairs that secrete an adhesive substance, present on the pulvilli or pads of the feet of some insects to walk on a smooth surface.

Teneral: A term applied to recently moulted, pale, soft bodied individual adult; incompletely hardened.

Tension receptors: Sensory receptors which respond to stretching or pulling of the cuticle, as for instance at the joints of insects (see also chordotonal receptors).

Tentaculum: A minute structure on the ventral side of the third abdominal segment which serves as a clap for the forcula (In Collembola).

Tent Caterpillar: *Malacosoma indicum* walker (Lepidoptera: Lasiocampidae). It is important pest of apple in north-western India, but more serious in Simla hills. It also causes damage to pear, apricot, walnut. The caterpillars feed gregariously on foliage, leaving behind only the midrib and other hard veins. The male moth is light reddish and the female moth is light brown.

Tenthredinidae: A family of the order Hymenoptera. It includes sawflies whose ovipositor resembles a tiny saw enabling the female to cut slits or pockets in plant parts where she lays her eggs. The larvae resemble caterpillars but have more abdominal prolegs. The larvae cause serious losses to various cultivated crops.

Tentorial pits: Pitlike depression on the surface of the head that mark the points of union of the arms of the tentorium with outer wall of the head. There are usually two tentorial pits in the epistomal suture and one at the lower end of each postoccipital suture.

Tentorium: The endokeletal brace of the carinum formed of united anterior and posterior pairs of arms, bearing primarily the origins of the ventral muscles of the gnathal appendages, and usually giving attachment secondarily, on a pair of dorsal arms, to the antennal muscles.

long, sharp ovipositor.

Tergite: (g has the 'J' sound). A sclerite or the tergum.

Tergum (Pl. Terga): The dorsal surface (cuticular plate) of any body segment.

Terminal: At the end; at the posterior end of the abdomen; the last of a series.

Terminal filament: The cellular end thread of an ovariole.

Termitarium: The large elaborately built nest of a colony of termites.

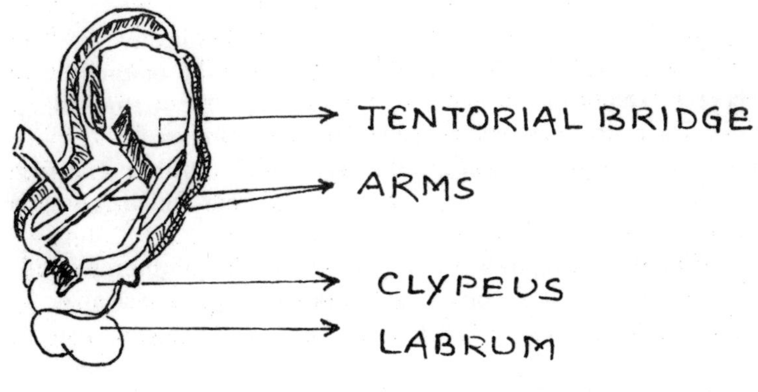

Tentorium

Tephrosin: It is a derivatives of rotenone and is found along with rotenone, in the Derris and cube roots. It is isolated from the leaves of *Tephrosia vogelii*.

Teratology: The study of structural abnormalities, especially monstrosities and malformations.

Terebrantia: A sub-order of Hymenoptera including parasitic wasps and gall wasps having a

Termite or white ant: This insect belongs to the order Isoptera and family Termitidae. Three main species of termites are found. They are: *Microtermes obesi* Holmgren, *Odontoterms obesus* (Rambur) and *Termes taprobanses*. They live under ground but make small earthen mounds or earthen passages that are visible above the ground. These are social insects and their colony organisation based on a caste system. In a colony, there are numerous workers, one queen and a king, and good number of complementary forms of true but young males and females. These feed on woodwork, fences, wooden poles, sugarcane, chillies, wheat and beans to get cellulose.

Termitidae: One of the principal families of termites including, among others, the genera *Termes* and *Nasutitermes*.

Termitophiles: A general name for many insects that live as inquilines in the nests of termites, scavenging or sharing the food of their hosts.

Terrestrial: Insects living on land; opposed to aquatic.

Testis: The male reproductive organs containing the primary germ cells, and in which the sperm cells undergo maturation.

Tetraethylpyrophosphate: TTEP; teppl bladden. It is a commercial compound contains 40% pure TEPP and is used for the control of aphids. It is highly toxic to warm blooded animals.

Tetranychid mites: These are commonly known as spider mites. They are soft bodied prostigmatid mites with long, recurved whip-like chelicerae inserted in a movable stylophore and the fourth palpal segment bears a strong claw. In recent years, these mites are of great concern to agriculture.

Tetanychus cinnabarinus (Boisduval): It is a common pest of vegetable and ornamental plants, and has been reported to be a serious pest of castor and soybean also. Its attack results in characteristic blotching of leaves.

Tetrathion: 0-Methyl, 0-Ethyl-S-(2-Ethylthioethyl dithiophosphate. It is an organophosphorus insecticide similar to thiometon and disulfotan in chemical properties.

Tettigonidae: Long horned grasshoppers or bush crickets. These are long legged jumping insects of the order Orthoptera, differing from other grasshoppers and lcusts (Acridiidae) by their long antennae.

Theca: A protective covering or case, e.g. the chitinous covering of a pupa; the strengthening plate of a fly's proboscis.

Therapsid reptiles: The reptillian group from which the mammals envolved.

Therapy (Physiotherapy and Chemotherapy): Physical or chemical curative measures involving the application of heat, cold or chemicals to destroy pests.

Thermal: Of or pertaining to heat.

Thigmotactic: Showing a tendency to press the body against a surface, or into crevices and corners.

Thigmotaxis: A locomotory response, sometimes an inhibition of movement, caused by the stimulus of touch.

Thimet: Phorate (0, 0-diethyl-S-(ethylthiomethyl dithio phosphate). It is a systemic organophosphorous insecticide of low water solubility. It has long residual activity in plants and at the same time, is highly toxic.

Thiocorn: Amidithion (0, 0-dimethyl-S-(N-2-Methyxoethyl carbamol methyl ditho-phosphate). It is similar to dimethoate and is a good contact insecticide with prolonged systemic action.

Thiolo: The redical-SH; the use of term 'Thiolo' commonly indicates the replacement of the oxygen in –OH group by sulphur.

Thiono: (The radical is = C8). The use of term 'Thiono' commonly indicates the replacement of oxygen in a carberyl group by sulphur.

Thiophos ME: Methyl ethyl parathion; Methyl ethyl thiophos (methyl-o-ethyl-0-4-nitrophenythiophosphate). It is an organophorus compound similar to parathion except one ethyl has been replaced by Methyl group.

Third axillary: The flexor sclerite of the wing base; the sclerite on which the flexor muscle is inserted.

Thoracic glands: Glands in the thorax of the insect, which when stimulated by other hormones from the brain, secrete the moulting hormone.

Thorax: The second division (middle part) of an insect's body between head and abdomen and is composed of three segments, known as the Pro,—meso,—and metathorax.

Threshold Level: The pests population level below which no damage is detectable.

Threshold limit value: The concentration of a contaminant or toxic substance in air, at which on the basis of present day knowledge, workers may be repeatedly exposed without any preceptable adverse effect on their health during their life time. This value should be used as a guide in the control of health hazards and should not be regarded as fine line between safe and dangerous concentrations.

Thrips: Any member of the order Thysanoptera. These are minute insects with short antennae and four narrow wings fringed with hairs, some species are however, wingless. Many of these major pests of cultivated crops, e.g. onion thrips (*Thrips tabaci*).

Thysaneura (thousanus, a fringe; *cura*, a tail): The spring tails. The medium tail has bristle like processes and on each side there are long cerci. The body is covered with scales; antennae are long; mouth parts are meant for biting.

Thysaneuriform: A name sometimes to describe *Campodeiform* or *oligopod* larvae on account of their resemblance to the bristle tails (Thysaneura). Such larvae are common in beetles, mayflies and dragonflies.

Thysanoptera: (*Thysanos*, a fringe; *pterron*, a wing). Thrips. The thrips have narrow wings, with long fringes. They are minute insects with 6-9 segmented antennae. The mouth parts are stylet like and are used for rasping and sucking plant sap. The tarsi are short, each ending in a vesicle.

Tibia: The fourth segment of the insects leg, between the femur and tarsus. It is usually long and thin, straight or gently curved, and often equipped with strong spines.

Tibial spur: A large spine on the tibia usually located at the distal end of the tibia.

Tick: A blood sucking arachnid parasite of the family Iodidae or Argasidae.

Til hawk moth: *Acherontia styx* Westwood (Lepidoptera: Sphingidae). This pest feeds on til in India, Indonesia, Sri Lanka, the Philippines and Burma. Besides til it has also been reported on potato, brinjal and lab lab. The moth also sucks honey from the honey combs in apiaries. It is also known as sphinx moth or death's head-moth. The adult is a large reddish brown, robust moth having forewings decorated with a mixture of dark brown and grey patterns with dark or black wavy markings and a prominent yellow spot on each wing.

Tillage: The opening and preparation of soil for sowing crops.

Dry tillage: Means the pulverisation and exposure of soil carried under dry condition.

Time sense: Insects, like most animals appear to have some internal clock which regulates their daily and seasonal activity. The rhythm of activity, although influences by the light, temperature, etc. is not dependent on these factors alone. The cockroach will become active between 6 or 7 o'clock each evening and even if it is kept in total darkness during day time. The rhythm appears to be controlled by the build up of particular hormones and in the case of cockroach it has been found that these hormones come from secreting cells in the suboesophageal ganglion, but other indications point to be the brain and the corpora allata.

Tingling: A peculiar pricking sensation.

Tiphiidae: A family of small, hairy wasps whose larvae are ectoparasites on the larvae of beetles. The wasp first paralyses its victim by stinging it and then lay its eggs on the surface of the body. *Tiphia popillivora* has been used to control the Japanese beetles (*Popilla Japonica*), a pest of cereals.

Tipulidae: A family of slender insects (Diptera) with long fragile legs, a tapering abdomen and two long wings. In the Tipulinae, the insects larvae are soil dwellers and do considerable damage to the roots of plants and have tough skin which gives them the name of *leather-jacket*.

Tissue culture: The cultivation of insect tissues in vitro is a technique which could lead to very important advances in our understanding of the modes of action of insect hormones and their various mechanisms.

Titillators: Terminal processes of the acdeagus.

Tobacco: It is a natural insecticide which is used as such and also extracted for its active insecticidal compounds like nicotine.

Tobacco Budworm moth: *Heliothis virescens,* a notuid moth (Lepidoptera). Whose larvae bore into the buds of the tobacco plants.

Tobacco caterpillar: *Spodoptera litura* (Fabricius) (Lepidoptera: Noctuidae). It is found throughout tropical and sub-tropical parts of the world. In India it feeds on castor, soybean, groundnut, tomato, cabbage, etc. The damage is done by the caterpillars which are velvety black with yellowish-green dorsal stripes and lateral white bands. The forewings of the adult moths have beautiful golden and greyish-brown patterns.

Tobacco-decoction: It has insecticidal properties and is prepared by boiling or soaking overnight tobacco leaves, stalks or wastes in water and by adding soap to the mixture.

Tolerance: The maximum amount of residue which may safely remain on a harvested crop, measured in parts per million (ppm); it is the ability of a variety of a crop to withstand the attack of a pest.

Tolerance limit, Tolerance level: It is the amount of a pesticide residue (ppm) that is permitted by regulation to remain in or on a commodity. It is set as low as possible.

Tongue: Usually applied in entomology to the proboscis of moths, butterflies and bees.

Tonofibrillae: Non-contractile tension fibrils such as those running vertically through the epidermal cells of insects and connecting the cuticle with the underlying muscles.

Tonus: The conditions of a muscle which remains in a continuous state of partial contraction enabling an animal to maintain its posture; a phenomenon most noticeable when insect remain for hours or even days without moving.

Topical application: The application to the top or to the upper surface of the plant.

Topochemical sense: The perception of sense along a path or on either side of its, enabling ants or other insects to follow a track previously used by themselves or by other insects and to return to it.

Topotaxis: Any reaction by which an insect orientates itself and moves in a definite direction in relation to stimulus.

Topotype: A specimen collected at the type locality.

Torgomen: The epidermal cell associated with a seta that forms the setal membrane or socket.

Tornus: The anal angle of an insect's wing; a distinct notch on the hind margin, dividing into two sections.

Tortoise beetles: Leaf beetles of a sub-family Cassidinae of family Chrysomelidae having the thorax and wing cases flattened and widened to cover the head and the greater part of the legs like the Carapace of a tortoise.

Tortricidae: A family of small moths (Microlepidoptera) whose larvae are generally leaf-rollers and do considerable damage to many trees.

Torus: A general name for a thickened part or pedicel on which an organ is borne, e.g. the base of an antenna.

Toxaphene: It is a chlorinated hydrocrobon insecticide, obtained by the chlorination of the naturally occurring tarpenes. A mixture of products formerly called chlorinated camphene, Hercules 3956 and now toxaphene are produced.

Toxic: Poisonous; injurious to animals and plants through contact or systemic action.

Toxicant: A poisonous substance that can injure or kill plants, animals or micro-organisms.

Toxicarl: A poison. It is a derivative of rotenone and occurs along with rotenone and deguelin in derris and cube roots.

Toxicity: Ability to poison. The toxicity after one exposure to a pesticide is called *Acute toxicity*, whereas after small and repeated doses over a period of time is called *chronic toxicity*. The toxicity value are expressed as *Acute oral/dermal* (a single dose received via mouth/skin/LD50 (lethal dose that will kill 50% of a large population of test animals) in terms of milligrams of the chemical per kilogram of body-weight (Mg/Kg); or inhalation data LC 50 (Lethal concentration of a pesticide in the air that will kill 50% of test animals) in terms of micrograms of mist or dust per litre of air (Mg/1); or parts per million (ppm) by volume of gas or vapour. The chronic toxicity values, on the other hand, are expressed as parts per million (ppm) of the daily diet for a specified number of days.

Toxicognath: A poison jaw; a modified leg (Centipedes).

Toxicological: Relating to poisons and their antidotes.

Toxin: A poinson produced by a plant or animal.

Tracer: Additive to facilities location of a deposit, by redioaction or fluorescent means.

Trachea: Pl. Tracheae. A tube of respiratory system, lined with taenidia, ending externally at a spiracle, and terminating internally in the tracheoles; the breathing tubes formed as multicellular invaginations of the ectoderm.

Tracheal commissures: Transverse tracheal trunks continuous from one side of body to the other.

Tracheal gills: Gills containing tracheae and tracheoles. These are respiratory organs of certain aquatic insects consisting of thin outgrowths of the body wall containing numerous closed tracheoles into which dissolved oxygen can only diffuse from the surrounding water.

Tracheal gills

Tracheal orifice: The primary opening at the point of formation of a trachea, whether exposed externally or concealed in a secondary atrial depressions of the body wall.

Tracheal system: The part of the respiratory system composed of the tracheae and tracheoles.

Trachein: A colloidal or jelly like material forming the walls of the tracheal air-sacs in the aquatic larvae of gnats (e.g. *Corethra*). It swalls up when absorbs the water and shrink on loosing it, thus an insect can control its buoyancy by causing a swelling or shrinkage of these bladders.

Tracheoles: The fine terminal branches of the respiratory tubes (tracheae) without spiral thickenings.

Trademark: Defined as "word, letter, device, or symbol used in connection with merchandise and pointing distinctly to the origin or ownership of the article to which it is applied."

Trade name: Name given to a product by its manufacturer or sold exclusively by that

Translaminar: A pesticide which passes through from one surface of a leaf to the other (from lamina to lamina) through the leaf tissue.

Translocation: The uptake of a pesticide into parts of a plant body and its subsequent dispersal to other parts of the plant body.

Translucent: Allowing light to pass through, but not necessarily transparent.

Transverse: Across, at right angles to the longitudinal axis.

Transverse costal vein: A cross vein in the costal cell (Hymeroptera).

Transverse cubital vein: A transverse vein connecting the marginal and cubital vein (Hymenoptera).

Transverse marginal vein: A cross vein in the marginal cell (Hymenoptera).

Transverse median vein: A cross vein between the median or discoidal vein and the anal vein (Hymenoptera).

Transverse radial vein: Same as transverse marginal vein.

Transverse suture: A suture across the mesonotum (Diptera).

Trap crop: The crop of plants grown especially to attract insect pests.

Trench fever: A human disease caused by *rickettesia*. It is nonfatal and chracterized by sudden onset of fever, headache, dizziness and pains in muscles and bones.

Triange: A small triangular cell or group of cells near the base of the wing (Odonata, Anisoptera).

Triazophos: Hostathion. It is a broad spectrum contact and stomach organophosphorus insecticide and acaricide. It has also ovicidal action against mite. It is used at the rate of 200 to 400 mg a.i./ha depending upon the crop and stage of the insect pest.

Trichlorphos: Dipterex. It is contact and stomach insecticide with penetrant action. It is recommended for use against flies, some bugs, some beetles, lepidopterous larvae, and ectoparasites of domestic animals. Its activity is attributed to its metabolic conversion to dichlorvos.

Tribe: A sub-division of a sub-family, containing a group of related genera. Names of tribes end in-*ini*; between genus and sub-family.

Trichobothria: A minute sensory hairs on the tarsi (spiders).

Trichogen cell: The epidermal cells from which a setae or scales develop on the body or wing of an insect.

Trichophore: The opening in the cuticula beneath a seta giving passage to the hair forming process of the trichogen.

Trichoptera: Caddis flies. Moth like insects whose wings are covered with hairs and mouth parts are reduced to take a little food only. The larvae are aquatic, hervivorous, mostly live in tubular cases formed of wood particles, sand, small cells etc.

Trigonal brace: A strong forked vein in the basal part of the wing, formed by fork of Cu_1 in the front wing and of M_{3+4} in the hind wing (Neuroptera, Ascalaphidae and Myrmelleontidae).

Trigonal vein: The vein forming the anterior fork of the trigonal brace (Neuroptera, Ascalaphidae and Myrmeleontidae).

Trinominal nomenclature: An extension of the binominal system of nomenclature to permit the designation of subspecies by three-word name.

Tripectinate: Having three rows of comblike branches.

Triploid: A cell or individual with three haploid chromosome sets; one of the forms of polyploidy.

Trithion: (Carbophenothion, 0, 0-diethyal-S-(4-Chlorophyenylthiomethyl) dithiophosphate). It is a powerful contact organophosphorus acaricide and insecticide.

Tritocerebrum (Metacerebrum): The third hind part of the insect brain, formed of the ganglia of the postoral second antennal somite.

Tritogenaphis: A bright red aphid from which a vermillion anthocyanin pigment can be extracted.

Triungulin larva: The active first instar larva of the certain beetles (Meloidae) that undergo hypermetamorphosis. The larva is minute, long legged and parasitic and attach itself to the bodies of the visiting bees. Thus, they are carried off to the bee's nest where they feed on the eggs. A similar type of larva is also found in Strepsiptera.

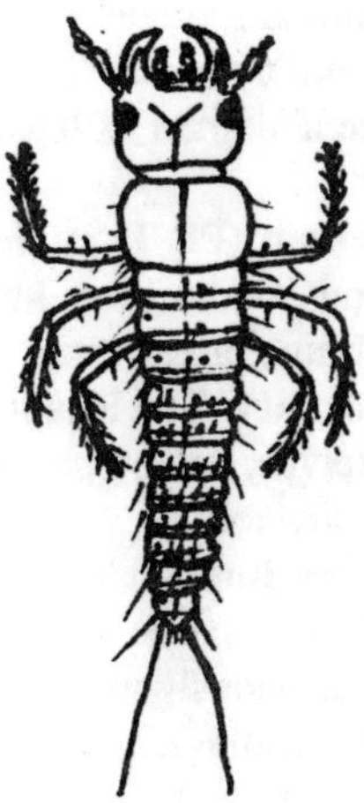

Triungulin larva

Trivial name: An obsolete designation by Linnaeus for the specific name, also a synonym for "Vernacular name".

Trochanter: The second segment of the leg uniting the coxa and femur, probably composed of two united trochanteral segments; in odonata a division between its component segments.

Trochantin: The precoxal sclerite of a thoracic pleuron derived from the anterior part of the primitive coxopleurite, usually articulated at its ventral end to the anterior margin of the coxa, and giving insertion to the tergal promotor muscle of the leg.

Any small sclerite interposed between two others, e.g. on the mouth parts.

Trogoderma: *Trogoderma granarium* (Dermestidae: Coleoptera). Commonly called Khapra beetle. It is a minute light brown insect with reduced wings and very long antennae. The larvae are flattened and hairy, and feed on stored cereals specially in Tropical countries.

Trophallaxis: The mutual exchange of food between insects of a colony, e.g. bees, wasps, ants; a behaviour pattern that helps to improve relationship and bind the colony together into a social unit.

Trophamnion: A sheet or membrane surrounding the eggs of some parasitic insects, especially those which exhibit polyembriony. It is able to absorb food materials from the tissues of the host.

Trophocytes: Cells that elaborate nutritive materials; the cells of the fat body having a tropic function in distinction from the urate cells; the nutritive cells or nurse cells of the ovary or testis.

Trophothylax: A food pocket in the first abdominal segment of the larva, in ants of the subfamily Pseudomyrminae.

Tropism: A reflox response whereby an insect turns in a particular direction in relation to a stimulus. The word is sometime wrongly used synonymously with *taxis*, the orientation of an animal with respect to stimulus, either positive (turning towards stimulus) or negative (turning away from stimulus).

Tropotaxis: A response involving the turning of an insect followed by movement in a definite direction in relation to stimulus. It thus combines a *Tropism* and a *Taxis*.

Truncate: Cutt off square at the end.

Trunk: The entire series of body segments of an orthoropod, including the cephalic, thoracic and abdominal section.

Truss cell: The cell immediately behind the point of fusion of Sc and R_1 (Neuroptera).

Tubercle: A small knob-like or rounded protuberance.

Turbidity: Suspended matter in water preventing light penetration.

Turret nose piece: This is a multiple mounting to carry two, or three pairs of objectives in microscopes, so that these can be quickly interchanged.

Twitch: Violent intermittent in voluntary contractions.

Tydeid mites: These are soft bodied yellow mites, commonly known as 'honey dew' mites. Most of the species are of no economic importance.

Tylus: A sclerotised plate in the sound producing organs of a Cicada.

Tymbal: The drum-like organ with which a male cicada is able to make a loud noise. Two such organs are situated at the base of the abdomen. Each consisting of a tightly stretched skin with an air space beneath it. Vibration of drum is brought about by minute but powerful muscles that cause a rapid in-and-out movements.

Tympanal hood: One of a pair of tubercles or rounded prominences on the dorsal surface of the first abdominal segment (Lepidoptera).

Tympanal organs: Auditory organs on the abdomen or legs of certain insects.

Tympanum (Pl. Tympana): A vibrating membrane; an auditory membrane or eardrum.

Type locality: The locality at which a holotype, lectotype, or neotype was collected.

Type method: The method by which the name for a taxon is unambig-uously associated with a definte zoological object belonging to the taxon.

Types: Forms designated when species or group is described to be most representative or typical, to serve as the reference if there is any question about what species or group includes.

Type selection: Type designation.

Type species: The species which was designated as type of a nominal genus.

Typhoid (fever): Enteric fever; an infectious disease due to *Bacillus*, characterised by fever, rose-red rast, enlargement of the spleen and ulceration of the intestine.

Typhus: A acute infection and contagious disease caused by rickettsia; marked by higher fever, great mental and physical depression and formation of pimples. It must not be confused with typhoid fever which is altogether different disease.

Typological thinking: A concept in which variation is disregarded and the members of a population are considered as replicas of the 'type', the Platonic *eidos*.

Tyrosine: An amino-acid, important in all living organisms. In insects it is the basis of the formation of melanin in cuticle.

U

Ultra-hazardous: A job or activity that is very dangerous.

Ultra violet vision: Most insects can perceive a wide range of light frequencies including all the coloure visible to the human eyes as well as infrared and ultraviolet which are beyond the range of humen eye vision. Moths are attracted to UV light. On the other hand ants will try to avoid it.

ULV: Ultra Low Volume.

Uncus: A process of the tenth abdominal tergum overhanging the anus in Lepidoptera.

Ungues: Lateral claws of the foot (pretarsus) secondarily developed from the base of the dactylopodite; characteristics of insects, but occurring in several arthropod group usually also called 'tarsal claws'.

Unguifer: The median dorsal process on the end of the tarsus to which the pretersal claws are articulated.

Unguitractor Plate : A ventral sclerite in the base of the pretarsus of insects upon which the depressor muscle of the pretarsus or retractor of the claws, is inserted by a long tendinous apodeme.

Uninominal nomenclature: The designation of a taxon by a scientific name consisting of a single word; required for taxa above specific rank.

Unintentionally: Did not mean to do it; done accidently.

Unisexual: Consisting of or involving only females.

Unsclerotized: Not hardened by the deposition of sclerotin.

Urate cells: The special cells of the fat body that become temporarily or permanently charged with excretory material (Uric acid in solid form).

Uric acid: The chief nitrogenous waste products of birds, reptiles and insects.

Urogomphi (Sing. Urogomphua): Fixed or movable cercus-like processes on the last segment of a beetle larva (also called pssudocerci or corniculi).

Urophora: A genus of gall-flies of the family Trypetidae; small two winged flies whose larvae make large round galls, at first green and soft but later becoming brown and hard. A number of larvae live in each gall.

Uroplata: A genus of the family Chrysomelidae having delicately ridged and scutplured bodies and elytra; wedge-shaped beetles. The tiny larvae are short legged and flattened, they tunnel inside the leaves of grass and other plants, making large irregular mines in which they pupate.

Uropod (Pl. Uropoda): One of the terminal pair of abdominal appendages, usually lobe-like (Crustacea).

Uterus: The term is strictly incorrect when applied to insects but is commonly used to denote a part analogous with the Uterus of mammals. A compartment of the genital chamber or vagina where the two oviduct join. In some insects such as tsetse flies the eggs hatch here and the larvae remain in the parent's body to be deposited only when they are nearly full grown and ready to pupate.

Utethesia: A genus of the moth of the family Arctiidae, generally brilliantly spotted with red and black (also see Sunnhemp hairy caterpillars).

Uv light: Radiation in the region of spectrum of lower wavelength (higher frequency) then the visible violet light extending from the end of visible light about 3900-4000 A° to down 200 A° light.

V

Vagina: A part of the definitive egg passage in many insects posterior to the true oviducted communis, derived from the genital chamber. A median duct chamber into which two oviducts open. It is formed by a folding of the body wall in the ninth abdominal segment.

Valid name: An available name that is not preoccupied by a valid senior synonym or homonym.

Valves of the heart: Internal valve-like lobes of the heart walls between the chambers, said to be present in certain dipterous larvae.

Valvifers: The basal plates of the ovipositor, probably derived from the exopodites of the gonopods, carrying the valvulae, including first valvifers.

Valvulae: The three pairs of processes forming the blades and ensheathing lobes of the ovipositor. The first and second valvulae are gonapophyses of the gonopods. The third valvulae are discal outgrowths of the coxopodites of the ninth abdominal segment.

Vamidothion: (Kilval, 0, 0-dimethyl-S-2-(1-N-methyl carbomyol ethyl mercapto)-ethyl thiophosphate). It is an organophosphorus insecticide and approaches the methyl demeton in insecticidal activity.

Vanessidae: A family of handsome, brightly coloured nymphalid butterflies of Lepidoptera order.

Vannal lobe: A lobe in the anal area of the wing, immediately distal to the jugal lobe when a jugal lobe is present (Hymenoptera).

Vannal Region (Vannus): The wing are containing the vannal veins, or veins directly associated with the third axillary; when large, usually separated from the remigium by plica vannalis, often forming an expanded fan-like area of the wing.

Vannal veins: The veins associated at their bases with the third axillary sclerite, and occupying vannal region of the wing (the annal veins except the first or postcubitus).

Vannus: The annal or porterior lobe of an insects' wing; a fan-like expansion separated from the rest of the wing by a furrow.

Vapour: Gas steam; mist; fog; fume.

Variability: The ability of an organism to change its character from one generation to another.

Variation ecophenotypic: Variation caused by nongenetic responses of the phenotype to local conditions of habitat, climate, etc.

Variety: An ambiguous term of classical (Linnaean) taxonomy for a heterogeneous group of phenomena including nongenetic variations of the phenotype, morphs, domestic breeds, and geographic races.

Vas deferens: Pl. = vasa deferentia. One of the lateral ducts of the male reproductive system.

Vas efferens: Pl. = vasa efferentia. One of the short ducts connecting the sperm tubes of the testis with the vas deferens, corresponding to the pedicel of an ovariole.

Vector: Any organism that can carry disease producing organisms from a diseased to a healthy plant or animal.

Vedalia: An alternative name for the lady bird beetle *(Rodolia)*.

Veins: The rib-like tubes that strengthen the wings; the tubular thickenings of the wings springing from the wing base and branching distally.

Veliidae: A family of predatory bugs of Hemiptera. It includes water crickets, water spiders or ripple bugs. These are similar to pondskaters but shorter and broader with thickened hind legs. Some species are wing less. They live on ponds and slow flowing streams, running on the surface and occasionally dive in search of prey.

Vena arcuata: The first jugal vein.

Vena cardinalis: The second jugal vein, usually opening as a basal branch of the vena arcuata.

Vena dividens: A secondary vein present in some Orthoptera lying in the fold between the remigium and vannus.

Vena spuria: A fase vein; a thickening resembling a vein in the wings of hoverflies (Syrphidae).

Venation: The arrangement of veins or nervures in the wings of insects. This helps in identification of species and genus. The generalised veins in the wings are: the costa—a thick unbranched vein forming the anterior margin of the wing; the subcosta—another unbranched vein immediately below the costa; the radius, media and cubitus which are usually branched; the anal veins—short and unbranched. All these veins run logitudinally and may be linked by a number of cross veins dividing the wing into the cells. The evolutionary trend seems to have been towards a reduction in their numbers.

Venter: The entire under surface of the animal between the two series of limb bases, or, when qualified by the designation of a segment, the corresponding surface of a single body segment.

Ventilation tracheae: Tracheae with collapsible walls, responding to varying sorrouding pressure.

Ventrad: Towards the ventral or underside of the body; downward.

Ventral: Lower or underneath; pertaining to the underside of the body.

Ventral diaphram: A membranous or muscular sheath present in some insects, stretched between the lateral edges of the abdominal sterna, sometimes extending into the thorax, separating the ventral sinus from the perivisceral sinus.

Ventral glands: The glands in the ventral part of the head of some insects (Thysanura, Ephemeroptera, Odonata, Isoptera, Dermaptera, Acrididae and Phasmida), corresponding to the prothoracic or ecdysial glands of other insects. These secrete a source of moulting hormone.

Ventral nerve cord: The chain of connected ventral ganglia, morphologically beginning with the tritocerebral ganglia of the brain.

Ventral sinus: The space of the definitive body cavity below the ventral diaphram, containing the nerve cord.

Ventral tracheae: The ventral segmental tracheae originating at a spiracle.

Ventral tracheal trunk: A longitudinal tracheal trunk uniting the series of ventral tracheae.

Ventral tube: A tube formed of two fused ventral appendages (sometimes called coxal vesicle) on the first abdominal segment in sprintails (Colembola order). It is everted by blood pressure, and retracted by muscles. Its function is not known but seems to help in adhesion, especially to a water film, and also sucking up water.

Ventriculus: (Midgut, Mesentron, Stomach). The stomach of the insects immediately behind the gizzard or proventriculus. The epithelial of this portion is the endodermal mesenteron of the embryo. In this part most digestion and absorption of food takes place.

Venules: The smaller cross veins of a insects' wing.

Vermiform larva: The larvae are maggot (worm) like, legs absent, without well developed head and body elongated, e.g. larvae of flies, and some Hymenoptera and Coleoptera.

Vermin: Pests; usually rats, mice or insects.

Vernacular name: The colloquial designation of a taxon.

Versonian cells or Verson's' cell: A specialized apical cell at the tip of the testis in some insects (also see apical cell).

Versons' gland: The moulting glands of caterpillars. These epidermal glands secrets a fluid having a property of liquifying the endocuticule and thus loosening the epicuticle prior to moulting.

Vertebrate: An animal with a bony spinal column. Examples: Mammals, fish, birds, snakes, frogs.

Vertex: The top of the cranium between and behind the compound eyes.

Vertical classification: Classification which stresses common descent and tends to unite ancestral and descendant groups of a phyletic line in a single higher taxon, separating them from contemporaneus taxa having reached a similar grade of evolutionary change.

Vesica: A terminal membranous part of the aedeagus.

Vesicle: A sac, bladder or cyst, often extensible.

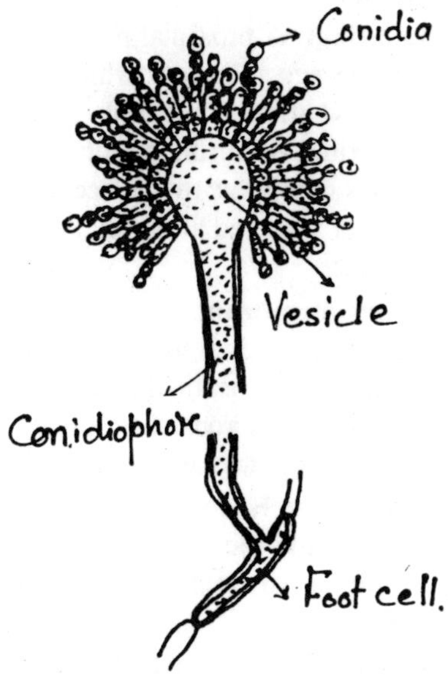

Vesicle

Eumenids. Most of these are black and yellow and fold their wings longitudinally when at rest.

Vesicula seminalis: A dilation of the vas deferens in which spermatozoa may be retained.

Vespa: A genus comprising large social wasps. Formerly the smaller species such as common wasps were also named Vespa but these are now placed into a separate genus *Vespula*.

Vespidae: The principal family of social wasps of Hymenoptera order. It comprises two genera, *Vespa* and *Vespula*.

Vespiformia: A group of Hymenoptera including both ants and wasps.

Vespoidea: A super family of Hymenoptera order comprising the social wasps and vespids and

Vespula: *Vespula vulgaris*, the common wasp; *V. germanica,* the German wasp and *V. sylvastris,* the tree wasp are very similar species all having black and yellow striped body. They live in spherical papery nestes in holes in the ground, hollow trees, out-houses, etc. and are commonest social wasps.

Vestibulum: An external genital cavity formed above the seventh abdominal sternum when the latter extends beyond the eighth.

Vestigial: Poorly developed; non-functional, or degenerated structure.

Vibrissae: Curved bristles located just above the mouth in some Diptera.

Viability: Being alive.

Vibrissae oral: A pair of stout bristles, one on each side of the face near or just above the oral margin, and larger than the other bristle on the vibrissal ridge (Diptera).

Vinculum: The entire coxosternal plate of the ninth abdominal segment (in Lepodoptera).

Virga: A terminal phallic spine, usually arising from endophallus.

Virus: A sub-microscopic pathogen that requires living cells for growth and is capable of causing disease in plants or animals. Plant viruses are often spread by insects.

Visceral: Pertaining to an internal

organ, especially one of the larger abdominal organs.

Visceral nervous system: See sympathetic nervous system.

Visceral trachea: The median segmental trachea originating at a spiracle, branching to the alimentary canal, the fat tissue, and the reproductive organs.

Visceral tracheal trunk: A longitudinal tracheal trunk closely associated with the walls of the alimentary canal.

Viscocity: It is the internal friction or resistance to flow shown by a liquid.

Viscous: Of the consistency of oil or treacle.

Vision: Two kinds of eyes, viz. compound eyes and ocelli are found in insects. Most adults have both kinds, but the larvae have only ocelli. These eyes are made up of lens, retina and nerve. Compound eyes contain numerous ommatidia, each containing a cuticular lens, a crystalline cone and several retinal cells. Ocelli have a cuticular lens and few retinal cells. In both cases the nerves from the retinal cells lead to optic lobes of the brain. Ocelli donot form an image at all but merely detect the changes of intensity of light. Compound eye can produce either a mosaic image composed of dots or a superimposition image in which the dots merge to form a blurred picture. Most insects can perceive colours as well as ultra violet and in some cases infra-red light. Many can also perceive patterns of polarization in the sky.

Vitellarium: A proximal part of an ovariole egg tube into which the occytes are liberated from germarium or distal region. The oocytes here grow by the accumulation of yolk and attain their mature size.

Vitelline membrane: The cell wall of the insect egg. A thin membrane lying beneath the chorion; undivided in meroblastic cleavage, surrounding the blastoderm.

Vitellophags: Endoderm cells proliferated into the yolk and accomplishing a partial digestion of the latter.

Vitrellae: The cells which secrete the crystalline cone in each ommatidium or optical unit of a compound eye.

Vitreous body: A name used to denote the crystalline cones in the eye of an insect.

Vitta (Pl. Vittae): A broad stripe.

Viviparity: The production of young ones, e.g. in aphids.

Viviparous insects: The insects who give birth to young ones, not egg laying; applied to insects in which the egg hatch internally, and either young larvae or fully fed ones are deposited, e.g. parthenogenetic females of aphids, tsetse fly.

Volatile: A compound is said to be volatile when it evaporates (changes from a liquid to a gas) at ordinary temperatures on exposure to air.

Volsellae: Ventral lateral processes of the phallobase (Hymenoptera).

Voltinism: The number of generations per year.

Volunteer: Crop plant growing accidently from shed seed; not deliberately cultivated.

Vomitous: Matter which is vomited.

Vulva: The external opening of the genital chamber or vagina serving in most cases for both copulation and the discharge of the eggs; sometimes on the eighth abdominal segment, sometimes on the ninth.

Vulvar lamina: The posterior margin (usually prolonged posteriorly) of the eighth abdominal sternite (in female Odonata).

W

Waiting Period: It is time limitation in respect of harvest of the treated crop or slaughter of the treated animal; Safety period.

Walking Stick: A colloquial name sometimes used for any of the stick insects like *Phasmida* and *Dixippus*. These come under family Phasmidae or Phasmatidae. The insects are thin legged and body generally match with the leaves of plants on which they live.

Walnut weevil: *Alcidodes porrectirostris* Marshall (Curculionidae: Coleoptera). It is a serious pest of walnuts in Kumaon, Kulu and Kashmir. The adults feed on buds, flowers, whereas grubs feed inside the fruits and cause premature dropping. The adult beetle is pitch black when young and turn dark brown with age. The grub is creamy white, legless, with a pale-brown head.

Wanting: Missing or lacking.

Warfarin: It is a rodenticide manufactured in India.

Warning Colouration: Conspicuous colours in insects frequently associated with qualities which render their possessor unpalatable or offensive to birds or other enemies.

Wasp: Winged insects of order Hymenoptera having a slender body with a narrow constricted waist and with the ovipositor modified to form a sting. These are omnivorous, social or solitary with biting or sucking mouth parts. They differ from bees as they donot store honey and pollen and donot secrete wax. They live in nests which are constructed by social wasps in the form of a sphere of cells of chewed wood palp; others live in holes in the ground or in trees. Also exhibit warning colouration.

Water Beetles: These beetles lead an aquatic life by remaining either wholly submerged or swimming on or just below the surface and come under the order Coleoptera. The hind or sometimes middle pair legs are flattened, fringed to enable to perform a rowing movement. They take supply of air either as bubble under the wing cases or trapped as a film by the hydrofuse hair under the body. The larvae are generally aquatic. The principal families of water beetles are; Dytiscidae (carnivorous beetles), Cyrinidae (whirligig beetles), Hydrophilidae (scavenger beetles).

Water bug: A colloquial name given to aquatic Hemiptera of the sub-order Heteroptera. Some swim under water, some on the surface. Some of them are carnivorous and others are hervivorous. The principal families are: Nepidae (water scorpions), Naucoridae (saucer bugs), Notonectidae, Crixidae, Gerridae, and Hydrometidae. Their legs are flattened or fringed for swimming.

Water Gnat: *Hydrometra stagno-*

rum (Heteroptera). A water bug that have long legs and it walks on the surface of water of ponds and streams.

Water Moth: A name sometimes loosely and colloquially used for a caddis fly (Trichopterous insect).

Water solubility: Capable of being homogenously mixed with water.

Water Tiger: A colloquial name sometimes used for the carnivorous larvae of Dytiscid water beetle.

Water Weevil: *Lissorhapterus simplex*, an insect whose aquatic larvae feed on the submerged roots of rice and other aquatic plants, and the adults feed on the leaves.

Wax: Most insects secrete a mixture of waxy substances which pass out in small quantities through pore canals in the epidermis and impregnate the epicuticle making it waterproof. Some other insects like honey bees produce wax by the glands present under the abdomen. The scale insects (*Ericerus* sp.) produce a China wax and Wooly aphis (*Eriosoma* sp.) produce 'wool' perhaps to protect the insects from natural enemies and weather conditions.

Wax Insects: A name for number of insects (plant bugs) of a family Coccidae that secrete a wax like substance. Chinese wax scale insect *(Ericeerus pela)* is bred commercially for making candles.

Wax moth: The larvae of *Galleria mellonella* feed on bee hives, hence known as wax worm.

Wave Moth: A name for number of Geometrid moths having grey or greenish wavy lines across the wings, e.g. *Delinia exanthemata*.

Weathering: The wearing away of pesticides from the surfaces they were applied to because of rain, snow, ice and heat.

Webbing Clothes moth: *Tineola bisselliella* is one of the commonest clothes moth. The moth is small, grey-brown with fringed wings and long antennae. The larvae feed on wool, fur, etc. and pupate in silken cocoons.

Web worm: A name given to gregarious caterpillars living in nests of silk threads produced by the labial glands of the larvae. The larvae of the cabbage web worm (*Crocidolomia binotalis*) of Lepidoptera order form webs of silken threads and feed on leaves remaining inside.

Weevils: These insects come under the order Coleoptera and family Curculionidae. The adults have long snout or rostrum which is used for boring into seeds, woods, stems, etc. Many of them are serious pests of crops, grains and stored products. e.g. Rice weevil (*Sitophilus oryzae* Linn.), Legume weevil (*Alcidodes collaris* P.).

Weighting: A method for determining the phyletic information content of a character; the evaluation of the probable contribution of a character to a sound classification.

Weismanns' Ring: A gland surrounding the main dorsal blood vessel just above the cerebral ganglia in insects such as house fly and blow fly. It is believed to secrete a hormone which induces the formation of puparium.

Wettability: The extent of which a solid is wetted by a liquid, measured by the force of adhesion between the solid and liquid phases.

Wettable Powder (W.P.): It comprises a finely ground powder and a wetting agent. In water, it forms a suspention that settles down soon unless the suspension is stirred.

Wetting agent: A compound which reduces surface tension and causes a liquid to contact plant surface more thoroughly.

Whame Fly: A local name (British) occasionally used for blood sucking horse flies.

Wheat Aphid: *Microsiphum miscanthi* (Takahashi) (Aphididae: Hemiptera). This is a minor pest of wheat, barley, oat, etc. and is widely distributed in India. The insects are green, louse-like and appear on young leaves and ears in large numbers specially during cold and cloudy wheather. The nymphs and adults suck the sap from the plants.

Wheat Bulb Fly: *Hylemia (Delia) platura, H. coarctata, H. cilicrura.* A small black fly of the family Anthomyidae (Diptera) is a serious pest of agricultural crops. It is smaller than house fly. It lays eggs in the soil and the ensuing maggot burrow into the shoots of wheat, barley, rye, soybean, etc.

Wheat Gall Nematode: *Anguina tritici* (Tylenchida; Tylenchidae). The 'earcockle' or mamni disease is caused by the nematode throughout the wheat growing parts of the world. If the black rounded mamni galls are soacked in water overnight the large number of larvae are set free. These are as wriggling thread-like creatures.

White Ants: A name commonly but incorrectly used for termites. These donot have any relationship with the ants except that both are having numerous castes, viz. queen, worker, soldier, etc.

White Fly: These are small Homopterous plant lice of the family Aleurodidae having the body covered with fine white powdery wax. They suck juice from the leaves, young shoots and fruits and smother the surface with a sticky secretion of honey dew. Green house white fly, *Trialeurodes vaporariorum* is spread all over the world from central America, attacks tomato, cucumbers, etc. Soybean white fly *(Bemesia tabaci)* is a serious pest of soybean, green gram, black gram in India and spreads 'yellow mosaic virus'.

White Grub: *Holotrichia consanguinea* (Blanchard) (Melolonthidae: Coleoptera). This pest attacks maize, chillies, okra, brinjal, pearl millet, sugarcane, etc. It is found in Haryana, Himachal Pradesh, Rajasthan and the Punjab. The grub eats away the fine rootless and also girdle the main root ultimately killing the plants. Beetles at night feed on the foliage of neem (*Azadirachta indica*) and banyan trees (*Ficus bengalensis*).

Whorl: The point where leaves or other plant parts unfold or are formed in a circular pattern.

Wide Field Eye Piece: These are designed to accept a bigger cone of light than standard eye pieces, and show you more of the object at one time. The lenses are bigger and consequently it is easier to bring ones' eye quickly into the line.

Wide range: Pesticides ability to kill several different kinds of pests.

Wilting: Drooping of leaves.

Wing Base: The proximal part of the base of the veins and the body, containing the humeral and axillary sclerites.

Wing Bud: A small fold of epidermis growing out from the thorax of a nymph or of a pupa, eventually to become the wing of the adult insect.

Wing Cell: The area of the membrane between the veins and cross veins.

Wing Cover (case): See elytron and tegmen.

Wing Muscles of the heart: See alary muscles.

Wing Pads: The undeveloped wings of the nymphs and naiads. They appear as two flat structures on each side.

Wing Region: The principal areas of the wings differentiated in the wing-flexing insects by the distinct lines of folding including the axillary, remegial, vannal and jugal regions.

Wings: Most insects have two pairs of wings, one pair on mesothorax and other on metathorax. However, in Diptera only fore wings (mesothoracic wings) are functional for flying, the latter being reduced to small halteres. Wings are formed from enlarged folds of epidermis and cuticle which grow out from the thorax. Wings may be membranous in bees, wasps or these may be thickened and hardened to a elytron or tegmen. The surface of the wings may be completely smooth (dragon fly) or covered with fine hairs (caddis fly); in mosquitoes these scales are only located along the lines of the veins and in butterflies and moths (Lepidoptera) the whole wing is densely covered with scales of various shapes, sizes and colours.

Wing scales: See scales.

Wing Veins: Narrow sclerotized tubes between the uper and lower epidermis of an insect's wing. Circulation of the blood or haemolymph takes place along these veins and in some cases have marked tracheal tubes. These veins are situated in furrows (concave veins) or on ridges (convex veins) or of both types in a single wing to stiffen the membrane like a fan. Primitive insects such as mayflies and dragonflies usually have a complex network of veins. Each species is characterised by a definite number and arrangement of veins.

Wire Worm: An elateriform larvae. A larva that is yellowish, slender, elongate, heavily sclerotized, with a few hairs on the body, with thoracic legs but without prolegs, e.g. the larva of the click beetle (family Elateridae, order Coleoptera). These feed on roots, tubers and young seedlings. They cause considerable damage to agricultural crops.

Wood Borers: Wood boring is caused by many insects either to obtain food or as a means of protecting the eggs and the developing larvae and pupae. Most of the insects belong to Coleoptera, some Hymenoptera and the larvae of some Lepidoptera. The boring action may be performed by the jaws (in beetles, weevils, larvae of moths) or by the ovipositor (in sawflies, wood wasps). Some insects have the enzymes *cellulase* which is necessary to digest the cellulose of the cell walls. Termites have an intestinal flora or fauna to digest cellulose.

Wood Naphtha: Useful as a relaxing agent. An impure distillate of wood, consisting of methyl alcohol with traces of esters, cresol, etc. It is probably these impurities that make it effective in relaxing the insects.

Wooly Aphid: *Eriosoma lanigerum;* It is a small plant louse, originally a native of America but now a world wide pest of apple trees. The adults and the nymphs suck the juice and reproduce parthenogenetically in large numbers. Unlike most aphids it exudes a white waxy substance forming a mass of wooly threads under which large numbers of the insects are protected.

Worker: They are sterile individuals among social insects such as bees, wasps, termites (females in ants, wasps and bees; both males and females in termites). They perform most of the work such as building of nests, food storage, etc. of the colony. They outnumber all other castes. In case of bees the worker develops from the fertilized egg, the larva being given a diet of nectar and pollen only. If given a more complete diet of 'Royal Jelly' which has been partly predigested and concentrated, the same larvae may become queen bee. Worker Hymenoptera are always sterile female but worker and soldier termites are sexually immature individuals of either sex.

WO Type: Out of two types of emulsions, in the WO type (water in oil type) oil is the internal and water is the external phase.

Wriggler: One that moves in a fashion of wavy track, e.g. the larvae of mosquitoes.

Wyeomia: A genus of mosquito whose larvae live in the liquid contained in pitcher plants in central and south America.

X

Xanthophyll: The yellow pigments of the leaves which sometimes absorbed by insects. It is present in the skin of caterpillar of the cabbage white butterfly and in the cocoons of the silk worms.

Xanthopterin: A yellow pigment derived from Pteridine found in the bodies of wasps and in the wings of certain butterflies (See pterin pigments).

Xenopsylla: *Xenopsylla cheopis*, the India rat flea that transmits bubonic plague.

Xixuthrus: A large brown long horned beetle from the Fiji islands; one of the worlds' largest species about 13 cm long. It is probably now extinct owing to the fact that the long, fat larvae are edible and were considered a great delicacy by the natives.

Xylophagus: A genus of flies (Diptera) with larvae living in decaying wood.

Xylota: A genus of hover-flies (Syrphidae) whose short tailed larvae live in moist decaying wood.

Y

Yellow Butterflies: A group of butterflies of the family Pieridae whose wings contain yellow Pterin pigments derived from uric acid, e.g. the Brimstone Butterflies, *Gonepteryx rhamni* and the clouded yellow, *Colias croceus.*

Yellow Fever: An acute infectious disease due the virus transmitted by certain mosquitoes and marked by fever, jaundice and albumin and globulin in the urine.

Yellow Fever Mosquito: *Aedes aegypti*, a tropical culicine mosquito the chief carrier of yellow fever virus in central Africa and central America. It is also, however, widely spread in areas where yellow fever does not yet occur. A few other species of *Aedes* can carry the disease if conditions are favourable.

Yellow Jacket: A colloquial name sometimes used for any of the yellow and black wasps.

Yield: The aggregate of crop procuction used by man. It does not include parts of the plants which are rejected as useless such as foliage and true roots of potato.

Yolk Cell: Cleavage cells remaining in the yolk and taking no part in the blastoderm formation.

Yolk Cleavage: The division of the yolk into masses containing from one to several cleavage nuclei.

Yolk: Deutoplasm: When the yolk is small in quantity or evenly distributed through the cytoplasm the egg is said to be *Alecithal* or *Hololecithal* when the yolk has central position, the egg is *Centrolecithal;* if the yolk lies at one end of the egg, the later is *Telolecithal.* Insect eggs are *Centrolecithal.*

Y-vein: Two adjacent veins fusing distally, forming a Y-shaped figure, e.g. the anal veins in the front wing.

Z

Zeugloptera: An alternative name for Micropterygidae. Minute primitive moths with biting mouth parts. The name is used when they are regarded as not being true Lepidoptera.

Zink Phosphide: It is a rodenticide now manufactured in India.

Zeographical Regions: The regions into which the world is divided for the purpose of study of animals.

Zootermopsis: A genus of termites that live in damp wood in the forests of western Canada and the United States.

Zoraptera: A small group of insects apparently related to book-lice and sometimes placed in the order Psocoptera. They live in decaying wood or sometimes in termite colonies. They comprise both winged and wingless individuals. Little is known about their life.

Zorotypus: The principal genus of the small and little known order Zoraptera.

Zunzler: A name taken from the German and sometimes used as a general term to denote various moths of the family Pyralidae, e.g. the Indian meal moth (*Plodia*), the flour moths (*Ephestia*).

Zygentoma: A suborder of the Thysanura (bristletails) including the silver fish (*Lepisma*) and firebrat (*Thermobia*).

Zygoptera: Damselflies. A group of dragonflies with slender bodies and narrow based wings which are held vertically above the abdomen when at rest. The hind wings are not different from the fore wings, as they are in true dragonflies (Anisoptera). The nymphs of these flies have three elongated caudal gills at hind-end of the abdomen.

Zygote: The fertilized egg or egg nucleus.

SECTION — II

PLANT PATHOLOGY

A

Abiotic diseases: Diseases caused by non-living agents, such as excess or deficiencies of minerals, herbicidal damage, frost damage, burning etc.

Abjection: Liberation or separation of spores from the sterigmata or tip of spore producing sporophore accomplished through the activity of the fungus.

Abjunction: Formation of spore due to cutting-off of the hypha as a result of septum formation.

Aboospore: Parthenogenetically formed oospore.

Abscess: A localized collection of pus.

Abscission: Shedding of any plant part, leaf, fruit.

Abscission layer: A zone of delicate, thin walled cells, surrounding a lesion on a leaf, the breakdown of which disjoins the affected area from the rest of the leaf.

Abstriction: A process in which separation of a cell or spore by septum formation is followed by abscission, resulting in liberation of that portion.

Acaricide: Chemical toxic to mites.

Accumbent: Resting against anything.

Acervulus (Pl. acervuli): A compact cluster of conidiophores and conidia present in the sub-cuticular or subepidermal layers which are not surrounded by fungal tissue, Characteristic of melanconiales.

Acetamides: A group of systemic fungicides which are basically amido derivatives of acetic acid. Two common fungicides of this group are "Curzate" and "Udonkor". They are effective against *Plasmopara viticola* and *Phytophthora infestanes*.

Acetate Malonate Pathway: A pathway for synthesis of phenols involving head to tail condensation of acetate units. Accumulation of Isocoumarin in carrot slices is an example of it.

Acetate-Mevalonate Pathway: A metabolic pathway of phenolic synthesis in which acetyl CoA units are condensed to form mevalonic acid which is then converted into cyclic terpenes or steroids.

Achlorophyllous Gr.: Having no chlorophyll.

Acid Curd: Coagulation of milk protein by acid.

Acid dye: A dye containing acidic atoms or anions which is actively staining part and has the affinity for cytoplasm.

Acid fastnen: Property of bacteria to retain the initial stain (dye) which is not washed off by acid alcohol.

Acid fast Staininig: A differential staining technique using the series of Carbol fuchsin stain (heated over flame), acid alcohol and methylene blue to stain bacterial smear. Those retaining red colour of carbol fushsin are acid fast and others are non-acid fast.

Acquired immunity: Resistance developed as a result of initial infection by the pathogen, or transfer of substances responsible for such resistence from infected host to other hert which confer a form of passive acquired immunity see also Active immunity.

Acquired Resistance to fungicides: Resistances gradually developed into a microbial population with the exposure to fungicides which are normally sensitive to fungicide. This developed resistance may have genetical basis.

Acquisition feeding: The feeding of a vector (insect) on a virus source for transmission tests.

Acrasin: A chemical secreted by few cells (myxamoeba) in members of acrasiomycetes which attracts other myxamoeba and stimulates the phenomenon of aggregation around such centres. This chemical is cyclic-AMP.

Acrogenous: Developing at the apical region (apex) of any structure (conidiophore. sporophore etc.).

Acropetal Gr.: Usually used for a chain of conidia, in which the youngest is at the apex.

Acropleurogenous: Developing at the end on the side portions.

Actidione (Cycloheximide): A glutarimide antibiotic produced as a by-product of streptomycin manufacture, by species of *streptomyces,* it is active against a wide range of fungi and yeasts but not against bacteria. Chemically it is beta (2-(3-5 dimethyl-2-oxychlorohexyl)-2 hydroxyethyl) glutarimide.

Actinomycetes: A group of microorganisms apparently, intermediate between bacteria and fungi and classified separately.

Actinomycosis: It is considered to be a polybacterial disease caused by actinomycetes.

Activated sludge: Material containing huge, active microbial population produced by vigorous aeration of sewage for its disposal.

Activator: Chemicals added to a pesticide to increase its toxicity.

Active immunity: Immunity acquired by an individual in response to the stimulus of pathogen or their metabolites.

Active Ingradient (A.I.): Active component (chemical) responsible for the desired action of a formulated product (fungicide, weedicide, insecticide).

Active Resistance: See Active Immunity.

Acylalanines: A group of systemic fungicides which are effective against oomycetous fungi. Some common examples of this group are metalaxyl, furalaxyl, miliforum etc.

Adaptation: Change in habits and other character of individual due to altered environmental factors.

Adaptive enzyme: An enzyme produced in responses to the presence of its substrate (also called induced enzyme).

Adelphogamy: Pseudomictic copulation of mother and daughter cells, e.g. yeasts.

Adenine: A purine component of nucleoside, nucleotide and nucleic acid.

Adenosine: A mononucleoside made up of adenine and D-ribose (Pentose) sugar which is produced by hydrolyses of adenosine monophosphate.

Adherence: Property of a fungicide to stick to surface.

Adhesive: Material added to increase the retention of chemical on plant surface e.g. Methy Cellulose preparations.

Adhesorium: Structure formed from the resting spores, which gets attached on host and penetrates the host, e.g. *Plasmodiophora* sp.

Adherent transmission: When pathogen remain attached on the surface of seed or vegetative parts and transmit the disease e.g. Bunt of wheat.

Adjuvant: An additive to spray to improve its properties of spreading, sticking adhering etc.

Adopted Name: See "Common name".

ADP (Adenosine diphosphate): A compound which upon phosphorylation (addition of phosphate and energy) forms high energy bonds as ATP (Adenosine triphosphate). These two chemicals play important role in energy reaction (respiratory) of organisms.

Aeciospores Gr.: A binucleate spore borne in an aecium.

Aecium (Pl. accia) Gr.: The first fruiting body (sorus) formed after plasmogamy bearing binucleate aecio spores (Stage I of Uredinales).

Aerial blight: A disease of soybean, affecting mainly foliage, caused by *Rhizoctonia solani*. Leaves show reddish brown spots, spreading very fast and covering the entire surface of leaves.

Aero-aquatic fungi: Fungi growing vegetatively inside water, but sporulate only when exposed to air.

Aerobe Gr.: An organism which essentially require free Oxygen for respiration.

Aethalium (Pl. aethalia) Gr.: One of the various types of fructification found in myxomycetes which are fairly large, generally cushion shaped, representing a stage of evolution of sporangia. It is a group of sporangia which has not separated into individual units, e.g. *Lycogala, Fuligo* (cf. pseudoaethalium, plasmodiocarp, sporangium).

Aerosel: A dispersion of spray droflets of diameter < 50 micron usually dispersed from a cannisten.

Aestivation: Dormancy during a hot or dry season.

Aetiology: See etiology.

Aflatoxin: A mycotoxin produced in grains, nuts and other food stuff as a result of infection with *Aspergillus flavus* and *A. parasiticus* and certain other fungi, making them unfit for animal and human consumption. Chemically, aflatoxins are bis-furanoisocoumarins and are highly heat stable; can be neutralized by alkali treatment only. They are highly poisonous and are related with cancer.

Agallol: Trade name of organomercurial fungicide (Methoxy ethyl mercury chloride); widely used for seed treatment especially as a dip for tubers and setts of vegetatively propagated crops.

Aganomycetes: Group of fungi which donot produces any conidia or other spore and remain sterile, may produce sclerotia, also called Mycelia sterilia.

Agar Agar: Dried polysaccharide extract made from red marine algae *(Gelidium sp.)* and is used for solidifying the culture media. Also extracted from other algae, *Gracilaria, Gigartina, Pterocladia* etc.

Agaric: A mushroom of Agaricacea with characteristic habit of gill bearing.

Agglutination: Clumping of cells forming precipitate.

Agglutinin: An antibody causing agglutination of bacteria or other living cells.

Aggressins: Metabolites that enhance the virulence of pathogens, but are not directly toxic to the host.

Agrimycin: A commercial antibiotic preparation having streptomycin sulphate and terramycin used for the control of bacterial diseases.

Agrosan G.N.: Organomercurial fungicide (Phenyl mercury acetate+ethyl mercury chloride) used as seed dresser.

Airspora: Spore population (spore flora) present on airborne particles of plant or animal origin.

Akaryote Gr.: A stage of plasmodiophorales during which the nucleoplasm losses its affinity for stain.

Alate: Having wings, especially for spores.

Algal Virus: Viruses infecting algae especially blue green algae. See cynophages.

Allergy: A condition of hypersensitivity to a specific chemical/organism.

Aleuriospore: A thick walled terminal spore formed from the blown out end of the sporogenous cell now such spore are called, terminal thalloconidia.

Aliphatics: A chemical group of systemic fungicides which includes organic compounds without ring structure and are member of "paraffin" hydrocarbons or their derivatives. They are effective specifically against pythiaceous fungi. Two important compounds are "Prothiocarb" and "Propamocarb".

Altenuene: A host selective toxin produced by *Alternaria tenuis* which induces chlorosis in leaves and is structurally similar to Aternariol.

Alternaric Acid: A non-specific phytotoxin produced by *Alternaria solani* producing wilting, necrosis and chlorosis on host.

Alternariol: A toxin produced by *A. tenuis* which is monomethyl ester and inhibits phenol metabolism of host plant.

Alternate host: Another hosts of different type on which a pathogen produces few stages of its life cycle and is essential for completion of pathogen's life cycle, e.g. *Barberis* for wheat rust, *Puccinia graminis.*

Alternation of Generation: Succession of gamete and spore bearing stages in the life cycle of an organisms usually having diploid (2n) and haploid (n) number of chromosomes.

Amastigomycota: A division of Kingdom Myceteae (fungi) consisting of fungi without centriole, having no motile cells, asexual reproduction by budding, fragmentation, sporangio-spores or conidia and haplobiontic-haploid life cycle with zygotic meiosis.

Amatoxin: A group of toxins present in poisonous mushrooms especially *Amanita phalloides.* They are cyclopeptides. Principal toxin of this group are alpha—amanitin and beta-amanitin. If eaten they produce adverse effects and symptoms include hypoglycemia, liver destrophy and kidney failure. (cf. pallotoxins).

Ambimobile: Referred for chemicals (esp. systemic fungicides), capable of being translocated both in upward and downward directions in plants.

Ambrosia fungi: Fungi growing in the tunnels formed by wood burrowing Scolytidae insects (ambrosia beetles). These fungi are a source of food for larva and adult insects.

Amerospore: Single celled spores (cf. **didymospore phragmospore,** dicty-ospore, scolecospore, heliospore and staurospore).

Amino acid: An organic compound containing amino (NH_2) and Carboxyl (COOH) groups.

Ammonification: Decomposition of organic nitrogenous compounds like protein, by microbes, resulting into release of NH_3.

Amphigenous: Referred for the hymenium of Aphyllophorales found all over the surface of sporophore (cf. Unilateral).

Amphigynous: Fungus having a state in which the oogonial initial grows through the antheridium.

Amphimixis: Sexual contact of two nuclei having no near relationship.

Amphitrichous: Having a single flagellum at each end of a cell.

Ampulliform: Swollen or flask shaped.

Amylase: A starch hydrolysing enzyme.

Amyloid reaction: A test in which basidiospores/basidia of Agaricales give a particular reaction with iodine. It is an estbalished test being used in classification of Agaricales.

Amylovorin: A non-specific, toxin produced by *Erwinia amylovora;* polypeptide antibiotic in nature, disrupting the cellular permeability and inducing shoot canker and necrotic spots.

Anaerobic: When no molecular oxygen is required for respiration.

Anamorph L: The conidial state (imperfect state) asexual part of the fungal life cycle.

Anastomosis: Fusion between various hyphal elements after coming in contact with each other.

Ancophylaxis: A state of increased sensitivity or hypersensitivity in animals following the injection of an antigen.

Androgynous: A condition of having both antheridium and oogonium on the same hyphae.

Androphore: A branch of hypha producing antheridia or male reproducing structure.

Angiocarpous: Fungi producing sporocarp, which remain closed atleast upto maturity of spores.

Angiospermic parasitis: Flowering plants living as parasites over another plants mainly belonging to families of Loranthaceae, Convolvulaceae, Scrophulariaceae Orobanchaceae, Hauraceae, Santalaceae and Balano-phoraceae.

Ang-Kak: Fermentation products from *Monoascus purpuren* grown on rice used for colouring the food in far east.

Angstrom (A): A small unit of measuring length equal to 1/10 milimicron (mμ) or 1/10,000 micron (μ).

Angular leaf spot: A disease of cotton caused by *Xanthomonos malvacearum* showing finger like lesion on leaves and then stem is also affected showing black arm symptom. It is also called "Black arm".

Anisogamous planogametes: Motile gametes of different sizes but identical morphology.

Anisogamy Gr.: Fusion of planogametes of different size but of similar morphology.

Annellide (Annellidic or annellophore) L: A sporogenous cell in which after the production of conidium the hyphal wall grows out through the conidial scar producing another conidium then forming a chain of conidia and sporogenous cell has a seris of rings at its apex.

Annulus (Pl. Annuli) L.: A membra-nous skirt surrounding the stipe of a Hymenomycetes which is remanant of the inner veil.

Antagonism: A phenomenon in which the activity of one organisms inhibit, limits or harms the activity of other participants of an ecological association of a particular habital.

Antagonistic crops: Non-host plants which are capable of killing the soil pathogens, esp. nematodes when planted as intercrop with a susceptible host crop. e.g. white mustard crop reduces the population of nematodes when grown with potato, as allyl-isothiocynate (alkaloid) produced by mustard is toxic to nematodes.

Antheridiol: A sex hormone (sterol) produced by somatic hyphae of female thallus of *Achlya bisexualis*, which induces the production of antheridia in the hyphae of male thallus growing in close proximity.

Antheridium (Pl. Antheridia) Gr.: A male gemetangium producing male gemeles.

Anthropochory: Dispersal of pathogens through human.

Anthrozoid Gr.: Motile male gametes of Monoblepharidales.

Antracnose: (i) Dark reddish brown lesions on stem, leaf or fruit usually caused by acervulus fungi. *Colletotrichum* and *Gleosporium*.
(ii) A plant disease/symtom having characteristic limited black lesions usually sunken, giving a fire or burned appearance of plant parts usually caused by one of the Melanconiales mainly *Colletotrichum* sp.

Antibiosis: (i) Antagonism between two organisms especially in soil.
(ii) Resistance of a plant to insect diseases attack e.g. thick cuticle, hairy leaves, toxic sap etc.
(iii) Production of some metabolite (antibiotics) which adversely affect other organisms.

Antibiotic: (i) Secondary metabolites of microorganism especially actiniomycetes and fungi, which inhibit the growth of other pathogenic organisms.
(ii) Chemical compounds of high molecular weight produced by microorganisms which have specific toxicity for other microorganisms sensitive to it.

Antibody: A complicated specific protein produced in the blood of animals (especially mammals) usually in response to stimulus, of any foreign materials (antigen). It act against any invading macromolecules and provides defence.

Antifeedant: Chemical inhibiting the feeding of certain insects.

Antigen: A substance (usually proteins, nucleic acid or polysaccharides) which after introduction into mammalion blood induces the antibody formation.

Antimycin: An antifungal antibiotic produced by species of streptomyces, effective against early blight of tomato, blight of oats and rice blast.

Antiseptics: Chemicals used commonly to control the growth of microorganisms.

Antiserum: Blood serum containing antibodies.

Antisporulant: Substances/chemicals preventing spore production without killing vegetative growth of a fungus.

Antitoxin: An antibody capable of uniting with or neutralizing a specific toxin.

Aphanoplasmodium (Pl. Aphanoplamodia) Gr.: A plasmodium (in early stage) of a network of very fine transparent strands not differentiated into ecto-and endoplasm and protoplasm is not coarsely granular e.g. *Stemonitis*. (c.f. "Phaneroplasmodium", "protoplasmodium")

Aphids: Insects belonging to order Hemiptera, having sucking type of mouth parts and act as vectors for transmission of many viral diseases.

Apical paraphyses: Paraphyses

originating at the top of a perithecial centrum which grows down in a palisade layer among the asci and have free tips.

Apiculate: Cells having one or more pointed ends like a lemon.

Aplanetic Gr.: Non-motile.

Aplanospore Gr.: Non-motile spore.

Aplerotic: A type of oospore not filling the oogonium; found in members of Pythiaceae (cf. plerotic).

Apomictic Gr.: Developing by phenomenon of Apomixis.

Apomixis: Produced normally by sexual reproduction but without having the stages of plasmogamy, Karyogamy and meiosis.

Apoplastic: Movement (esp. of systemic fungicides) in the non-living vessels and tracheids of xylem in a passive manner from root to leaves. Most of the systemic fungicides have apoplastic, upward and outward movement and usually deposit at leaf margins.

Apothecium (Pl. Apothecia) Gr.: The fruiting body of Ascomycetes in which the hymenium is open and not covered by other fungal tissues at maturity stage.

Appressorium (Pl. Appressoria) L: Flattened enlargement of hyphal tip or germ tube which attaches itself to the host before actual penetration. Infection peg usually grows from it.

Apron: See "Metalaxyl".

Arbuscules: Minute branches, especially of fungal hyphae, forming the mycorrhizae, VAM.

Arbutin: A phenolic glucoside already present in tissues of pear tree which is supposed to be hydrolysed by beta-glucosidase of host origin into glucose and an antimicrobial semiquinone (phenol) which provides resistance against fire blight caused by *Erwinia amylovora*.

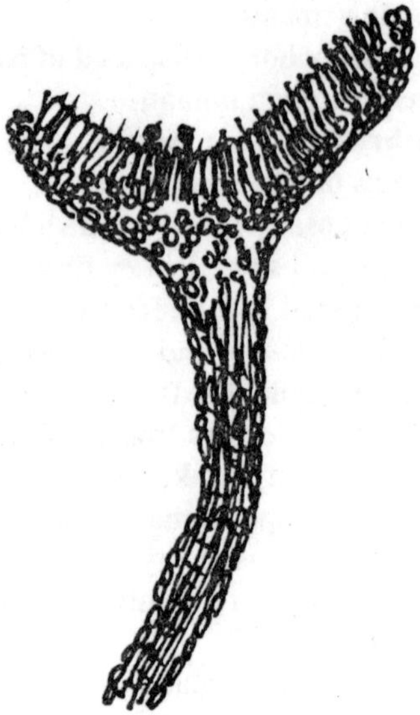

Apothecium

Archicarp Gr.: The initial stage in development of female reproductive organs and supporting cells (in Ascomycetes).

Aretan: Organomercurial fungicide similar to Agallol.

Arthroconidium: See arthrospore.

Arthrospore Gr.: Spore formed as a result of fragmentation of hypha.

Ascigerous Gr.: The ascus stage of ascomycetes.

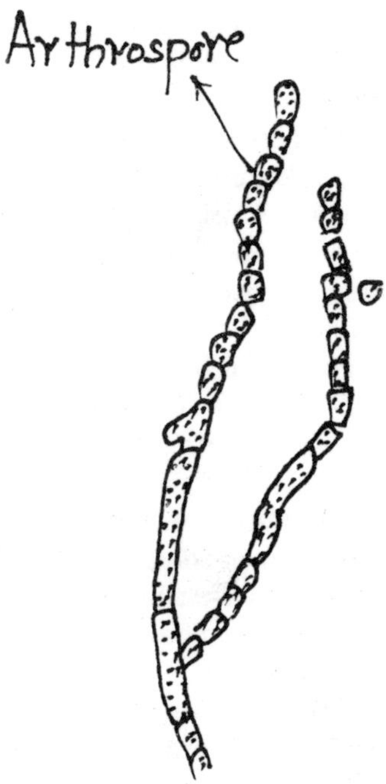

Arthrospore

Ascocarp Gr.: A multicellular well defined fructification formed in Ascomycetes bearing ascus with ascospores.

Ascogenous hypha Gr.: A specialized hypha giving rise to one or more asci.

Ascogonium (Pl. Ascogonia) Gr.: Female gametangium of Ascomycetes, participating in Sexual reproduction.

Ascoma: Ascus bearing structure-see ascoparp.

Ascomycetes: A class of fungi under subdivision Ascomycotina. Its members reproduce asexually mainly by conidia and sexually by ascospores found in sac like ascus.

Ascomycotina: A subdivision of fungi under Amastignomycota division. Its members produce typically meiospores (ascospores) in sac like structure (asci) by free cell formation. It has a single class Ascomycetes.

Ascophore: An erect, septate structure produced by the germination of zygote, found in Endomycetales of Hemiasomycetidae, which bear the asci.

Ascospore Gr.: Spore formed after meiosis inside an ascus by free cell formation.

Ascostroma (Pl. Ascostromata) Gr.: A stromatic ascocarp bearing asci directly in locules of stroma.

Ascus (Pl. Asci) Gr.: A reproductive sac like cell in which definite number of endogenous ascospores are produced after meiosis, by free cell formation.

Ascus mother cell: The binucleate crook cell in Ascomycetes which develops into ascus after Karyogamy.

Aseptate L.: Lacking septa or cross walls.

Aseptic: Free of microorganism capable of causing infection.

Asexual L: Reproduction not involving the union of two nuclei.

Aspergillosis: A group of diseases of animals and human being caused by species of genus *Aspergillus.*

Asporogenous Gr.: Non-spore forming.

Assay media: Media employed for

assaying the production of various metabolites, like amino acids, antibiotics, enzymes, toxins etc. *in vitro*.

Assembly: Integration of nucleic acid and protein coat of a virus particle during replication of virus.

Assimilation: The uptake of nutrients and their utilization by organisms.

Astomate (Astmous): Having no ostiole.

Astomous: See "Astomate".

Asymptole: The point at which the growth of a population reaches numerical stability.

Atomiser: Apparatus for breaking a liquid stream into a very fine droplets by a stream of air.

Atomize: Making a very fine spray (small droplets) of a solution by releasing through a nozzle under pressure.

Atrophied: An organ which has not developed properly or has withered reduced in size, rudimentry, vestigial.

Attenuation: Weakening or reducing the virulence of pathogen.

Aureofungin: A broad spectrum antifungal antibiotic produced in submerged culture of *Streptoverticillium cinnamomeum var terricola.* It is systemically translocated in plant system.

Autocide: Substances produced by an organism which is lethal for itself. Usually produced due to induced genetic change, a method for pest control.

Autoclave: An apparatus using high pressure steam for sterilizing equipments, media etc.

Autoecious Gr.: Capable of completing their life cycle on a single host (cf. heteroecious).

Autogamy: A process in which the pair of nuclei fuse inside the female sex organ without fusion of cells.

Autogenous vaccine: A vaccine prepared from the bacteria isolated from the suffering patient (host) to be vaccinated.

Autolysis: Disintegration of cells by their own metabolites (enzymes).

Automixis: Self fertilisation by the fusion of two closely related sex cells or nuclei.

Autonomous transmission: Transmission of pathogens to short distances by the active growth of hyphae e.g. wood rotting fungi *Armillaria, Fomes, Ganoderma* etc.

Autotrophic Gr.: Organisms which do not require the external supply of organic food for their growth (cf. heterotrophic).

Autotrophs: Microbes which use only inorganic materials as a source of nutrient, CO_2 is the sole Carbon source. (cf. Heterotrophs, Necrotrophs).

Auxotrophs: Mutant or strain of a microorganism which require specific nutrient which are not needed by its parent strain (prototroph).

Auxins: (Indole-3-acetic acid). An endogenous plant growth regulator which promotes growth along the longitudinal axis and is responsible for many disease symptoms like galls, malformation, elongation, witches broom in infected plants due to unusually high concentration in diseased tissue.

Auxotrophic: Microorganisms or their mutants which can be grown on culture media only on addition of certain specific nutrients.

Avirulent: Lacking virulence.

Avoidance: Method of disease control based on growing the crop in localities/seasons where/when a disease is not active.

Axenic Gr.: Without presence of another organism (used for purity of culture etc.) (cf. gnotobiontic).

Axenic culture: Pure culture of a single species free from other living organisms.

Axeny: Passive (resistance of plant to a pathogen).

Axoneme: The main core of fungal flagellum consisting of 2 central microtubules surrounded by 9 double microtubules.

Azotodesmic Gr.: Nitrogen fixing; referred for Cynophycophilous lichens having nitrogen fixing blue-green algae as phycobiont.

Azygospores: Structure morphologically similar to a Zygospore (fungal spore) but is non-sexual in origin (in Zygomycetes), developed partheno-genetically.

B

Bacillus: Generic name of a rod shaped bacterium.

Bacteremia: Condition in which bacteria are found in the blood stream.

Bacteria: A group of organism different from fungi.

Bacterial leaf blight (BLB): Most destructive disease of rice, caused by bacterium, *Xanthomonas oryzae*, in which yellow, straw coloured lesions with wavy margins are formed on both margins of leaves, leading into drying or blighting of foliage.

Bacterial leaf streak (BLS): Bacterial disease of rice caused by *Xanthomonas translucens* f. sp. *oryzincola* in which translucent streaks are formed between veins of lamina which become orangish red. These streaks coalesce and leaves are dried.

Bacterial pustule: A common disease of soybean caused by *Xanthomonas phaseoli* var. *sojensis*. Small, pin head size, reddish brown, raised pustules are formed on both the surface of leaf, which are surrounded by a yellow halo.

Bactericide: Substance/chemical capable of killing bacteria.

Bacteriocin: Factors inducing sharp disorders in bacterial metabolism.

Bacteriolysin: An agent causing lysis/disintegration of bacteria.

Bacteriolytic amboceptors: The cytolytic antibodies in serum that lyse the bacteria in complement fixation test.

Bacteriophage: Virus which infect bacteria and cause lysis of bacterial cells.

Bacteriostatic: Checking the bacteria growth without killing them.

Ballistospore: A basidiospore which is forcibly abjected.

Bakane disease: A disease of rice caused by *Gibberella fujikuroi (Fusarium monillforme)* in which plants become etiolated, abnormally tall, thin, yellowish in colour, which lateron dry and are killed. Foot rot symptoms are also seen in this disease.

Basal body: See "Blepharoplast".

Base Plate: A hexagonal plate at the end of the helical tail of some bacteriophage to which tail fibers are attached.

Basic dye: A dye (stain) containing a basic organic grouping of atoms (Cation) which is actively staining part combining with an acid usually inorganic. It has affinity for nucleic acid.

Basidiocarp: A fructification bearing basidiospores produced in Basidiomycetes. See "Basidioma".

Basidiole: A structure similar to basidium in morphology without sterigmata present in hymenium of basidiomycetes. They may be immature basidium or sterile structure (cf. Cystidium).

Basidioma (Basidiome): A basidium producing structure or basidiocarp. Also see "basidioc-

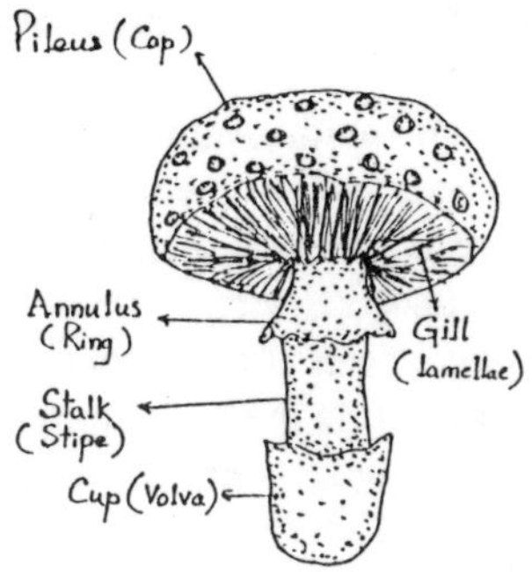

Basidiocarp

arp".

Basidiomycotina: A subdivision of fungi under Amastigomycota division. They have septate mycellium; reproduce typically by basidiospores on the surface of basidia. It has a single class-Basidiomycetes.

Basionym: The name or epithet bringing synonym on which a new transfer or combination is based.

Basidiospore: A sexual spore produced exogenously on a basidium after karyogamy and meiosis.

Basidium: A characteristic reproductive organ of basidiomycetes. In larger forms it is a short cell, wider at the top than at base with 4 small sterigmata each of which bears a basidiospore. It is the site of Karyogamy and meiosis.

Basipetal: A succession of conidia with the youngest at the base of the chain.

Bavistin: A systemic fungicide of benzimidazole group 2-(methoxy carbomyl) benzimidazole, effective against ascomycetes and fungi imperfecti.

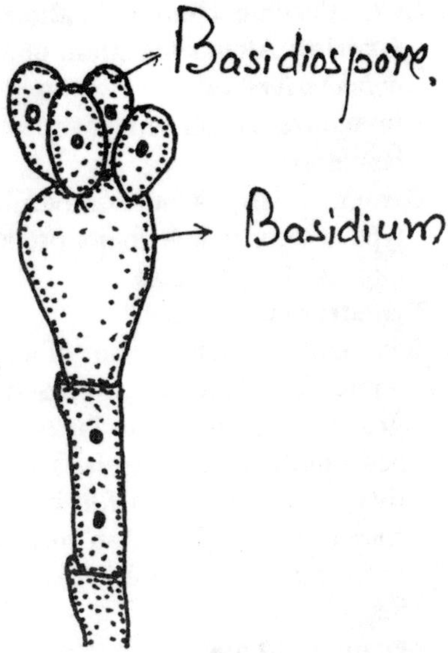

Basidium

Baycor: See "Biloxazol".

Bayleton: Trade name of systemic fungicide of triazol group with common name "triadimefon". Chemically it is 1-(4-chlorophenoxy)-3, 3-dimethyl-1-1H-(1, 2, 4-triazol-1-yl) butane-2-one. It is effective against *Botrytis cinerea, Clados-porium cucumerium, Penicillium expansum* and *Hatilago maydis*. It is translocated both upward and downward directions. It is used as seed dresser to control *Ustilago nuda, Helminthosporium* sp. powdery mildew in barley; smut, bunt, rust, powdery mildew in wheat and as foliar spray for powdery mildew.

Beta-Carotene: A precursor of vitamin A produced commercially by fermentation with a fungus *Blakeslea trispora*.

BCG Vaccine (Bacillus calmette-Guerin): Attenuated strain of *Mycobacterium tuberculosis* used to immunize human being against tuberculosis.

Benign (Virus): A mild or avirulent type of virus which do not produce any visible symptoms.

Benlate: See "benomyl".

Benodanil: Common name of a systemic fungicide or carboxamide group. Chemically it is 2-iodobenzonic acid anilide. It is effective against rusts and *Rhizoctonia solani* and is better than mebenil. Trade name is "Celirus". Not marketed in India.

Benomyl (Benlate): Systemic fungicide of benzimidazol group, chemically it is methyl-N (1-butyl carbamoyl) 2-benzimidazole carbamate.

Benthos: Term used collectively for all organisms living along the bottom of ocean or lakes.

Benzanilide: A group of systemic fungicides (mainly mebnil and benodanil) which are carboxylic acid anilides and are closely related to non-systemic salicylanilide.

Benzene compounds: Organic compounds having benzene group, being used as fungidies for the control of diseases of fruits and ornamentals. e.g. PCNB, deoxon, dinocap.

Benzimidazoles: A group of systemic fungicides having a broad spectrum of activity. They have two types of derivatives, both being substituted at 2-carbon atom of heterocyclic ring. e.g. thiabendazole, MBC.

Benzoic acid: Antimicrobial substances widely used as food preservative also as sodium benzoate.

Biloxazol (Bitertanol): Systemic fungicide of triazol group. Chemically it is p-(1, 1-biphenyl)-4-yloxyl-a-(1, 1-dimethylethyl)-1H-1, 2, 4,-triazol-1-ethanol. It showed excellent activity against powdery mildew, apple scab, Cercospora and rusts. Trade name is "Baycor" as 25 or 50 wp.

Binal: Bacteriophage having two different symmetryes e.g. cuboidal head and helical tail.

Binapacryl: A fungicide, effective against powdery mildew and some rusts. Chemically it is 2-sec. butyl-4, 6, di-nitro phenyl-3-methyl-2-butenoate. It is available under trade name "Morocide" as 50 wp or 40 EC.

Binding hyphae: One of the types of hyphae present in basidiocarp which are thick walled, aseptate, without clamp connection, highly branched, functioning to weave other hyphae together. They are sterile produced from the generative hyphae and do not produce basidia. (cf. generative, skeletal types).

Binomial L: A scientific name of an organism, which consist of two names the first for genus and second for species. This system of naming was first started by Linneus.

Bio-assay: Quantitative estimation

of biologically active substances in terms of their efficacy on living organisms under standard condition.

Biochemical defence: Resistance imparted to the host against a particular pathogen due to certain chemical(s), present before infection or produced as a result of infection (post-infection).

Biochemical media: Media used for the differentiation of microorganisms on the basis of their biochemical activity.

Biodeterioration: Any undesirable change in the properties of a material of economic importance as a result of vital activities of organisms.

Biological control: Control of disease through the activity of some organism other than host. These biological agents may be predators or parasites.

Biological form or Race: See physiological race.

Biological specialization: Manifestation by members of a species of physiological specialization of infecting on a particular host species.

Bioluminescence: Emission of light by living organisms.

Bionomics: The study of organisms in relation to their environment i.e. ecology.

Biotin: Vitamin, occurring as alpha and beta isomers, required by certain organisms.

Biotrophic parasite: A fungus parasitising on other fungi which feeds on living cells but does not kill its host.

Biotrophs: Organisms which obtain their food from the living tissues on which they complete their life cycle regardless of their ability to be cultured e.g. rusts, smuts, mildews. Also see "Parasite".

Biotype: (i) A subdivision of physiological race which are morphologically identical.

(ii) In bacteria a subdivision of species or serotype distinguished by some special physiological character.

Biparental: Progeny having two different parental types.

Bipolar sexuality (Bipolarity): The sexual compatibility in basidiomycetes controlled by multiple allelic genes at a single locus. In this two basidiospore in a basidium are of one strain and two of another (cf. tetrapolar sexuality).

Bitunicate: Double walled-used for ascus, inner wall is elastic and expands greately beyond outer wall (separating it) at the time of spore liberation (cf. unitunicate). They are also called "Jack-in-the box" ascus or "Fissitunicate ascus".

Bivoltine: Having two broods or generation in a year.

Black Arm: A bacterial disease of cotton showing dark brown to black lesion on stem and fruiting branches.

Aso see: "Angular leaf spot".

Black heart of potato: A non-parasitic disease of potato tuber showing discolouration of iner tissues of tuber; caused by high temperature and lack of oxygen during storage and transit.

Black leg: A disease symptom where basal part of stem becomes necrotic.

Black mold: Commonly referred to *Aspergillus* sp. (cf. green mold).

Black rot of pineapple: A rotting disease caused by *Ceratocystis paradoxa.*

Black shank: A tobacco disease caused by *Phytophthora parasitica* var. *nicotinae*. First symptom is soft rot of stem near ground level and is followed by root and stem blackening and mortality of young plants.

Black tip of mango: A non-parasitic disease of mango caused by injury from atmospheric pollutant like smoak from brick klin which disturbs the boron metabolism leading to necrosis of fruit tips.

Blastic: A type of conidial development in which conidiospore arises from only a part of the pre-existing cell and is significantly enlarged before it is delimited from the conidiogenous cell. (cf. thallic).

Blasticidin: An antibiotic produced from *Streptomyces griseochromogenes;* active against both bacteria and fungi. It is specially used for the control of rice blast fungus at 5-10 ppm.

Blastimycin: See blasticidin.

Blast of rice: A severe and important disease of paddy caused by *Pyricularia oryzae* forming spindle shaped spots on foliage, leaf sheath, nodes, seeds etc.

Blastogenous: A method of conidial development in which conidium originates as a bud from the conidiophore wall (cf. murogenous, phialogenous, porogenous).

Blastomycetidae: A subclass of fungi under Deuteromycetes in which vegetative thallus is yeast cell with or without psedomycelium and true mecelium (if present) not well developed.

Blastospore: An asexual spore produced by budding.

BLB: See "Bacterial leaf blight".

Blepharoplast: A chromatic body of nuclear origin at the base of a flagellum. It is attached to the nucleus by a rhizoplast. It is also called "Kinetosome: or "basal body".

Blewitt: Common name used for fungus *Trichoderma nudum* used as edible food.

Blight: A disease symptom on leaves showing an unlimited, irregular necrosis leading to sudden and serious damage to foliage.

Blimix 4: Trade name or fungicide, copper oxy chloride 4% dust.

Blitox 50: Trade name of fungicide, copper oxy chloride 50% wettable powder.

Blood serum: The fluid portion from clotted blood or blood plasma.

Blotch: A necrotic spot having fibrillose margins of visible mycelial strands.

Blue copper: Trade name for copper oxy choloride.

Blue mold rot: Rotting diseases caused by *Penicillium* sp.

BNPRA medium: A selective medium used for isolation of *Phytophthora* spp. from soil and plants.

B.O.D. (Biological oxygen demand): During oxidation (decomposition) of organic matter of sewage, aerobic bacteria require oxygen, and amount of oxygen utilized (B.O.D.) is proportional to the amount of organic matter present in the sample.

Boletes: Common name for a group of fungi belonging to Agaricales in which basidia are not borne on gills, but line the inside or deep tubes or sometimes shallow bits formed on the undersurface of the fleshy basidiocarp.

Boom: Horizontal or vertical light frame on which several spray nozzles are mounted.

Bordeaux coefficient: A measure for standard and uniform comparison of fungicides for their efficacy by comparing the fungicidal value of a particular material with standard Bordeaux. It is achieved by dividing the LD_{50} for Boreaux mixture (in ppm) with that of test chemical.

Bordeaux mixture: Most ancient and widely used chemical mixture used for the control of plant diseases, contains varying ratios of copper sulphate, hydrated lime and water standard being 5:5.50. The solution of the two chemicals are prepared separately and are mixed just prior to use.

Bordeaux paste: A mixture of copper sulphate, hydrated lime and small amount of water to form a paste, usually in ratio of 1 lb:1 lb:1 gallon. It is used as fungicide for tree wound dressing.

Bortan: See "Dicloran".

Botulism: Food poisoning due to toxins of *Clostridium botulinum* (a bacterium).

Bracket fungi: Referred for the members of "Polyporales" which produce bracket shape fruiting body on tree logs etc. e.g. *Ganoderma applanatum.*

Breed count: A microscopic method of counting bacteria in a dried, stained film or milk.

Brestan: See "Triphenyl tin acetate".

Brestanol: See "Triphenyl tin chloride".

Bromatium (Pl. Bromatia): An enlarged hyphal ending induced and used as food by fungi cultivating ants.

Bromothymol Blue: A pH indicator (dye) with range of 6.0-7.6.

Broom rape: See Orobanche.

Broth: A liquid nutrient culture medium, used to grow the microorganisms.

Brownian movement: A peculiar circular motion exhibited by colloidal particle and bacteria in suspension due to bombardment by the molecules of fluids.

Brown plaster mould: Common term used for the fungus *Papulospora byssina* found in the commercial beds of mushroom harming their production.

Brown spot of rice: An important disease of paddy, producing eye shaped, brown coloured spots on leaves, coleoptile, leaf sheath, glumes seeds. It is caused by *Helminthosporium oryzae* or *Cochliobolus miyabeanus.*

Bud blight: A disease of soybean, probably of viral origin, exhibiting curving and necrosis of apical buds, coupled with necrotic specs and ring spots on leaves. Tobacoo ring spot virus, necrotic strains of Soybean mosaic virus, Tobacco streak virus and few others have been supposed to be associated with the disease.

Budding: Phenomenon of producing new cells from a small protrusion or outgrowth of the mother cell (found in yeasts) (cf. fission).

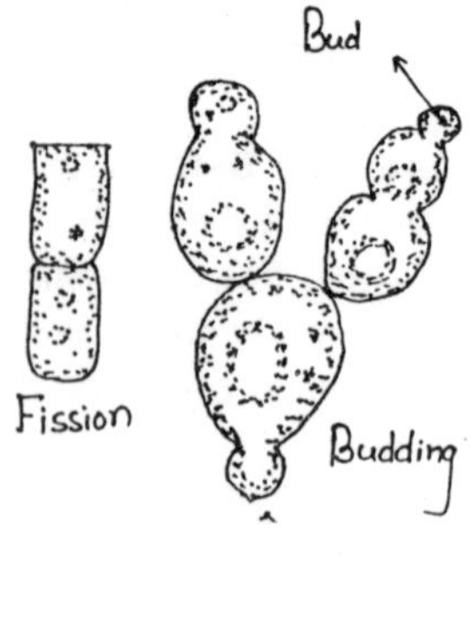

Budding

Bud necrosis: A viral disease of groundnut, caused by Tomato spotted wilt virus, characterized by flat, hard, curved and necrotic apical bud.

Bud rot of Palm: A disease of palm caused by *Phylophthora palmivora* exhibiting discoloured spots on leaves leading to drying of foliage.

Buffer: Any substance in solution which resist the change in pH due to addition of acid or alkali.

Bulbiformin: An antifungal, polypeptide antibiotic produced by *Bacillus subtilis,* reduced pigeon pea wilt incidence.

Bunchy top of banana: A virus disease of banana in which the leaves become bunched together at the apex of the pseudostem to form a rosette or bunchy top.

Burgundy mixture: A modification of bordeaux mixture in which lime is replaced by sodium carbonate. The ratio is $CuSO_4$ (4.536 kg); Na_2Co_3 (5.666 kg); Water 50 Gallons.

Butyl isocyanate: A volatile fungitoxic substance produced due to degradation of "benomyl" after moistening. It is effective against *Ustilago maydis* and *Saccharomyces cerevesiae.*

C

Caeoma: The form genus of aecial stage of rust in which aecium lacks a peridium.

Caldariomycin: ($C_5H_8O_2$ Cl_2) A chemical compound produced by the fungus *Caldariomyces fumago,* containing the chlorine metabolized by the fungus from medium. The fungus utilizes 95% chlorine in the medium.

wounding. (ii) Undifferentiated mass of tissues grown from one type of cell/tissue in tissue culture studies.

Canker: A necrotic symptom of woody parts (in trees) spreading wound in bark, is usually surrounded by raised and swollen lumer like margin.

Capillitium (Pl. Capillitia): A mass of sterile fibers within a fructification interspersed among spores, found in myxomycetes and Gasteromycetes.

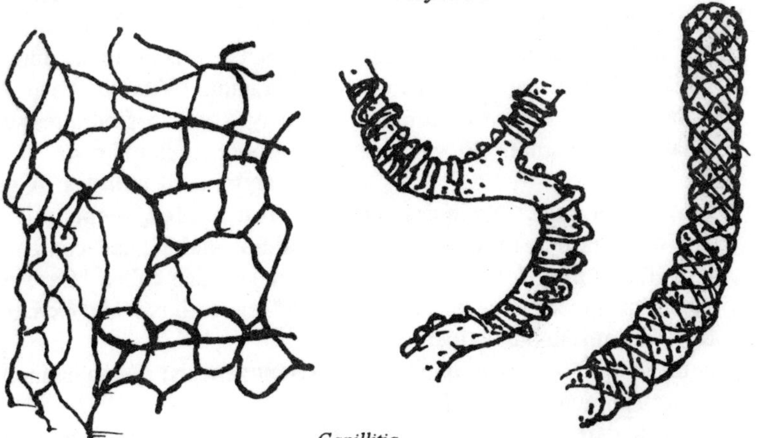

Capillitia

Calixin: Trade name of systemic fungicide with excellent prophylactic and curative action against powdery mildew of cereals, sigatoka disease of banana. *(Mycosphaerella musicola)* and other ascomycetes. Chemically it is N-tridecyl-2, 6 dimethyl, morpholine and commonly known as *Tridemorph.*

Callus: (i) Parenchymatous tissue produced by the cambial activity for providing protection after

Capitate hyphopodia: A short, more or less lobed outgrowth of Meliolates hyphae consisting of a short stalk and a terminal "head cell:. Also see "Hyphopodium". (cf. Mucronate hyphopodia).

Capsidiol: A phytoalexins diffusing from pepper fruits inoculated with a non-pathogenic fungi.

Capsid: Protein coat of virus particle.

Capsomere: A subunit of protein

coat (Capsid).

Capsule: A gelatinous covering or slime layer surrounding the cell wall of bacteria.

Captafol: Heterocyclic nitrogenous fungicide, Cis N-(1, 1, 2, 2-tetra chloro ethyl thio) -4-cyclohexene-1, 2-dicarboximide, being used as seed dresser. Its use as spray has been banned in India. Available as "Foltaf 80 wp". Earlier it was marketed under the trade name "Difolatan 80 w".

Captan: A commonly used heterocyclic nitrogenous fungicide having a wide spectrum. Chemically it is N-trichloromethyl-thio-4 cyclohexene 1, 2 dicarboximide. Available as "Captaf" 50 wp and 75D.

Carbamate fungicide: An important group of organic sulphur fungicides. They are derivatives of dithiocarba-mic acid, e.g. ziram, ferbam, thiram, maneb, zineb etc.

Carbendazim: Common name of systemic fungicide, methyl-2-benzimidazole carbamate (MBC) having a broad spectrum of activity, except Phycomycetous and dematious fungi and bacteria. It is marketed as "Bavistin 50 wp", "Derosal 50 wp", "Agrozim 50 wp" and a specific seed dressing formulation "Subeej 25 DS".

Carbohydrase: Enzyme hydrolying carbohydrates.

Carbon di sulphide (CS_2): Antimicrobial chemical used as soil fumigant for checking nematodes and also in citrus orchards against *Armillaria mellea.*

Carboxamides: A group of systemic fungicides which are carboxylic acid anilides or carboxanilides or oxathiins. They are substituted derivatives of salicylamide. They are highly effective against smuts. Two compounds of this group are carboxin and oxycarboxin. Also see "oxathiins".

Carboxin: Common name of a systemic fungicide of oxathiin group. Chemically it is 5, 6-dihydro-2-methyl-1-1, 4-oxathiin-3-carboxanilide; abbreviated as "DMOC". It is available under the trade name "Vitavax 75wp". It is widely used as seed dresser for control of smut and rusts.

Carboxylic acid anilides: A group of fungicides, especially of oxathiin and related compounds. They are also called "Carboxanilids". eg. carboxin and oxycarboxin.

Carcinogenesis: Establishment of cancer.

Carrier: An organism harbouring pathogenic organisms and transfer-ing them to healthy host.

Carriers: (i) A finely divided inert powder on which active ingredient remains attached in a dust preparation.

(ii) An agent especially insect which disseminates the pathogen from one place to other or diseased to healthy plants.

Cassia oil: An essential oil having antimicrobial property and act as food preservative.

Catabolism: Activities of an organism pertaining to degradation or breaking of chemical compounds mainly for deriving energy.

Catechol: A phenolic compound present in many plant tissues and supposed to be responsible for resistance.

Catenate or Catenulate L: Spores borne in a chain with end to end.

Cauliflower fungus: Common name for *Sparasis* belonging to Aphyllophorales. Its fruiting body consists of a multiple flattened branches arising from a thick stalk.

Causal agent: Factor mainly responsible for making a plant diseased. It may be animate or inanimate.

Cavity slides: Microscopic slides having a concave depression, which is used for putting a hanging drop of spore suspension on cover glass to study the germination of spores.

Cell Theory: A theory proposed by Schleiden and Schwann (1838) defining the cell as structural and metabolic unit of plants and animal.

Cellular slime molds: Common name for the members of class Acrasio-mycetes of Gymnomycota. (cf. True slime mold).

Cellulase: An enzyme hydrolysing the cellulose (present in the plant cell wall) into cellobiose; it is produced by a number of plant pathogenic fungi and bacteria during the course of pathogenesis and cell wall degradation. This enzyme has two components C_1 and C_x, C_1 is responsible for degrading native cellulose into linear chains and C_x converts it into cellobiose and ultimately into glucose.

Cellulose: A linear polymer of glucose which is an integral component of secondary walls of plant cells.

Cell volume assay: A method used to determine the inhibition of spore germination by fungistates, based on principle that volume of spore cell increased during germination. Increase in cell volume is measured by centrifugation in hematocrit tubes.

Centric: Referred to a type of nuclear division involving centriole.

Centrifugal: Away from the centre outwardly. Principle of centrifugal force is used in centrifuge to separate solids or higher molecular weight components in a solution.

Centrifuge: An apparatus in which solutions (colloidal or otherwise) are rotated at high speed resulting into separation of different components according to their molecular weights.

Centripetal: Towards the centre.

Centrum (Pl. Centra): The totality of structures (asci and hamathecium) enclosed by the ascocarp wall.

Cephalexin: A semi synthetic cephalosporin (Antibiotic).

Cephalodium: A swollen, delimited region or warty structure within or on surface of a diphycophilous lichen thallus containing an alga different to that characteristic of the rest of thallus. They are reproducing structures of lichen and usually contain blue-green algae.

Cephaloridin: A semi synthetic antibiotic of Cephalosporin group.

Cephalosporins: A group of antibiotics produced by *Cephalospor-ium* spp. which are closely related to Penicillin in structure.

Ceraceous (Cereous): Wax like.

Ceresan dry: An organomercurial fungicide used as seed dresser, chemically it is Phenyl mercury acetate.

Ceresan M: Ethyl mercury-p toluene sulphanamide.

Ceresan wet: Methoxy ethyl mercury chloride.

Charcoal rot: A common disease of many crops caused by *Macrophomina phaseolina-Rhizoctonia bataticola*, affecting roots and stem. Due to production of large number of black coloured scleretia and pycnidia, the diseased portion becomes dark black, just like charcoal.

Chaubattia paste: An antifungal paste prepared by mixing Copper carbonate (800g), red lead (800g) in one litre of lanolin or linseed oil; used as wound dressing fungicide against diseases of trees.

Chemotaxis: Movement (positive or negative) in response to some chemical, mainly referred to attracting or repelling of leucocytes by the microorganism.

Chemotherapy: Treatment of disease by the use of chemical.

Chemotrophs: Organisms incapable of utilizing radient energy and relying upon oxidation of chemical compounds for their energy requirements.

Chemotropism: Growth in a particular direction in response to stimulus of a particular chemical (cf. Geotropism phototropism).

Chestnut blight: A serious fungus disease of Chestnut in America, caused by *Endothia parasitica*.

Cheshunt compound: An antifungal compound having two parts of Copper sulphate and 11 parts of Ammonium Carbonate. One ounce of this mixture is dissolved in 2 gallons of waters for spray on disease affected crop. Useful against damping-off.

Chinese yeast: Concentrated fungal mycelium of *Amylomyces rouxili* used as starter in fermentation of food stuff in Indonesia and other eastern countries.

Chinomethionat (Oxythioquinox): Common name of a fungicide also acting as an acaricide as well as insecticide. Chemically it is 6-methyl quinoxaline-2, 3-dithiol-cyclic carbonate. It is very effective against powdery mildew of fruits, cucurbits and ornamentals @ 0.03 to 0.05% and also controls the eggs and larval stages and adults or spider mites and *Psylla piricola*, white fly and aphids. It is available under trade name of "Morestan" as 25 wp. It is not compatible with EC formulations, alkaline products like Bordeaux mixture, lime sulphur, thiram and oil based products.

Chitin: A nitrogenous polysacicharide found in the cell wall of fungi and outer layer of insects.

Chitosome: A small, microvesicular structure much smaller than vesicle found at the hyphal apex. It is supposed to be conveyor of chitin synthetase having destination at cell surface.

Chlamydospore: A thick walled hyphal cell transformed into a nonsexual resting spores.

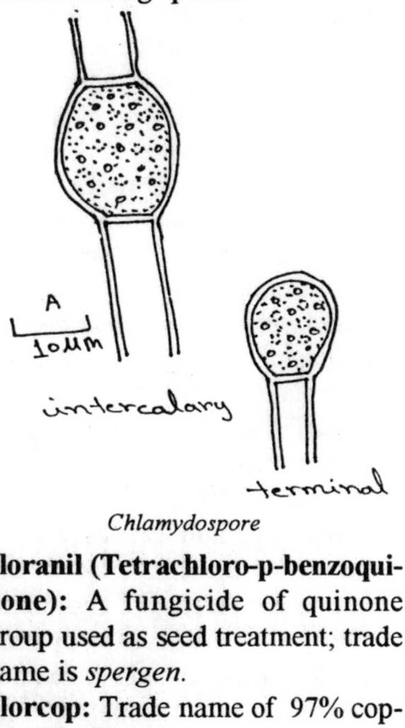

Chlamydospore

Chloranil (Tetrachloro-p-benzoquinone): A fungicide of quinone group used as seed treatment; trade name is *spergen.*

Chlorcop: Trade name of 97% copper oxy chlorite.

Choloroneb: Common name of a phenol derivative, acting as systemic fungicide, which is chemically 1, 4-dichloro-2, 5-dimethoxy benzene. It is available under trade name of "Demosan 65W" and is used for seed and soil treatment alongwith other conventional seed dressers like thiram for the control of seedling blight by *Rhizoctonia* spp. *Pythium* spp., *Sclerotim rolfsii* etc.

Chloronets: (1, 4-dichloro-2, 5-dimethoxy benzene): A systemic fungicide of benzene group, active against, *Rhizoctonia, Sclerotium rolfssi* and *Phytophthora.*

Chlorophycophilous: A category of lichens in which the phycobiont is green algae, usually belonging to Chlorophyceae (cf. Cyanophyco-philous, diphycophilous).

Chloropicirin: A soil fumigant act as fungicide and larvicide.

Chlorosis: Symptom of chlorophyll deficiency showing yellowing of plants.

Chlorothalonil: Broad spectrum, non-systemic fungicide, effective mainly against tikka and rust of groundnut, early and late blight of potato, apple scab and blister blight of tea, Chemically it is tetrachloro iso-phthalonitrile. Marketed as "Kavach 75 WP%.

Chorioallantoic membrane: In chick embryo outer covering of embryo sae (chorion) is fused with a sac like outgrowth on ventral side of gut (allantois) forming a double membrane which is used for cultivating viruses in laboratory.

Chromatophore: A pigment containing body especially chlorophyll containing granules found in bacteria.

Chromogenesis: Production of pigments by microorganisms.

Chromosome: A filamentous structure found in the nucleus of cell which contains gene.

Chytrid: Member of chytridiomycetes.

Chytridiomycetes: A class of fungi under Mastigomycota, having the forms producing posteriorly, uniflagellate motile cells, with whiplash type flagellum.

Cicerin: A phytoalexin associated with resistance in *Ascochyta blight* of Gram.

Cilium: A hair like appendage on certain cells.

Cinnamon: An essential oil used for food preservation due to its antimicrobial property.

Cirrhus (Pl. Cirrhi): A ribbon like cylinder of spores held together by mucus after release from an ostiole.

Cistron: A region of any genome, capable of accomplishing a single function; term coined by Benzer.

Citrus greening: A widely prevalent mycoplasma disease of citrus showing yellowing of midrib and lateral veins and short, upright weak shoots with bushy growth.

Clamp cônnection: A hook like curved outgrowth of a cell formed during cell division in basidiomycetes. It provides the passage for moving the daughter nuclei into the penultimate cell and ensures the dikaryotic conditions.

Clavate: Club shaved.

Cleistothecium (Pl. Cleistothecia): An ascocarp in which the asci are surrounded by fungal tissue in which there are no regularly formed openings.

Climatograph: A polygonal diagram drawn by plotting temperature means against relative humidity means.

Clone: The aggregate of individual organisms produced from one through asexual means.

Club root: A disease of crucifers caused by *Plasmodiophora brassicae* in which roots become swollen and club shaped due to hypertrophy and hyperplasia.

Clypeus (Pl. Clypei) L: A shield shaped, stromatic tissue around the mouth of a perithecium, usually found in members of Phyllachoracea.

Coagulase: An enzyme produced by pathogenic *staphylococci* causing coagulation of blood plasma.

Coccus: A spherical bacterium.

Coelomycetes: A group of fungi imperfecti having a definite sporulating apparatus produced within the cavity of the substrate or even within the cavity of fungal tissue. Conidia are produced in pycnidia, pycnothyria, fungal stroma or acervulus.

Coelomycetidae: A subclass of fungi under Deuteromycetes. Its members have true, septate mycelium and produce conidia in pycnidia or acervuli.

Coenocytic: Non-septate multinucleate condition in lower fungi in which nuclei lie freely in cyloplasm and are not separated by cross walls.

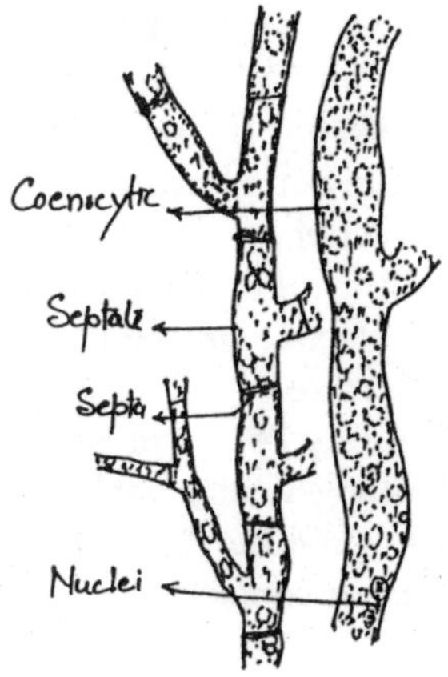

Coenocytic (non-septate) and Septate hyphae

Coenzymes: Non-proteinaceous, freely existing catalysts mediating the biochemical reactions.

Coffe rust: A serious disease of coffee plantation in Sri Lanka, caused by the fungus, *Hemillia vastatrix.*

Coined name: See "common name".

Cold sterilization: Method(s) of disinfecting especially the culture media without heating them to avoid the risk of heat damage; use of membrane filter and propylene oxide are common methods of sterilization.

Colicin: A cytoplasmic genetic determinant, characterized by a narrow range of inhibitory substances (protein) in bacteria, *Escherichia coli.*

Collar: (i) A thin disc on which head and tail of a binal virus are attached.

(ii) The region where stem and root join.

Collar rot: Rotting of the stem at or above the soil level.

Collarette: A cup shaped structure at the apex of a phillide through which a succession of conidia is produced or a frill left at the apex of a sporangiophore after rupture of the sporangium found in some mucorales.

Collateral host: Referred to grasses which act as hosts for the rusts or smuts of cerels and help in the built up of inoculum and secondary spread of disease.

Colletotin: A nonspecific phytotoxin produced by *colletotrichum fuscum* which is similar to alternaric acid and disrupts cell permeability or affect pectic enzyme.

Colonization: The invasion of a substratum or habitat by a fungus.

Colony: A discrete mass of cells or hyphae developing from a single cell, hyphal fragment or spore. Group of individuals of same type living in close association.

Colostrum: First milk secreted by a mother after giving birth to infant.

Columella (Pl. Columellae): A central sterile column of fruting body (Sporangiophore) after an extension of supporting stalk found in some Mucorales.

Commensal: A relationship between

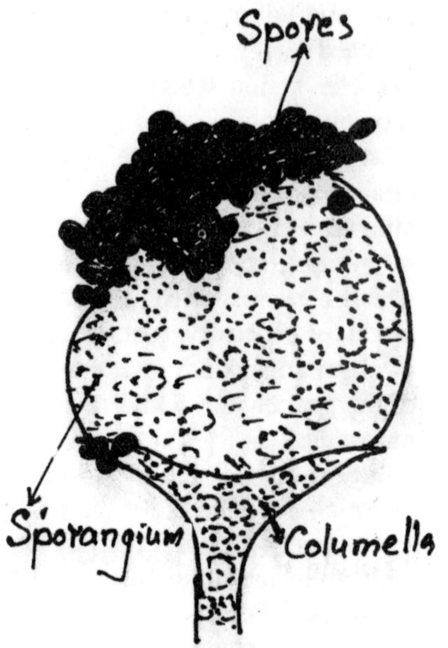

Columella

members of different species living in proximity (the same culture) without harming each other.

Commensalism: A kind of symbiosis, in which one of the partners lives at the expense of the other with out causing any harm.

Common name: Term used for pesticides. A name given to any fungicide/pesticide, other than its chemical name, which is uniform and universally accepted and helps in communication. There may be different trade names for the same fungicide, but it will have only single common name. It is also called "Adopted name" or "Coined name".

Communicable: Disease whose pathogens are easily transferred from one host to other.

Compatible: (i) Fungal strains which are able to act as partner in sexual reproduction and produce spores.
(ii) Able to act with other component/partner like mixing of one fungicide with other pesticide without any deleterious effect.

Compatibility: (i) Ability of fungal strains to act as partner in sexual reproduction.
(ii) Ability of two pesticides to remain effective on the target after getting mixed.
(iii) Ability of a strain/isolate of the pathogen and variety of the host to produce the disease after coming in contact with each other.

Competition: An active demand of two organisms for the same commodity (space, food etc.) which is not sufficient for all the organisms of that habitate.

Competetive Saprophytic ability: Relative capability of successfully competing with other saprophytes.

Complex virus: Virus having complex architecture and having a specific mode of development e.g. pox-virus.

Compliment: A normal thermolabile protein component of blood serum which participates in antigen-antibody reaction.

Complimentary strand: A newly synthesized strand of DNA using the already present strand as template.

Complementation: (i) Successful multiplication of a functionally in-

complete bacteriophage when it is associated with another functionally complete phage which acts as helper.
(ii) Compensation by a nucleus for deficiencies of other nuclear type(s) in a heterokaryon.

Compliment fixation: The combination of compliment with an antigen-antibody complex.

Composting: Degrading the complex organic wastes from town, agriculture etc. with microbial activities to such forms which can be used by plants.

Compound interest disease: A disease whose increase is mathematically analogus to compound interest money. There are several generations of the pathogen in the life of a crop causing secondary infection, Late light of potato, stem rust of wheat and powdery mildew are example of such diseases.

Compound oosphere Gr.: An oosphere with many functional gamete nuclei.

Conidiogenous cell: The cell of hyphe which gives rise to conidium.

Conidioma: A specialised multi-hyphal, conidia bearing structure.

Conidiophores Gr.: A specialized structure (hypha) bearing conidia.

Conidium (Pl. Conidia): Thin walled non-sexual spore borne terminally and exogenously on a condidiop-hore becoming deciduous at maturity, which are not formed by cytoplasmic cleavage.

Conjugate nuclear division: Simultaneous division of the two nuclei in a dikaryon, giving rise to four daughter nuclei. These generally become separated by a septum into two cells, the sister nucli migrating into different daughter cells.

Conjugation: The fusion of the genetic material of two related species or types of bacteria. It resembles a reduced sexual process.

Contact poison: Chemicals killing pest by contact action.

Contaf: See : "Hexaconazole".

Contagium vivum fluidum: A latin term coined by M.W. Beijernick for describing a diffusible, filtrable and infectious suspension of tobacco leaves infected with TMV.

Contamination: Entry of some undesirable microorganism in some material or object especially culture medium.

Context L: The fibrous tissue which make up the body of the pileus in basidiomycetes.

Control: Prevention or reduction in the development of disease.

Coombs test: Test used for detecting incomplete agglutinins in rhesus negative mothers.

Copper carbonate: Inorganic copper salt used in preparing fungicidal paste (Chaubatia paste) for applying on trees. Also see "Chaubatia paste".

Copper fungicides: Fungicides having copper as an active ingredient. They are basically used against downy mildew and many foliar diseases and in diseases of tress. Basically they include, copper sulphate, copper carbonate, copper oxy chloride, cuprous oxide preparations. e.g. Bordeaux mixture, Burgundy mixture, Cheshunt compound, Chaubatia paste, fytolan, blitox, peronox etc.

Copper oxy chloride: Inorganic form of copper, used as fungicide against many foliar diseases. It is available as Fytolan and Blitox 50 WP.

Copper sulphate: Inorganic salt of copper which is used as an active ingredient in preparing many fungicidal mixtures. e.g. Bordeaux mixture, Burgandy mixture, Cheshunt compound etc.

Coprophilous Gr.: Living on dung (Fimicolous).

Copy choice: Mechanism of genetic recombination in bacteriophages in which exchange of fragments of one viral genome with that of another viral genone both simultaneously infecting the same host cell.

This exchange takes place during replication of the nucleic acid molecules of both the viruses taking place at the same time in the host cell.

Coral fungi: Commonly referred to members of family Clavariaceae of Polyporales (Aphyllophorales). They have an upright coral shaped fruiting body with amphigenous hymenium.

Core: The central nucleic acid portion of a virus particle.

Coremium Gr.: (Synnema). An erect compact cluster (Fascicles) of fused conidiophores of different lengths.

Cornmeal agar: A common medium used for cultivation of fungi. It contains corn meal extract and agar-agar.

Corticolous: Forms (esp. lichens) growing on bark in tropics and subtropics.

Cortina L: A veil extending from the margin of the pileus in some hymenomycetes.

Cosmopolitan: A species occurring very widely throughout the major regions of the world.

Cottony leak: A fruit rot disease of cuccumber caused by *Pythium aphanidermatum*.

Coverage: Distribution of fungicide over a discontinuous area such as leaves of a tree.

Covered smut of wheat: See "Hill bunt".

Creosote: An organic chemical used to protect the wood from fungal attack; some fungi like *Cladosporium resinae* is resistent to creosote and coaltar.

Crook cell: During sexual reproduction of Ascomycetes the curved cell at the apical region of ascogenous hyphae which becomes binucleate as a result of laying of septa and serves as ascus mother cell.

Crop rotation: A cultural method of controlling the soil borne plant diseases in which the pathogen living in the soil are kept deprived of a susceptible host for some period by growing a non-host crop.

Cross-protection: Inoculation of host with mild or avirulent strain of pathogen which provides resistance against the attack of virulent strain of same pathogen (found in case of viruses and fusaria).

Crown Gall: A bacterial disease of fruit trees caused by *Agrobacterium tumefaciens*) in which outgrowth (gall) appear on the stem which are knobby and knotty and become cleft and woody.

Crozier: A recurved hook shaped hypha at the tip of an ascogenous hypha which becomes the ascus later on.

Cruciate: In the form of a cross.

Cruciform L: Intranuclear division in which the chromosomes are arranged in a ring around a dumbell shaped nucleolus found in plasmodiophoromycetes and some protozoa.

Crustose thallus L: A flat crust like Thallus that adheres closely to the substratum usually in lichens (cf. foliose, fruticose).

Cryogenic storage: A method of storing cultures at ultra low temperature (–196°C) by keeping them in a suspending medium like 10% glycerol and immersing in liquid nitrogen.

Cryptogram: A descriptive code summarizing some of the important properties of a virus for its identification.

Cuboidal: A category of virus having a cube like morphology.

Cultivar (cv): A cultivated variety of a plant.

Culture L: To grow an organism and resulting growth. Usually for microbes on artificial media for experimental or industrial use.

Culture media: Media (synthetic or natural) prepared for growing the microbes in the laboratory. It contains most of the substances required for the growth of microbes. Also see "Medium".

Cuman-L: Trade name of Ziram, available as 30% EC. Also see "Ziram".

Cupramar: Trade name for copper oxy chloride, available as 50% wettable powder.

Cupravit: Trade name of 73.5% WP copper oxy chloride.

Cuprous oxide: Inorganic copper used as fungicide against various foliar diseases. Available as dust and WP formulations under the name "Peronox".

Cuproxol: Trade name of copper fungicide having 50% copper oxy chloride.

Curative: The ability to cure a diseased plant.

Cuscuta: A phanerogamic parasite attacking trees, shrubs and crop plants. It is non-chlorophyllus drawing nourishment from its host through haustroria, also called "Dodder".

Cutaneous: Pertaining to cutin layer of skin.

Cyanophage: Virus infecting cyanophycean (bluegreen) algae.

Cyanophycophilous: Lichen thallus having blue-green algae of Cyanophyceae as phycobiont. (cf. chlorophycophilous, diphycophilous).

Cyathin complex: A complex of

metabolites produced by fungus *Cyathus helenae* of *Nidulariales.* It has antimicrobial properties and inhibits the growth of actinomycetes, bacteria and some fungi.

Cycloheximide: See Actidione.

Cyphella: A break in usually lower cortex of a lichen thallus lined with a layer of loosely connected cells formed from the medulla.

Cyprex: See "Dodine".

Cyst: A sac, especially a resting spore or sporangium like structure.

Cystidium (Pl. Cystidia): A sterile cell found among the basidia often projecting beyond the hymenium. It is morphologically different from the basidium.

in diseased tissues.

Cytolysis: Dissolution of a cell.

Cytoplasm: Colloidal solution present inside cells in which proteins, carbohydrates, lipids and minerals are dispresed in water.

Cytoplasmic inheritance: Transmission of characters in the off springs through cytoplasm and not the chromosomes. Such characters of the Fl are found to be invariably associated with the female parent.

Cytoplasmic membrane: A semi permeable membrane lying just beneath the inner surface of the cell wall. It is highly organised, specialized, complex structure

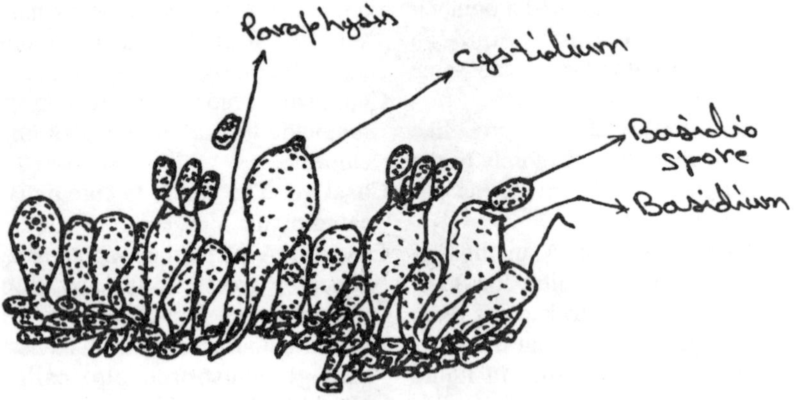

Cystidium

Cytochrome: A chemical participating in reversible Oxidation reduction reaction of respiration.

Cytokinins : Growth regulating substances necessary for cell growth and differentiation. They are associated with the development of any disorders due to altered quantities

made up of three layers, i.e. lipid, protein and carbohydrate and is also called "triple membrane".

Czapek-Dox agar: A common synthetic culture medium used for growing the fungi and bacteria in laboratory.

D

Dalmau technique: A method of observing the pseudomycelium and mycelium on culture plate.

Dalton: A unit of measuring the molecular weight of complex macromolecules. It is equivalent to 10^{-24} of molecular weight of oxygen.

Damping off: Toppling down symptom in young seedlings due to rotting of stem tissue at ground level.

Dark field microscopy: A technique of microscopic examination in which field remain dark and the object is brightly illuminated.

Dazomet: (Tetrahydro 3, 5 dimethyl 2N-1, 3, 5 thiudiazene 2-thione)-1 wettable powder used for soil treatment; releasing methyl isothiocynate in soil which controls nematodes and some soil borne fungi.

DCMOD: See " Oxycarboxin".

DCNA: See "Dicloran".

DD mixture: A mixture of dichloropropene, and dichloropropane used as soil fumigant for control of nematodes.

Deaminase: An enzyme removing amino group from a compound and releasing ammonia.

Death cup: Common name for deadly poisonous mushroom, *Amanita phalloides.*

Decarboxylase: An enzyme liberating carbon dioxide from the carboxyl group of a molecule.

Decomposition (Decay): Separation of a material into its components like breakdown of organic matter by microbes.

Defective phage: Functionally incomplete, bacteriophage which can not complete the replication cycle.

Deflocculating agent: Inert material, capable of keeping the particles away from each other in the solution to prevent flocculation to ensure dispersion and retard setting of a solid. eg. geletin, plant gum, milk products etc.

Defoliant: Chemical spray inducing premature leaf fall.

Dehydrogenase: An enzyme oxidizing a substrate by removing hydrogen from it.

Dematiaceous: Dark coloured, or pigmented (usually used for dark hyphomycetes, dematiaceae).

Demicyclic: Used for rusts which during their life cycle do not have the uredinial stage. (cf. macrocyclic, microcyclic).

Demosan: See "Chloroneb".

Denitrification: Reduction of nitrates to free nitrogen.

Denticle: Small tooth like projection which bear spores.

Deoxyribonucleic Acid (DNA): Nucleic acid found in nucleus of cell which contains phosphoric acid, D-oxyribose sugar and any of the 4 bases.

Deposit: A quantity of dry fungicide deposited on unit area of material treated.

Deposition velocity: Velocity at which the spray impinges on the target plant.

Dermatomycosis: A fungus infection of animal or human skin like, ringworm, tinea etc.

Dermatophyte: A parasitic fungus attacking and causing disease on skin or its appendages (hair, skin, nails).

Dermatotrophic: Having a selective affinity for the skin.

Dessication: Removal of water for preservation.

Detergent: A substances used for cleansing.

Deuteromycetes: A class of fungi under Amastigomycota division, with characters of Deuteromycotina subdivision. Commonly referred as "Fungi imperfecti".

Deuteromycotina: A subdivision of Amastigomycota division. They do not have sexual reproduction, but parasexual cycle may be found; reproduce asexually by conidia from various types of condiogenous cells. They are commonly referred as "Fungi imperfecti". They have a single form class Deuteromycetes.

Dexon: A fungicide of diazo compound, acting specifically against Phycomycetous fungi attacking seedlings etc., like *Pythium, Aphanomyces, Phytophthora*. Chemically it is Sodium p-dimethyl aminobenzene-diazo sulphonate. It is degraded in sun light. Commercially available as "Dexon 70 WP" and 5G.

Dextran: A polysaccharide of glucose formed outside the cell by enzymatic action of certain microorganisms.

Dialysis: Separation of soluble substances from colloids by diffusion through a semipermeable membrane.

Diaspore: Referred to the propagules of lichens consisting of both the symbionts.

Dichlone: (2, 3, dichloro, 1, 4, napthoquinone), a fungicide of quinone group, used as seed dresser and spray and more effective than chloranil. Trade name "Phygon".

Dichotomous: Branching into two arm of more or less equal size.

Diaporthin: A phytotoxin produced by *Endothia parasitica* in chestnut which causes necrosis of conducting vessels.

Diclinous: Form having male (antheridium) and female oogonium) reproductive organs on different hyphae.

Diclobutrazol: New systemic fungicide of triazol group, effective against rusts, powdery mildews etc. Chemically it is (2R, 3R) and 2S-3S)-1-(2, 4, dichlorophenyl)-4, 4-dimethyl-2-(1, 2, 4-triazol-1-yl) pentan 3-ol. It also shows growth retarding activity in plants by shortening the internode and is more effective when applied to soil. Trade name is "Vigil" 50% Suspension. Not marketed in India.

Dicloran: A fungicide also called "DCNA" active against *Botrytis* fruit rot, *Monilinia, Rhizopus, Sclerotium cepivorum, Sclerotinia sclerotiorum* etc. Chemically it is 2, 6-dichloro-4-nitroaniline (DCNA), and is available under trade name of "Bortan 50 or 75 WP".

Dictyosomes: Spherical vesicles associated with the edge of the membrane bound sacs (cisternae) which constitute the golgi apparatus in Oomycetous and few other fungi.

Dictyospore: Spores having both longitudinal and transverse septa eg. *Alternaria* (cf. amerospore, phragmo, didymo, scoleco, helio and stauro spores).

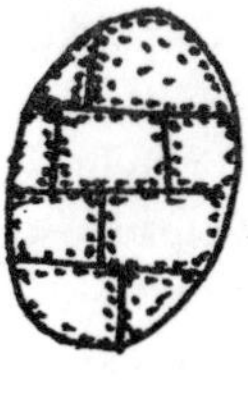

Dictyospore

Didymospore: Spore having one septa i.e. bicelled spore (cf. amerospore, phragmo, dictyo, scoleco, helio and stauro).

Didymospore

Die-back: A disease which starts from apical portion, resulting in browning and withering of affected branch. Attack spreads downward. Commonly caused by *Colletotrichum* sp.

Differential media: Culture media in which certain reagents/chemicals are incorporated which exhibit a specific change in appearance of media after growth of a particular species and thus help in differentiating the microorganism. eg. hemolytic, bacterial growth show a clear zone around colony on blood agar medium due to hemolysis of RBC.

Difolatan: A heterocyclic nitrogenous fungicide with properties similar to captan, chemically it is (N(1, 1, 2,2, tetrachloro-ethyl sulpenyl) Cis-4 cyclohexane 1, 2-dicarboximide).

Dihydropyran: A group of systemic fungicides, closely related to oxathiins, which are effective against rusts, smuts and *Rhizoctonia* spp. Pyracarbolid is one of the compounds of this group.

Dikaryon: A cell or hypha in which two haploid nuclei of genetically different nature are found closely associated in pairs after plasmogamy (cf. heterokaryon, monokaryon).

Dikaryotic: Fungal hyphae having dikaryons.

Diluent: Any inert material added to a fungicide to reduce its concentration.

Dilution end point: The extent to which sap from virus infected plant can be diluted without loosing its ability to infect fresh host.

Dimethirimol: Systemic fungicide of pyrimidine group. Chemically it is 5-n-butyl-2-dimethyl amino-4-hydroxy-6-methyl pyrimidine. It is available commercially as "Milcurb 10 or 12.5% EC". It is effective against powdery mildew of cucumber, melons, marrow and squash, when applied to roots.

Dimidiate: Structures lacking one half or having one half very small than the other.

Dimitic: A system in which generative hyphae are combined either with "skeletal" or "binding" types of hyphae to form the basidiocarp. (cf. monomitic, trimitic).

Dimorphic: Producing two morphologically different forms not essentially required to complete the life cycle (usually in animal parasities single celled on host and mycelial forms in culture).

Dimorphism: Existence of two morphologically different forms in a dimorphic organism.

Dinocap: A benzene group systemic fungicide highly effective against powdery mildew and mites, chemically it is 2-(1-methyl-heptyl), 4, 5 dinitro phenyl crotonate and sold as Karathane.

Dioecious: Having male and female reproductive structure on separate thalli.

Diphycophilous: Lichen thallus having both green (Chlorophyceae), as well as blue-green (chynophyceae) algae as phycobionts. (cf. Chlorophycophilous, Cyanophycophilous).

Diplanetic L: Having a succession of two morphologically different zoospore stages separated by a resting stage (in oomycetes) (cf. monoplanetic).

Diplanetism: The succession of two zoospore stages in a diplanetic organism.

Diplobiontic Gr.: Type of fungal life cycle where both haploid and diploid thalli are found, alternating with each other. e.g. Allomyces (cf. haplobiontic).

Diplococci: Coccoid bacteria occurring in pairs.

Diploid Gr.: Having 2 N (double) number of chromosome within a single nucleus.

Diploidization: Karyogamy of haploid nuclei to form a diploid nucleus especially in a heterokaryon.

Diplomastigomycotina: Subdivision of fungi under Mastigomycota division. Its members produce typically biflagellate zoospores, having one whiplash and other tinsel type flagellum and have haplobiontic-diploid life cycle. It has a single class-Oomycetes.

Disaccharide: Sugar composed of two monosaccharide.

Discomycetes: A group of ascomycetous fungi in which ascocarp is a flat dish or cup like body, apothecium.

Disease: Malfunctioning of an organism, caused by external agents resulting into abnormalities.

Disease cycle: Series of events in disease development, including various stages of the pathogen and its perpetuation from one season to other.

Disease development: Series of events from infection to development of disease symptoms.

Disease frequency: See Disease Intensity.

Disease Intensity: The intensity of a disease, may be the number of diseased host plants (or organs) in 100 plants, counted irrespective of the severity on each of them (or with a threshhold), which is expressed as 0-100% and sometimes, also called the 'disease frequency'.

Disease severity: It is a measure of assessing the seriousness of a disease with respect to the losses to the host plants. It is a balance between the qualities and weaknesses of the host and the pathogen, and depends upon the inoculum potential and susceptibility of the host.

Disinfect: To kill/remove the pathogen, present inside plant tissue.

Disinfest: To kill/remove the organisms present on the surface of plants.

Disjunction: Dissociation of the components of virus particles.

Disjunctor: Sterite cells found in between two conidium to keep them apart in a conidial chain.

Disorder: A harmful nonpathogenic deviation from the normal growth and metabolism.

Dispersal: Movement of individual out of a population (emigration) or into a population, the movement of plant disease propagule.

Dissimilation: Metabolic decomposition within an organism.

Dissemination: Spread of pathogens propagule from its source to surrounding through various agencies like insect, man, mites, air, water etc.

Distosepta = Pseudosepta: A cell appearing to be septate but not having true septum e.g. *Helminthospo-rium Drechslera*.

Dithane: Trade name for a number of bisdithiocarbamate fungicides like Zineb as Dithane Z-78 and Maneb + Zn as Dithane M-45.

Dithane M 22: Trade name for Maneb 80 WP. Also see "Maneb".

Dithane M-45: See Maneb.

Dithane Z-78: See Zineb.

Dithane A-40: See Nabam.

Dithianon: Broad spectrum, protectant, organic fungicide, especially effective against scab, downy mildew of grapes, coffe rust, leaf spots etc. Chemically it is 5, 10-dioxanaphtho (2, 3-b)-1, 4 dithiin-2, 3-dicarbonitrile. Trade name is Shan 75 WP. Under testing in India.

Diurinal: Active during day time,

daily cycle.

Divaricate: Types which are divergent (branching) at right angles.

DMOC: See "carboxin".

Dodder: See Cuscuta.

Dodemorph: Systemic fungicide of morpholine group. Chemically it is 2, 6-dimethyl-4-cyclo-dodecyl morpholine. Its spectrum of activity is similar to tridemorph. Trade name is "Meltatox" 40 EC. Not marketed in India.

Dodine: Common name of guanidine derivative, acting as potent fungicide against apple scab and has been reported to inhibit Tobacco Mosaic Virus also. Chemically it is n-dodecyl guanidine acetate and is sold under trade name "Cyprex" as 65 WP.

Dolipore Septum L: A septum which flares out in middle portion of a hypha. It forms barrel shaped structure with an open end. Found in dikaryotic hyphae of baidiomycetes.

Dormancy: A period of rest or minimum metabolic activity, especially in spores. (Constitutive dormancy is imposed by the organisms while exogenous dormancy is imposed by external conditions.

Dormant: Alive but not in growing state, resting stage, e.g. buds, seeds, spore.

Dosage response curve (DR curve): When inhibition units correlated to units of standard deviation of the normal distribution (probits) are plotted against the logarthims of fungicide doses, a straight line is usually obtained, which is called DR curve. From these curves, effective doses (ED) are estimated usually at 50% response level.

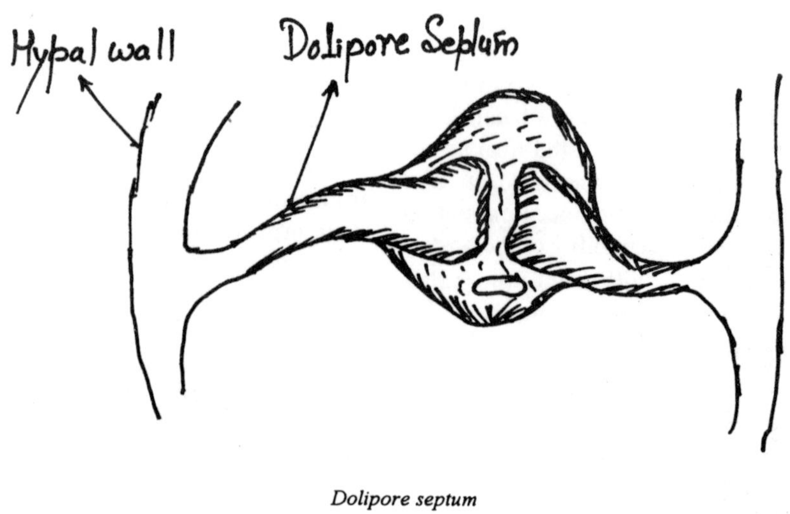

Dolipore septum

Dose/dosage: Quantitiy of pesticide applied per unit area or per unit volume or per unit of weight.

Downy mildew: A disease of a number of crops caused by member of peronosporales. The affected leaves show the growth of sporangiophores of fungus on their lower surface.

DR curve: See "Dosage response curve".

Dressing: Covering with fine coating of chemical, eg. seed dressing with fungicide.

Drift: Spray or dust blown away by wind currents away from the target area.

Dual host virus: Viruses which can multiply within cells of two taxonomically different unrelated host plants eg. Kyasanur forest disease virus.

Duster: Equipment for applying pesticide (dusts) to a crop.

Dutch elm: A disease of elm tree caused by *Ceratocystis ulmi*. Affected plant show defoliation, wilting and browning of Xylem tissue.

Duter: See "Fentrin hydroxide".

Dwarf mistletoe: Common name of an angiospermic parasite *Arecuthobium* attacking conifers.

E

Ear Cockle: A disease of wheat caused by nematodes *Anguina tritici*, exhibiting dwarfing, crinkling of leaves and formation of dark coloured hard structure in place of grain.

Early Proteins: Proteins, particularly enzymes which are synthesized first during the multiplication of bacteriophage and are involved in various biochemical activities of phage multiplication.

Earthstars: Common name for basidiomycetous fungi of family Geastracea, which are "puffballs" in which the outer peridial layers split along radial fissures and when wet, open out in shape of a star. e.g. *Geastrum*.

EC: See "Emulsifiable concentrate".

Eclipse: The time interval between the injection of nucleic acid into host cell and formation of first complete phage particle.

Ecoclimate: Climate within a crop or one type of plant community.

Ecology: Study of the interactions between an organism and its environment.

Economic damage: The injury made to a crop which will justify the cost of artificial control measure.

Economic threshhold: The pest population level at which control measures should be started to prevent the pest population from reaching the economic injury level.

Eco-system: The interacting system of the living organisms of an area with their environment and amongst themselves.

Ectal Excipulum Gr.: The outer layer of apothecium.

Ectotrophic Mycorrhiza: A symbiotic association of fungi and roots of higher plants on the surface of root. (cf. Endotrophic mycorrhiza).

Edelfale: The noble rot-*Botrytis cinerea* rot of grapes is called Edelfal in Germany as it evaporates the water of grapes thereby increasing the sugar content.

Edema: The excessive accumulation of fluid in body tissue.

Ediphenphos: Common name of an organo-phosphorus fungicide specifically and highly effective against paddy blast. Chemically it is 0-ethyl-S-S-diphenyl-dithiophosphate (EDP), and is available under the trade name "Hinosan 50 EC". It also has side effect against *Pellicularia sasaki* and *Cohiliobolus miyabeanus* on paddy as well as some insecticidal activity against leaf and plant hopper.

EDP: See "Edifenphos".

Efficiency: A fixed reduction in pest population regardless of a number of pest involved.

Egg: Female gamete.

Electrotaxis: Attraction and movement of Zoospores towards some electrical current.

Emulsifiable concentrates (E.C.): Liquid formulation of pesticides which spontaneously disperse in water as fine droplets to form an emulsion.

Emulsion: A stable dispersion of oil droplets in aqueous solution or vice-versa.

Enations: Outgrowth on leaves infected with virus,—a symptom of certain viral diseases.

Encapsulation: Covering the pesticide with a non-volatile envelops of gelatine of minute size for delayed and slow release.

Endemic: A disease established in a particular area, country, state.

Endoascus: The inner wall of a bitunicate ascus which is elastic and extensible. Also see "Endotunica".

Endobiotic Gr.: Growing entirely within a living host (cf. epibiotic).

Endoconidium (Pl. Endoconidia): An asexual spore formed inside a hypha (endogenously) and extruded out.

Endogenous: Produced within a structure.

Endomycin: An antibiotic produced by *Streptomyces endus*, is effective against wheat rust and *Botrytis* fruit rot of straw berry.

Endospore: The innermost layer of a spore wall (cf. epispore, perispore).

Endotoxin: A toxin produced in an organism and liberated only at the disintegration of organism.

Endotrophic mycorrhiza: Symbiotic fungus present inside the root tissues (cf. Ectotrophic mycorrhiza).

Endotunica: The inner layer of the ascus wall; also called "Endoascus".

Enriched media: Culture media in which special components like blood, serum, or extract of plant or animal tissues is added to provide additional nutrient for supporting growth of fastidious heterotrophs.

Enteroblastic conidia: In this type of conidial development, conidium is produced from within the conidiogenous cell either including only inner part of cell wall or no part of cell wall.

Enterothallic: A type of thallic conidial development in which outer wall of the conidiogenous cell does not take part in the formation of conidial wall. (cf. holothallic).

Entner-Doudoroff Pathway: A biochemical pathway different from Embden-Meyerhoff-Paranas and Hexose monophosphate, pathway for conversion of glucose to pyruvic acid (glycolysis) which is specific only to certain fungi and bacteria.

Entomogenous: Usually for fungi growing on or inside an insect.

Entomogenous fungi: Fungi growing on the body of insects e.g. *Entomophthora.*

Entomophagous: Animal/Plant/fungus feeding upon insects.

Entomophthorales: An order of Zygomycetes forming asexual spores as conidia which are forcibly shot-away at maturity. Most of the member are found on the body of insects.

Envelope: The outer membranous

lipo-protein covering around the nucleocapsid particles of most animal virsus.

Epibasidium Gr.: A part of a heterobasidium which is borne by the hypobasidium and in turn bears a sterigma and basidiospore.

Epibiotic: Growing on the outside of a living host.

Epidemic: Widespread and destructive occurrence of a disease in a large population.

Epidemiology: The study of epidemic, spread of disease in relation to environmental and other conditions existing in the area.

Epigean Gr.: Above the ground surface.

Epinasty: Downward curvature of leaves.

Epiphragm: Membrane found over the fruiting body of Nidulariaceae.

Epiphyte: A plant (usually non-parasite) attached to another plant for mechanical support e.g. Orchids, Lichen, Mosses.

Epiphytology: Study of epiphytotics-See epidemiology.

Epiphytotic Gr.: See Epidemic, epiphytotic is for plants where as epidemic is basically for human population.

Epiplasm: Portions of the cytoplasm left in the ascus, outside the ascospore walls which possibly nourishes the developing ascospores and also deposits the various external markings/striations on ascospore walls.

Episome: Small genetic elements which may occur independently in Cytoplasm or attached to chromosome of bacterial cell and multiply with it and act as F+ factor controlling the ability of the cell to act as gene doner. e.g. temperate phage causing lysogenic conversion.

Epispore: The outermost layer of the spore wall which is often ornamented (cf. endospore, perispore).

Epithecium (Pl. Epithecia): A layer at the surface of hymenium of apothecium formed by the union of tips of paraphyses over the asci.

Epixylous Gr.: Occurring on wood.

Epizootic Gr.: Widespread occurrence of an animal disease.

Eradicant: A chemical/fungicide used to eradicate the pathogen from a surface on which it has already established. e.g. organo mercurials.

Erdin: A metabolic product of *Aspergillus terreus* produced in culture which contains most of the chlorine utilized by the fungus from culture, chemically it is $C_{16} H_{10} O_7 Cl_2$.

Ergosterol: A precursor of vitamin D and is synthesized by a number of fungi and yeasts.

Ergot: A disease of some cereal crops in which infected grain is converted into hard structure "Sclerotium".

Ergotin: A medicinal alkaloid obtained from the sclerotia of *Claviceps purpurea.*

Ergotism: A disease of animals and humans caused as a result of eating the sclerotia of fungus *Claviceps purpurea.* Symptoms include gangrenous legs, hoofs and tails and abortion of pregnent females.

Erumpent: Bursting out (of spores etc.) through the surface of the substratum.

Erwinia amylovora: First plant pathogenic bacterium described by J.J. Burrill (1839-1916), causing fire blight of apples and pears.

Essential elements: Elements essentially required for the growth and development of a particular organism and are indispensable for it.

Esterase: Group of enzyme hydrolysing esters.

Ethazol: Common name of a fungicide having systemic action. Chemically it is 5-ethoxy-3-(trichloro methyl)-1, 2, 4-thiadiazol. Available under trade name "terrazole 30 WP" and in mixed formulation with PCNB. It has wide spectrum of activity against smuts, bunts, *Rhizoctonia, Pythium, Phytopht-hora and Fusarium* and is used for seed and soil treatment. It is also called "Turban".

Ethirimol: Common name of systemic fungicide of pyrimidine group, chemically being, 5-butyl-2-ethylamino-4-hydroxy-6-methyl pyrimidine. Trade names are "Milstem" (seed dresser), "Milgo" etc. Its combination with captafol is available as "Milcap:. Effective against powdery mildews.

Ethyl mercury chloride: An organo mercurial,seed dressing fungicide. Trade names, "Ceresan", "Tafasan" etc.

Etiolation: Yellowing of tissue and elongation of stems caused by reduced light or darkness.

Etiology: Study of causal agent of a disease and its relationship with susceptible host.

Ethylene: A highly active plant growth regulator which is produced in higher amounts in diseased tissues and is responsible for many symptoms like, epinasty, abscission, premature senescence, early ripening and shedding of leaves.

Ethylene oxide: A common volatile biocide used for gaseous sterilization of material, chamber etc.

Eucarpic Gr.: A condition in which only a part of an assimilative thallus is converted into a reproductive structure(s).

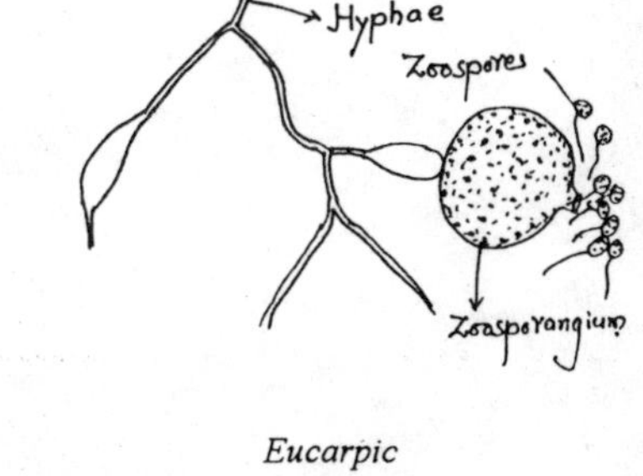

Eucarpic

Eukaryotes: Organisms with well developed nucleus and chromosomes.

Eukaryote: Other than primitive organisms having a well defined nucleus surrounded by a nuclear membrane and other complex organelle like mitochondria.

Eurithermophile: An organism capable of growth at 60°C and also at temperature near 30°C.

Excipulum (Pl. Excipula) L: In discomycetes the tissue beneath the subhymenium which also forms the outermost sterile margin of the apothecium.

Exclusion: A measure of pest/disease control in which pathogen is kept away from an area, state, or country, which is free from that disease.

Exergonic: Energy yielding as in a chemical reaction.

Exigynous: Forms or conditions when antheridial stalk arises directly from the oogonial cells above the basal septum.

Exoascus: See "Exotunica".

Exoconidium (Pl. Excoconidia) Gr. : An asexual spore formed on the surface of a hypha (exogenously).

Exogenous L: Produced on the outside of a structure.

Exomycology: Mycology of outer space/region.

Exotic: Foreign, not native to the area.

Exotoxin: Toxin produced in surrounding medium.

Exotunica: The outer layer of the ascus wall. It is also called the "Exoascus". (cf. Endotunica).

Exploitation: An ecological interrelationship in which one organism inflicts harm on another for its own benefit (parasitism or predation).

Extracellular virion: Complete infective virus particle as they exist prior to infection.

Extramatrical: Living on or near the surface of matrix or substratum.

Exudate: Droplets of liquid usually containing chemicals discharged from diseased or healthy plant tissue.

F

Facultative saprophyte: An organism which normally lives as a parasite but can also live saprophytically (cf. Facultative parasite).

Facultative parasite: An organism which normally lives as a saprophyte but may become parasite under certain condition (cf. Facultative saprophyte).

Falcate: Sickle shaped (usually for shape of conidia).

Fallowing: The practice of leaving fields without any crop for certain period to control soil-borne diseases, especially, nematodes.

False morels: Commonly referred for genus *Gyromitra* of "Helvellaceae" under Discomycetes.

False smut: A disease of paddy caused by *Ustilaginoidea virens*, in which grains are converted into large, velvety green mass of balls.

False truffles: Commonly referred to the basidiomycetous fungi of order Hymenogastrales having hypogeous basidiocarps.

Fassicle(s): In groups; used for spore bearing structure Conidiophoeres etc.

Fassiculate: Fungi producing erect bundles fassicle of conidiophores (Stipes).

Fastidious organism: An organism which can not be easily isolated or cultivated on ordinary culture media.

Fastidious heterotrophs: Heterotrophic microorganism which require specific nutrients like vitamins and growth promoting substances for their growth.

Fenfuran: Systemic fungicide of carboxamide group used as selective seed dresser for the control of wheat bunt, barley smut, loose smut of wheat, barley, oat smut etc. Chemically it is 2-methyl-furan-3-carboxylic acid anilide. Trade name is "Panoram". Not marketed in India.

Fentin hydroxide: Organic tin compound acting as fungicide for the control of *Cercospora, Septoria, Helminthosporium, Alternaria, Pythium, Phytophthora, Rhizoctonia* etc. Chemically it is tri-phenyl tin hydroxide (TPTH) and is available under trade name "Duter" 20 WP or 50 WP.

Ferbam: Iron containing fungicide of dithiocarbamate group (Ferric dimethyl dithiocarbamate) used as foliar and fruit spray. Marketed as 'Hexaferb' 'Karbam' etc.

Fermentation: Anaerobic oxidation of carbohydrates or allied compounds by the action of microbial enzymes without involving gaseous oxygen.

Fertilization: Sexual union of both cytoplasm (Plasmogamy) and nuclei (Karyogamy) of gametes.

Fertilization: The sexual union of two protoplasts resulting in double (2 n) number of chromosome.

Fertilization tube: An extension of antheridium which penetrates the wall of the oogonium to reach the oosphere and through which the male protoplast (gametes) passes (in oomycetes).

Fibrinolysin: A substance produced by hemolytic Streptococci, which can liquify clotted blood plasma or fibrin clots.

Field immune: Plants showing resistance under field condition although may sccumb under inoculated conditions.

Field resistance: Resistance observed under natural field condition but not under artificial inoculated conditions, also referred to horizontal resistance.

Filamentous: Thread like, filiform.

Filiform: See filamentous.

Filler: A dilutent in powder form used for dust preparation of fungicides, or inert component of a pesticide.

Filterable agents: Disease causing agents which can pass through bacteria proof filters. Originally it was a criteria of defining virus.

Fibriae: Very small, short and numerous hair like appendages (other than flagella) found over the surface of bacteria, supposed to be functioning as adhering or attachment organs, can be seen only under electron microscope.

Fimicolous: See "Coprophilous".

Fire blight: Bacterial disease of Apple and Pears caused by *Erwinia amylovora* in which flowers become brown and black and drop followed by browning and blotching of leaves and twigs.

Fission: Cleavage or splitting of cell into two parts (cf. budding), a mode of asexual reproducing in bacteria and yeast.

Fissitunicate: See "bitunicate".

Flagellate L: Having a flagellum or flagella.

Flagellum (Pl. Flagella): A hair like structure which propels a swimming cell.

Flagging: Loss of rigidity and drooping of leaves and tender shoots preceding the wilting of a plant.

Fleck: A minute spot.

Flexuous hyphae L: Hyphae in a pycnidium which fuse with pycniospores by plasmogamy (in the uredinales).

Flimmers: See "Mastigonemes".

Floccose L: A type of fungal (colony) growth (in culture) in which mycellium appears fluffy.

Flocculent L: An adherent aggregate of microorganisms or other material floating on a liquid.

Flodding: A method of controlling some disease like *Macrophomina phaseolina* and some nematodes by filling the fields with irrigation water.

Fluorescein: A fluorescent substance absorbed to antibiotics in order to facilitate their detection and location on a tissue.

Fluorescence microscopy: A microscopic technique in which microorganisms are stained with a fluorescent dye and observed by illumination with UV light.

Flying pin: Mechanism of transmission of virus causing myxometosis of rabbits by mosquitoes.

Foliicolous: Growing or living on leaves.

Foliose thallus L: A flat leaf like thallus spreading over the substratum attached only at localized points in lichens (cf. crustose, fruticose).

Folpet: A fungicide of heterocyclic nitrogenous group effective against few pathogens. Chemically it is M (trichloro methyl thio) phthalinide.

Foltaf: See "Captafol".

Fomites: Inaniamate objects carrying viable pathogens.

Foot cell: Basal cell some time found as a support to stipe of *Aspergillus*.

Foot rot: A disease showing rotting of lower portion of stem root axis.

Form: A subdivision of a species below the rank of a variety.

Formalin: A liquid chemical used for fumigating soil.

Forma specialis: A subdivision of species based mainly upon physiological character, especially specifically for hosts and very rarely on morphological characters. Special forms or variety, group of biotypes specific for a particular host species. Several formal sp. (varieties) of identical morphology constitute a fungal species and they differ from each other in their ability to attack and infect a particular host species.

Form genus (or Species): A genus or species for imperfect state.

Formulations: Physical state in which a chemical is marketed for fungicidal/insecticidal use, such as wettable powder, emulsion, dusts etc.

Forssman antigens: Heterophile antigens widely distributed in nature.

Fouling: Impairing the physical appearance of matured due to activity of fungi like bridging the gaps of electrical equipment and causing short circuits, etching the glass lines by the fungal metabolic, spoiling the paint film etc.

Fox fire: Phenomenon of bioluminescence, exhibited by the mycelium of certain Agaricales (e.g. *Armillariella mellea*) growing deep in the decaying wood.

Fractional sterilization: See tyndalization.

Fragmentation L: A method of asexual reproduction in which thallus is divided into many segments, each of which is capable of growing into a new individual.

Free living: An organism living freely, unattached or a pathogen living in soil outside its host.

Freeze drying (lyophilization): A process for maintaining the cultures of sporulating forms. Spore suspension is prepared in a suspending medium (like 10% skimmed milk and 5% inositol in distilled water), distributed in small glass ampule, frozen and dried under vacuum, ampules are sealed and stored at low temperature. The process is not suitable for mycelial culture or forms with large fragile spores.

Freezing injury: A non-parasitic

disease caused due to excessive low temperature showing burning and blighting of leaves.

Friesian classification: System of classifying the Aphyllophorales, based on the gross morphology of the basidiocarp with particular emphasis on hymenial configuration (cantharelloid, toothed, lamellate, porus etc.). This is named after great Swedish taxonomist Elias Fries.

Frog eye leaf spot: A foliar disease of soybean, caused by *Cercospora sojina,* producing round, grey coloured sports surrounded by reddish margin, which resembles the eye of frog.

Fructification: A discrete macroscopic or microscopic structure bearing sexual or non-sexual spores. It includes all accessory tissues or organs associated with this structure.

Fruticose thallus L: An upright or pendulous thallus in lichen (cf. crustose, foliose).

Fuchs-Rosenthal blood counting cell: See "Haemocytometer".

Fuel deterioration: Petrolium product especially high quality fuel is deteriorated due to attack of a fungus *Cladosporium residua* which grows in between water and petrolium layer and also blocks the supply line central valves etc. in the fuel tank of air craft.

Fumigant: A volatile substance forming toxic gas which is used to disinfest certain areas from pests.

Fumigation: Application of a fumigant for disinfection of an area.

Fungicide L: A chemical or physical agent that kills fungi (c.f. fungistasis).

Fungi imperfecti: Forms of fungi lacking perfect (sexual) stages. Also see "Deuteromycotina".

Fungimar: Trade name for 50% cuprous oxide, a fungicide.

Fungistasis: Inhibition of fungal growth which is not lethal.

Fungistat: A chemical or other factor inhibiting the growth of the fungus till it remains present in contact with fungus but does not kill it.

Fungus (Pl. Fungi): A heterocarpic, a chlorophyllous organism whose usually walled, thread like cells absorb nutrients in the form of a watery solution, and reproduce by sexual or asexual spores.

Funiculose: Fungi shows rope like growth of mycelium in culture colony.

Funiculous (Pl. funiculi): A hyphal cord which attaches a peridiole to the peridium (in Nidulariales).

Fusaric acid: A toxin produced by the wilt pathogen *Fusarium oxysporum.*

Furalaxyl: Systemic fungicide of acylalamine group, effective against Pythium damping-off and some downy mildews. Chemically it is methyl-N-(2, 6-dimethyl-phenyl)-N-furoyl)-2-alaninate. Marketed as "Fongarid 50 WP". Not available in India.

Furcarbanil: Systemic fungicide; a derivative of oxathiin group which is a furan; chemical name being, 2, 5-dimethyl-3-curanilide. Has wide spectrum of activity particularly against loose and stinking smut, *Helminthosporium, Fusarium, Macrophomina* etc. Trade name is "Campogran". Not marketed in India.

Fusicoccin: A phytotoxin produced by *Fusicoccum amygolali,* which has growth regulatory property and causes stomatal opening.

Fusidic acid: An antibiotic used mainly against penicillin resistant staphyloccoceal infections, isolated from *Fusidium occineum, Mucor ramannianus* and *Cephalospo-rium* sp.

Fusiform: Conidia having spindle shape or with both narrowing ends.

Fuziwan: See "Isoprothiolane".

Fytolan: Trade name of 88% copper oxy chloride WP.

G

Gall: A tumor or outgrowth formed due to hyperplastic growth of unorganised cells as a result of microbial, insect attack.

G-Antigen: A particle of viral nucleic acid, haemagglutinn and lipid which serve as precursor to complete virus particle during multiplication of certain animal virus.

Gametangial contact: A condition in sexual reproduction when two gametangia come in contact but donot fuse. The male nucleus migrates into female gametangium through a pore or fertilization tube. Also called "Gametangy".

Gametangial copulation: Fusion of two gametangia or their protoplast resulting into a zygote and resting spore. Also called "Gametangiogamy".

Gametangium (Pl. Gametangia): A differentiated specialized cell producing discrete gametes or whose undifferentiated protoplast functions in the place of discrete gamete.

Gamete Gr.: A differentaited reproductive cell which is capable of undergoing plasmogamy with another reproductive cell.

Gametothallus (Pl. Gametothalli): A thallus producing gametes (cf. sporothallus).

Gamma irradiation: Gamma rays produced by a cobalt 60 source apart from being a mutagenic agent are being used for strilization (disinfection) of materials which are not easily penetrated by ultraviolet radiation like materials packed in cardboard, plastic goods, lab ware, media food stuff etc.

Gangliform: Having knots.

Gas atmosphere: Perishable material like grain or fruits are stored in vesseles containing inert gases like Co_2, nitrogen etc. which protects it against attack of microbial deterioration and may be stored either at normal temperature or at some degree of cooling only.

Gaseous sterilization: Fumigation of material in closed chamber with some volatile biocidal gases like ethylene oxide, propylene oxide, methyl bromide, ozone sterilises it.

Gel: A jelly like conidial mass.

Gelatin: A protein from skin, hair, bones etc. which is used in culture media for determining the proteolytic activity of microorganisms and also for peptone preparation.

Gelatinase: Gelatine degrading enzyme.

Gemma (Pl. Gemmae): A thick walled cell similar to chlamydospore, especially in lower fungi.

Gene: A material substance in the chromosome which determines or conditions one or more characters in the off-springs. The smallest functional unit of the genetic material.

Gene for gene hypothesis: An hypothesis proposed by Flor to explain the inheritance of resistance in case of flax rust, according to which for every gene of resistance in host there is a complimentry gene of virulene in pathogen.

Generative hyphae: One of the 3 types of hyphae found in sporophores of basidiomycetes, which are the only one, capable of producing basidia. They are usually thin walled, profusely branched and septate and generally have clamp connection. They are essential part of the basidiocarp. (cf. skeletal, binding).

Genotype: A particular set of genes (genetic construction) present in the cells of an individual.

Genus (Pl. Genera L): A taxonomic group of a number of species. The first name (generic name) of bionomial is that of genera.

Geodin: A complex produced ($C_{17}H_{12}O_7Ci_2$) as a metabolite of *Asprgillus terreus* in culture which contains most of the chlorine utilized by fungus.

Geotropism: Movement (bending) in response to stimullus of gravitational force of earth.

Germicide: An agent which can kill germs (usually pathogenic microorganisms).

Germination L: The emergence of protoplast (usually vegetative) from the spore.

Germinative transmission: Transmission of pathogen through seeds of host which on germination facilitates the spread of pathogen also.

Germ pore: A differentiated frequently apical area or hollow portion found on the wall of spores, through which germ tube comes out during germination of spores.

Germ tube: At the time of germination, the fungal spore gives rise to one or more tubular structure called germ tube which develop further into hyphae.

Giant cell: A multinucleate mass of protoplasm formed by coalescene of several adjacent plant cells; also called syncytium, usually found in plants infected by certain nematodes.

Gibberellins: A group of growth regulating substances which play a vital role in development of many diseases eg. 'bakanae' disease of rice, caused by *Gibberella fujikuroi*.

This plant hormone was first discovered as a metabolite of pathogenic fungus *Gibberella fujikuroi (= Fusarium moniliforme).*

Giemsa staining: A stain used to demonstrate the presence of rickettsia within host cell and also for examination of protozoa in blood film.

Gill: A hymenium covered, plate like appendage which hangs from the underneath surface of the basidiocarp of some hymenomycetes.

Gleba: Spore bearing inner fertile tissue of fruiting body which is enclosed by a peridium throughout development (in Gasteromycetes).

Gliotoxin: An antibiotic substance secreted by *Trichoderma viride*, a strong antagonistic fungus.

Globose: Spherical in shape.

Globulin: Water insoluble protein which can be dissolved in dilute neutral salt solution. Most of the antibiotics are globulins.

Glucomannan: A hemicellulose made up of a hetropolymer of D-glucopyranose and D-manno-pyranose linked at beta 1, 4 and galacto glucomannan, linked at 1, 6 with D-glucopyranose side branches.

Gluconic acid: An organic acid produced from sugars by the action of *Penicillium (P. purpurogenum* sons) and *Aspergillus (A. niger* group). It is used in medicinal and food preparations as Calcium salt.

Glucosylation: Addition of a glucose molecule to nucleic acid of certain bacteriophage mutant enabling them to infect hitherto resistant strain of bacterium as the nucleic acid becomes resistant to bacterial nucleases.

Glycogen: A carbohydrate stored by animals which releases glucose on hydrolysis.

Glycolysis: First step of respiration in which glucose molecule is broken into a 3-carbon compound-pyruvic acid and lactic acid involving several steps.

Glyodin: A derivative of glyoxalidine acting as heterocyclic, systemic fungicide for the control of *Monilia, Stemphylium. Alternaria and Glomerella.* Available under trade names of "Glyoxalidine", "Glyox-ide", "Glyodin" as 30% liquid or 70 WP. It also show activity against Tobacco Mosaic virus.

Gnotobiotic Growth: (Pure culture growth). Growth of an microorganism in the absence of any other organism in its vicinity.

Gonotocont: Organs in which meiosis takes place.

Graft: Artificial union between tissues which is a common practice in horticulture and serves as a mean of transmission of diseases especially viruses.

Grafting: A method of plant propagation in which a bud or a scion of one plant is transplanted on another plant. It can be the joining of cut surfaces of two plants so as to form a living union.

Gram positive and Gram Negative Bacteria: Bacteria in which retain the crystal violet colour of Gram staining technique are Gram positive and others which become discolorize and take red stain are called Gram negative.

Gram Stain: A differential stain by which bacteria are differentiated into Gram positive or Gram negative depending upon their ability to retain or lose the primary stain of Crystal violet after being subjected with decolorizing agent.

Grassy shoot: A mycoplasma disease of sugarcane. Important symptoms are production of large number of small, thin, yellow or albinow tillers at ground level giving grassy appearance. Diseased plants produce no canes.

Grease band: Adhesive material (resin in caster oil or sticktile) applied as a band around a tree to trap or repele ascending wingless female moth or ants.

Green ear disease: A disease of bajra (pearl millet) caused by *Sclerospora graminicola,* exhibiting downy mildew symptoms on foliage and malformation of ears. Malformed ears produce bunch of green, leafy structures, giving appearance of phyllody symptoms.

Green mold: Commonly referred to the fungus *Penicillium* spp. (cf. black mold).

Gregarious: Found in large groups, but not joined together.

Grey mold: Commonly referred to the fungus itself or the diseases caused by *Botrytis cineria,* especially on chickpea.

Griseofulvin: A chlorine containing metabolic product of *Penicillium greisofulvum.* Also acting as strong systemic antibiotic against fungal infection of skin, hair and nails.

Growth Curve: Graphic representation of the growth of bacteria in phases in culture medium.

Growth inhibitor: A natural substance which inhibits the growth of a plant.

Growth regulator: A naturally produced chemical controlling the enlargement, division or activation of plant cell.

Gossypiol: An antifugal substances (Phytoalexin) accumulated in xylem vessels of *Gossypium* stem as a result of introduction of conidia of *Verticillium alboatrum, Rhizopus nigricans* and also heavy metal ion.

Gum: Complex polysaccharidal substances formed by cells in reaction to wounding or infection.

Gummosis: A disease of citrus caused by *Phytophthora* sp. exhibiting water soaking and oozing of dark gummy substances at basal portion of stem.

Gummosis: Production of gum by or in a plant tissue; usually in response to attack of pathogen forming a characteristic symptom of disease.

Guttate: Having tear like drops.

Guttation: Exudation of water from plants particularly along the leaf margins through special structure hydathodes.

Gymnocarpous: A type of basidiocarp development in which the hymenium remains naked and is never covered. It is initially formed on the outside of basidiocarp. (cf. hemiangiocarpous).

Gymnomycota: A division of kingdom Myceteae (fungi) consisting of phagotrophic organisms having vegetative thallus without cell wall.

Gymnothecium (Pl. Gymnothecia) Gr.: A naked ascocarp which is a cluster of asci, surrounded by a lax network of hyphae forming a loosely interwoven ascoma with or without appendages. It is found in order Onygenales of Plectomycetidae.

H

Haemaglutinin: Agent which causes agglutination of red blood corpuscles. Viruses act as agglutinating agents.

Habitate: The natural place of occurrence of an organism.

Haemocytometer: An instrument having a microscopic slide with marked squares of different size and a cover glass at a specific distance over the slide which is used to measure the blood cells or spore density in a given solution. It is also called "Fuchs Rosenthal blood counting cells".

Halmophagous: In ectotrophic mycorrhiza, forms having a mantle and a Hartig net.

Halophile: Microorganisms which grew luxuriently at high salt concentrations.

Hanging drop technique: A technique for observing the movement or other activities of living microorganisms in a suspended drop of fluid using a cavity slide.

Hantigen: Substance found in flagella of certain bacteria which act as antigen.

Haplobiontic Gr.: The type of fungal life cycle in which their is only one free living thallus which may be haploid or diploid. (cf. "Diplobiontic").

Haploid Gr.: Having a single set (n) of reduced number of chromosomes in a nucleus.

Haploid incompatibility: When male and female reproductive organs developed on the same individual, can not participate in sexual reproduction and require sex organs of another individual for it. Phenomenon is also called "Self infertile" or "physiological heterothallism".

Haploidization: In a heterokaryon, the transformation of a diploid nucleus to a haploid nucleus through a seris of mitotic divisions in which aneuploid nuclei are formed.

Haplomastigomycotina: A subdivision in fungi having members with uni-or biflagellate (both flagella whiplash type) zoospores and haplobiontic-hapliod or diplobiontic life cycle.

Hapteren Gr.: An organ of attachment in some lichens or Nidulariales. In Nidulariales, it is a mass of adhesive hyphae at the base of funiculus.

Hartig net: The intercellular hyphal network formed by mycorrhizal fungi on the surface of roots.

Hatching factor: A material produced by the roots of certain plants that is believed to increase the hatching of eggs of certain nematodes.

Haustrorium (Pl. Haustroria): A specialized hyphal extension in some parasitic fungi which penetrates the host cell and absorbs nutrients; it is common in obligate parasite.

Head: Cuboidal component of the binal viruses.

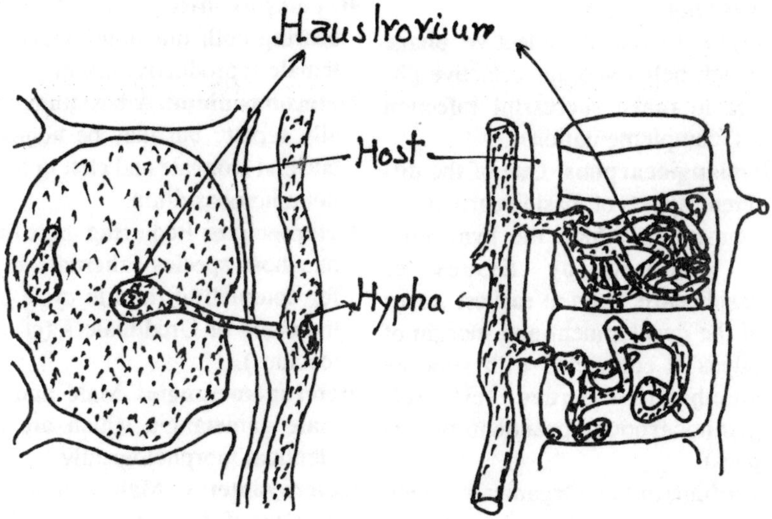

Haustroria

Heald-Pool Effect: Many fungi do not grow and sprout properly alone in culture but when grown as mixed culture with other forms they start sporulating which is supposed to be due to the specific requirement of certain chemical which are produced by other fungi in excess of their requirement. This effect is called Heald-Pool effect.

Heat therapy: Treatment of disease (eradication of the pathogen especially viruses) by application of heat. It is used mainly for treating planting material like, sugarcane or soil treatment for nurseries.

Helical: Spiral arrangement of Capsid subunits surrounding the nucleic acid core of the viruses.

Helicospore: A coiled or helical spore (cf. amero, didymo, phragmo, didyo and Scoleosporo).

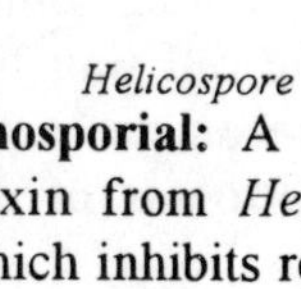

Helicospore

Helminthosporial: A non-specific phytotoxin from *Helminthosporium* which inhibits respiration in roots.

Helminthosporiosis: A brown leaf spot disease usually of rice) caused by *Helminthosporium.*

Helotism: A relationship of algae and fungi in lichen.

Helper Virus: A defective phage which helps another defective phage to make successful infection (cf. complementation).

Hemiangiocarpous: One of the different types of basidiocarp development in which the hymenium is enclosed by tissues of basidiocarp, even at earliest stages of the development and margin of pileus is connected with stipe by membranous inner veil. (cf. gymnocarpous, pseudoangiocarpous).

Hemibiotrophs: Organisms which attack living tissues just like biotrophs but continue to develop and sporulate after the tissue is dead.

Hemagglutination: Clumping of red blood cells.

Hemicellulase: A group of enzymes which degrades hemicelluslose like araban, galactan, xylan etc. forming a layer in the cell wall. These enzymes play a role in the pathogenesis and penetration of the pathogen.

Hemicellulose: A group of polysaccharide resembling cellulose, and found in the cell wall like xylan, galactan, araban etc.

Hemolysis: Process of dissolving RBC.

Herbaceous plants: A higher plant which does not develops woody tissue.

Hermophroditic: Special producing both male and female organs on a single individual.

Hermophrodite: An individual bearing both functional male and female reproductive organs.

Heterobasidium: A basidium usually septate but may be non-septate and elongate and cleft in form; not a holobasidium.

Heteroecious: Requiring more than one host species (alternate host) for completing the life cycle e.g. members of Uredinialees (cf. autoecious).

Heterogametangia: Male and female gametangia which are not identical morphologically.

Heterogametes: Male and female gametes of the same species which can be distinguished morphologically.

Heterogametes

Heterogamy: Sexual union (fusion) of two morphologically different gametes.

Heterokaryon: A cell (or thallus of such cell) having genetically different nuclei (cf. dikaryon, homokaryon).

Heterokaryosis: The condition of fungus when it contains genetically different type of nuclei in a single cell.

Heterokaryotic: Condition of being heterokaryon.

Heterokont: Having different types (one tinsel and one whiplash) of flagellum eg. zoospores of *Saprolegnia*.

Heteromerous Gr.: Type of lichen thalli in which algal cells forms a distinct layer within the thallus. Most of lichens belong to this category. (cf. homoisomerous).

Heterothallic: Self sterile condition. Requiring a partner of sexually different strain for reproduction though, it may be identical morphologically (cf. homothallic).

Heterothallism: Occurrence of two different type of self incompatible strain looking identical morphologically eg. *Mucor* sp.

Heterotophy: A hyperplastic state when an agents grows at an unusual pace (eg. development of ear in the tissue of Corn plant).

Heterotroph: Microorganisms which can not utilize molecular CO_2 as sole carbon source and require organic compound for their growth. (cf. Autotroph, Necrotroph).

Heterotrophic: Organisms which require organic matter for their growth and development (cf. autotrophs).

Hexaconazole: Systemic fungicide of triazole group having wide range of activity against *Uncinula, Hemileia, Cercospora, Venturia, Erysiphe, Alternaria, Colletotrichum, Phoma, Pyricularia, Septoria, Puccinia, Rhizoctonia, Sclerotium* etc. Chemically it is RS-2 (2, 4-dichlorophenyl)-1-(1 H-1, 2, 4-triazol-1-yl) hexan-2-ol. Trade name is "Contaf 5 EC". Not marketed in India.

Hexamer: The number of components protein subunits forming a Capsomore.

Higher fungi: Loosely used common term for the members of Ascomycotina, Basidiomycotina and sometimes also Deuteromycotina.

High Volume Spray: Spray of pesticides in which over 1000 1 of liquid per ha is to be sprayed on bushes and 700 1 per ha on ground crops.

Hilar appendix: A minute projection of basidiospore, close to the point of attachment to sterigmata of basidia. A small bubble formed at this appendix is supposed to be involved in the discharge of basidiospore at maturity.

Hill bunt: A disease of wheat caused by *Tilletia caries* and *T. foetida* in which grains are affected becoming brownish in colour and give foul smell, like rotten fish. It is also called "covered smut" or "stinking smut".

Hinosan: See "Ediphenphos".

Hircinol: An antifungal phenanthrene substance produced in *Laroolessum hircinum* tissues in response to infection of *Rhizoctonia repens*.

Holdfast: Basal projection of thallus used for attachment on substratum.

Hollow cone: Spray jet with a core of air breaking to give drops in an annular pattern.

Holobasidiomycetidae: A subclass of fungi under class Basidiomycetes. They have a non-septate basidia (holobasidia) formed on a persistent hymenium on various types of sporophores opening only after maturity of spores.

Holobasidium: An undivided, usually club shaped simple basidium, not a heterobasidium.

Holoblastic: A blastic conidium in which both inner and outer walls of conidiogenous cell swells out to form the wall of new conidium. (cf. enteroblastic).

Holocarpic: The entire assimilative thallus being converted into reproductive structure (cf. eucarpic).

Holomorph: The fungus exhibiting all the stages (forms) i.e. imperfect and perfect in its life cycle. (cf. Teleomorph and Anamorph).

Holothallic: A type of thallic development of conidia in which the layers of the wall of the conidiogenous cell are involved in the formation of conidium wall. (cf. enterotha-llic).

Holotype: The type specimen of plants, fungi (culture) or diseased specimen (herbarium) preserved in herbari on which the original description of the species by the author is based.

Holozoic: Organism which injest the solid food particles.

Homoisomerous Gr.: Referred to the thallus of lichen in which algal components are more or less evenly distributed throughout the thallus. (cf. heteromerous).

Homokaryon: A cell having genet-

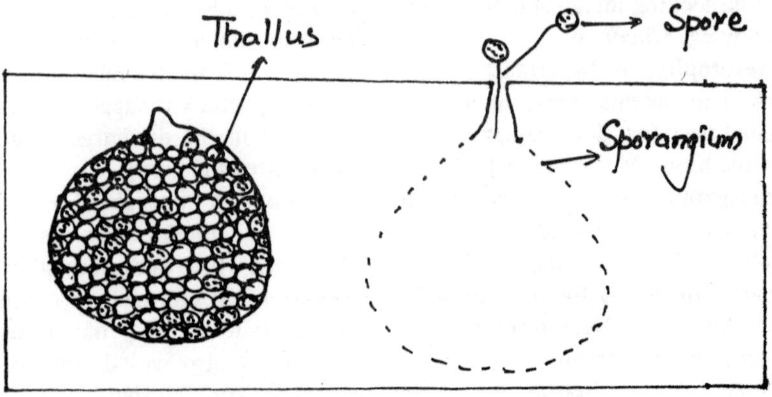

Holocarpic

ically similar nuclei or a thallus of such cells (cf. heterokaryon).

Homokaryotic: A condition of being homokaryon.

Homologous: Forms which resemble each other morphologically but not necessarily in their function and physiology.

Homonyms: Several different taxa (species) which have been given the same name.

Homothallic: Self-fertile conditions in which the single strain of a species can reproduce sexually without requiring another strain (cf. heterothallic).

Homothallism: Condition of being homothallic or self fertile.

Horizontal resistance: An evenly distributed resistance in the host plant against most of the existing races of the pathogen and is usually controlled by a number of genes (Oligogenic or polygenic resistance). It is also called "Durable resistance" or "Field Resistance".

Hormone: A growth regulator, usually referred to a particular auxin.

Host indexing: To, screen the plants for their being the carrier of any pathogen, usually done by growing them in isolation and observation e.g. as done for potato virus.

Host: A living organism providing nourishment and possible protection to microbes (fungus, bacteria etc.) living in or on it.

Host induced mutation: Changes in the nature of virus induced by its host e.g. conversion of non-virulent type into virulent type after glucolysation inside the host cell.

Host range: Various host plants which may be attacked by a pathogen.

Hot Water Treatment: A method of disinfecting the seed and planting materials by dipping in warm water for specific temperature and specific period.

Hülle cells: Rounded thick walled sterile cells surrounding the cleistothecia of many *Aspergilli*.

Humectant: Material added to a spray to delay its evaporation from the plant surface.

Humus: The dark organic layer of soil consisting of partially decayed plant material.

Hyaline: Colourless, more or less transparent, used for fungal hyphae or spores.

Hybernation: Dormancy during summers or winter when an organism has minimum metabolic activity.

Hybrid: The off-spring of two individuals differing in one or more heritable characteristics from its parents.

Hydathodes: A structure with one or more opening usually found at the margins of the leaves which discharges water from internal tissue to the outside in the form of fine droplets.

Hydrochory: Dispersal of the pathogen propagules through the agency of water. (cf. Zoochory, antropochory).

Hydrolysis: Enzymatic breakdown of a chemical compound utilizing the molecule of water.

Hydrophillic: Substances/chemicals which are miscible or soluble in water.

Hydrophobic: Substances/chemicals which are immiscible with or insoluble in water.

Hydrolytic enzymes: Enzymes which split the substrate molecule utilizing a water molecule. Many plant pathogenic fungi and bacteria attack the component of plant cell wall through these enzymes.

Hymenium (Pl. Hymenia): A palisade like layer of asci or basidia including some sterile cells like basidiole, paraphyses or cystidia.

Hyperparasite: A parasite parasitising another parasite.

Hyperpalasia: Abnormal rate of cell division or execessive multiplication resulting into abnormal growth of plant tissue which is a symptom of many infectious plant disease eg. wart of potato (cf. Hypoplasia).

Hypersensitive plant: A plant showing extreme resistance against a pathogen by quick and sudden death of cells around the point of attack and thus preventing further spread.

Hypertrophy: Abnormal development of a tissue or cell or components of cell in size which creats the symptom of abnormal growth or malformation.

Hypha (Pl. Hyphae): Filamentous part of a fungus, usually septate and consisting of several cells in linear succession.

Hyphal body: A hyphal segment which becomes separated and reproduce by budding or fission, as found in Entomophthorales.

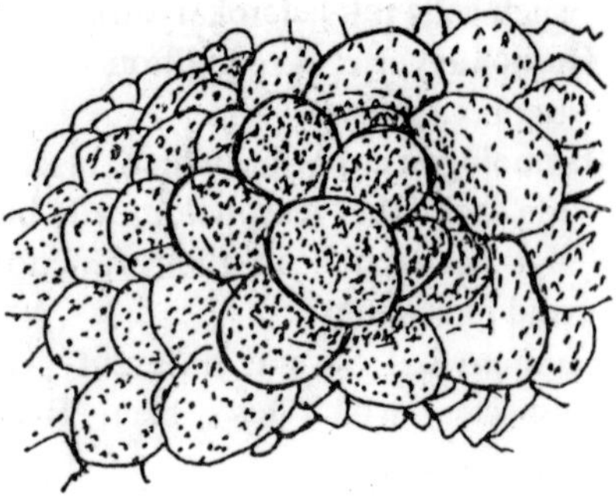

Hypertrophy

Hyphomycetales: A class of fungi-impertecti (Deuteromycotina) which does not form pycnidia or stromata and produce conidiophores openly on any part of the mycelium.

Hyphomycetidae: A subclass of Deuteromycetes, whose members have true, septate mycelium; produce conidia on conidiophores arising in various ways, but not inside pycnidia or acervuli.

Hyphopodium (Pl. Hyphopodia) Gr.: A thread like, more or less lobed, one to two ceiled structure on the superficial mycelium of Meliolales of Pyrenomycetes, which serves as an organ of attachment and absorption. They are of 2 types (i) capitate, (ii) mucronate.

Hypobasidium: Enlarged basal portion of a heterobasidium which may bear epibasidia.

Hypochytridiomycetes: A class of fungi under Mastigomycota division. It is a very small group of aquatic fungi having anteriorly uniflagellate, motile cells with tin-

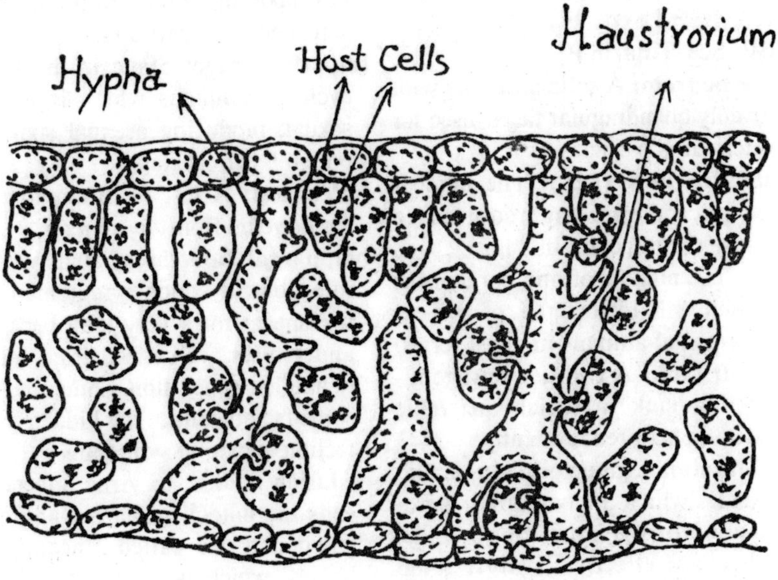

Hypha

sel type flagellum.

Hypodermis: Cellular layer below the cuticle or epidermis.

Hypogean: Organ /organism growing below the ground level.

Hypogeous Fungi: Fungi which produce their fruiting structures within the soil.

Hypoplasid: Underdevelopment of a tissue or plant due to decreased cell division.

Hypoplasia: Reduced rate of cell division resulting into the dwarfing which may be pathological in nature.

Hypopodium (Pl. Hypopodia): A small appendages on a hypha which is characteristic of Meliolales.

Hypostatic gene: A gene which can not express itself or is not effective in the presence of another gene called 'Epistatic gene'.

Hypothallus: A thin transparent deposit at the base of the fructification of some myxomycetes.

Hypothecium (Pl. Hypotheci): A thin layer of loosely interwoven hyphae in an apothecium just below the hymenium.

Hypotrophy: Underdevelopment of a tissue or plant due to abnormally reduced cell enlargement, which maybe pathological in nature.

Hystothecium: An elongated fruiting body of "Hysteriales" which remains closed at first but opens at maturity by a long slit following a line of dehiscence.

I

IBP: See "Kitazin P".

Icesahedrom: A solid structure with twenty quandragular faces, used for such virus particles.

Infective Dose (I.D.): The number of microorganisms required to cause the infection in a host.

ID_{50}: The number of microorganisms which will infect 50% of the experimental animals in a test series.

Imhoff tank: A two tier sewage digestion tank in which anaerobic bacteria act predominantly.

Imidazoles: A group of systemic fungicides which are basically heterocyclic nitrogenous compounds. They are effective against fungi which have become resistant to benzimidazoles and have wide spectrum. Two important compounds are imazalil and fenapanil. None of them are marketed in India.

Immigration: Movement of individuals into a population.

Immune : A host plant which by virtue of its inherited genetical qualities does not permit the establishment of a particular pathogen and remains free from that disease.

Immunization: Method of imparting immunity by injecting an immunogen (usually a weak strain or killed preparation of the same pathogen) into a susceptible host organism.

Immunogen: An agent capable of inducing immunity in a susceptible host upon injection into it e.g. inactivated virus particles.

Imperfect stage: Stage in the life cycle of a fungus which is non-sexual; producing asexual spores only (*cf.* perfect stage).

ImVic test: A test to differentiate *Escherichia* from *Aerobacter*.

Inapparent Infections: Such viral infections in which symptoms do not appear for a very long period and viruses are carried from one to another generation. Sometimes, some factors induce the hidden infection to exhibit symptoms.

Inclusion bodies: In virus infected cells the mitochondria got aggregated and are called 'Inclusion bodies' which is a typical symptom of virus infected plant cells.

Incompatible: Inability of the two partners to react and exhibit the desired results e.g.

(i) Ability of the two strains to participate in sexual reproduction and produce sexual spores.
(ii) Interaction of a host variety and pathogen strain to produce the disease symptoms.
(iii) Mixing of the two pesticidal chemicals to improve their efficacy without any harmful effect on their properties.

Incubation period: The time period between the penetration of a host by the pathogen and the first appearance of the disease symptom, on the host.

Incubator: Double walled, air tight chambers used for growing the microorganisms. The temperature of this instrument is automatically and continuously maintained at the desired level by means of an electrical thermostat so as to provide the optimum temperature for the growth of the particular organism.

Indar: See "Triazol".

Indehiscent: Usually referred for such fruiting body which do not have special mechanism for opening.

Indexing: A procedure of determining the presence of virus infection inside a given plant. In this method, bud, scion, sap etc. is applied from one plant to virus sensitive plant (indicator plants) and symptom appearance is observed.

Indicator plant: A plant exhibiting characteristic reaction when infected with a virus or exposed to a specific toxic chemical. It is used for the identification of these factors.

Indigenous: Native or of common occurrence in that particular area.

Indole-3-acetic acid (IAA): A growth regulating substance whose quantity is disturbed in diseased tissue producing various types of abnormalities. Also see Auxins.

Induced Mutation: Development of mutation in response to some artificial factors like radiations, mutagenic chemical etc. Mutation may be useful for disease resistance.

Induction: Conversion of temperate phage to lytic phage, one which may be spontaneous or induced.

Indusium (Pl. Inusia): A hanging skirt like structure from the recapticle of expanded fruiting body of Phallates. e.g. *Dictyophora*.

Inert: Materials having no chemical or biological action. Usually used for material added to the pesticides which themselves do not have any pesticidal activity.

Infect: Phenomenon of invading or attacking the host by the pathogen.

Infection: Establishment of the parasitic relationship between two organisms following the penetration (*cf.* disinfection).

Infection Court: The site of infection where first infection of the parasite takes place.

Infectious: A disease in which the casual agent may be transmitted from diseased to healthy plants, causing disease on it.

Infective: An organism which is capable of making infection.

Infestation: Introduction of the pathogen in the environment usually surface of the host (*cf.* disinfestation). Term commonly used for insects.

Inflamation: Reaction of any tissue due to the irritation by any foreign material, causing migration of leucocytes and increased flow of blood to the area, producing swelling, reddening, heat, pain and tenderness.

Inhibition: Prevention of the growth and multiplication of the microorganisms due to some factor.

Injector: An instrument used for introducing the pesticide below the soil surface or into the transport system of a tree.

Innate: Immersed.

Inner veil: The hyphal membrane covering the gill of a young mushroom (Agaricales).

Inoculate: Artificial introduction of microorganisms (usually pathogen) within or upon an organism (host) or a substratum or culture media.

Inoculating needle: An specialized needle with a long handle and a loop of wire at the end which is used for inoculating the cultures of fungi, bacteria etc. on host tissue or on culture media. It can be sterilized by heating over a flame.

Inoculum: (i) The portion of microbes which is used for inoculation.

(ii) The portion of the pathogen capable of dissimination and initiation of disease.

(iii) The portion of the microbial culture used for inoculating the fresh culture media.

Inoculum potential: The energy of growth or amount of the microorganism available to cause infection at the infection court.

Inoperculate: Asci or sporangia opening through a pore lacking any lid or operculum (*cf.* oporculate).

Integrated control: Combined use of biological, chemical and cultural control methods (compatible to each other) for the control of the pathogen. It is a very carefully reasoned use of several methods in conjugation with a minimum disturbance to the natural balance.

Intercalary: Usually referred to the position of the spores or other reproductive structures in the hypha; which are formed along and in the middle of the hypha and not at its tip. (*cf.* terminal).

Intercellular: The position of any object/organism etc. when found in between the cells i.e. in the space between cells.

Interference: Intervention of the successful viral infection due to the variety of reasons like failure of infecting particle to get absorbed to the specific receptor on the host cell wall.

Interferon: A substance (non-antigen portion of 63,000 mol. weight) produced in virus infected cells which moves to neigh-bouring cells and make them resistant against that virus.

Interferon: A specific proteinaceous substance which interfere with the process of multiplication of virus in host cell. They are host specific and interfere with the protein synthesis of viral particle.

Intracellular: Position of any object or organ inside a cell or through the cells.

Invasion: Infection and spread of pathogen through the host tissue after penetration.

In vitro: Growth or activity of any organism in culture or outside the host tissue.

In vivo: Growth or activity of any organism inside the host tissue, especially living tissue.

Ipomeamarone: An antifungal substance (phytoaleixin) produced in sweet protato tissue infected with black rot fungus *Ceratocystis fimbriata*.

Irradiation therapy: Irradiation with X-ray, gamma rays or ultra-violet rays sterilizes the planting material or area. It is used mainly for the working area of laboratory etc.

Isdia: Small coral like branched out-growth on the upper surface (cortex) of lichen thallus mainly containing algal component which serve as a mean of asexual reproduction. (*cf.* Soredia).

Isocoumarine: A phytoalexin isolated from carrot tissue inoculated with a non-pathogenic fungus *Ceratocystis fimbriata.*

Isoelectric precipitation: The precipitation from the suspension of a virus or a protein when the pH is at the isoelectric point i.e. when there is no positive or negative surface charge on the virus or the protein

Isogametangia: Morphologically similar gametangia usually of opposite sex.

Isogametes: Morphologically similar gametes of opposite sex.

Isogametes

Isogamy: Sexual union of morphologically identical gamates (*cf.* Heterogamy).

Isokont: Flagella or cilia of equal length.

Isolate: A single spore culture and subcultures derived from it. Also used to indicate collection of a pathogen made at different times from different places or different hosts.

Isolation: The separation of pathogen from its host on a nutrient culture media.

Isomorphic: Forms (especially zoospores/gametes) which are similar in morphology but not in size.

Isoplanogametes: Motile gametes of usually opposite sex which are identical morphologically but may fuse with each other for sexual reproduction.

Isoprothiolane: organo-sulphur, systemic fungicide having curative, protective and eradicant properties against blast as well as *Helminthosporium sigmoideum* and *Fusarium nivale* spots of rice. Chemically it is diisopropyl 1, 3, dithiolan-2-ylidenemalonate. Trade name is "Fuziwan 40 EC or 12 G".

Isthmus: Thickened medial perforated septum of a polarilocular ascospore.

Itaconic acid: An organic acid produced by *Aspergillus itaconicus* and *A. terreus* (commercial production) and used as copolymer with acrylic resins to bond printing ink.

J

Jack-in-the box: See bitunicate.

Jack-0' lantern mushroon: Common name of *Omphalotus olearius* fungus of Agaricales. Its basidiocarp have luminescent gills.

K

Kahn test: A flocculation test for the diagnosis of syphelis.

Kappa (Cuppa) Particle: Cytoplasmic particles which act for cytoplasmic inheritance in *Paramecium.* They are episomeic in origin.

Karathane: See Dinocap.

Karnal bunt: A disease of wheat seeds caused by *Neovassia indica* found in northern plains of India. Affected ears produce discoloured seeds filled with black powder but remains covered with seed coat.

Karyollagy: Fusion of sexually undifferentiated cells.

Karyochorisis: Somatic (mitotic) nuclear division taking place as a result of constriction of the nuclear membrane.

Karyogamy: Fusion of two nuclei usually of opposite sex or strain, during the course of sexual reproduction.

Kasunin: An antibiotic, very effective against the rice blast disease.

Kavach: See "Chlorothalonil".

Keratomycosis: Fungus infection of the cornea of eye.

Khaira disease: A mineral deficiency disease of rice caused due to the non-availability of Zinc.

Kinetins: A growth regulating substance (cytokinins) which supports the growth of rust culture on excised leaves.

Kinetosome: See "Blepharoplast".

Kitazin P: Systemic fungicide of organophosphorus group also called IBP. Chemically it is 0, 0-diisopropyl-S-benzyl-thiophorate, abbreviated as IBP. Available under trade name "Kitazin P as 48 EC, 17 G or 2% dust. It is effective against rice blast, stem rot (*Helminthosporium sigmoideum*) and sheath blight (*Pellicularia saskii*). Besides it also has insecticidal activity against rice leaf and plant hoppers.

Kledusity: Quality of a host variety to escape the infection under the conditions which are suitable for the establishment of the infection on other varieties.

Kline test: A microscopic test (using flocculation) for the diagnosis of Syphilis.

Koch's postulates: Four postulates given by Robert Koch which must be fulfilled for the pathogeneicity confirmation of any host-pathogen combination.

Koch's Sterilizer: An steam autoclave used to sterilize the media etc. under pressure and temperature of the steam.

Kojic acid: An acid produced by *Aspergillus flavus* group which gives characteristic red blood colour with aquous ferric chloride and is a useful test for the diagnosis of fungus.

Koleroga: A disease of Areca plants also called "Mahli" caused by *Phytophthora arecae*, exhibiting water soaking from the base of nut, development of fungus over it and utlimately shedding of nuts.

Kreb's cycle (Tricarboxilic acid Cycle): A cyclic seris of chemical reactions in the respiration, in which pyruvate is oxidised in several steps through several organic acids and energy is released either as ATP of $NADPH_2$.

L

Lacques: Pesticide incorporated into a varnish to achieve slow release for a long period of time.

Lactiferous hyphae: Hyphae having milky fluid inside it (*cf.* generative hyphae and skeletal hyphae).

Lactofuschin: A staining-cum-mounting medium used for preparing the temporary slide of fungi. Prepared by dissolving 0.1% acid fuschin in pure lactic acid. It rapidly stains even the spores and colour fades by exposure to strong light.

Lactophenol: It is a mounting fluid for preparing the slides of fungi for microscopical examination. It is prepared by mixing phenol (10 g); lactic acid (10 g), glycerol (20 g) and distilled water (10 ml). Its refractive index is 1.45.

Lactose: A disaccharide carbohydrate called 'milk sugar' which releases glucose and galactose upon hydrolysis.

Lamella (Pl. Lamellae): See Gill.

La pourriture noble: See Edelfaule.

Larva: The life stage of a nematode between the embryo and the adult.

Latency: A state of inapparent infection of a virus without expression of any symptom.

Late Proteins: Proteins synthesized during the later stages of the virus replication which are usually structure proteins of the Capsid component of the virus.

Latent infection: The state in which a host is infected with a pathogen but does not show any visible symptoms but may cause infection to a healthy host.

Latent period: (i) The penultimate stage in replication of bacteriophage. This period begins with the formation of the first new phage particle in the host cell and ends with the lysis of it.

(ii) Time required between the infection and the appearance of the first symptom.

(iii) The period required by a vector to become infective after acquiring the virus by feeding on an infected host.

Latent virus: A virus which does not produce any visible symptoms on its host.

LD (Lethal Dose): The number of microorganisms required to cause the death of an individual plant or animal species, after causing the infection.

LD_{50}: (i) Number of microorganisms that will kill 50% of the animal in a pathological test.

(ii) Lethal dose of toxicant required to kill 50 per cent of large group of individuals of one species.

Leaf rot of Pan: A pan (betlevine) disease showing the brown-black spots and rotting of leaves, caused by *Phytophthora parasitica* var. *piperina*.

Leaf spot: A self limiting lesion on the leaves (usually caused by a pathogen).

Leather fungi: Commonly used

term for members of Thelephoraceae.

lenticel: A structure of bark or some fruits which permits the inward or outward movement of gases, and also serves as a source of entry for plant pathogens.

Lesion: A localized area of discoloured diseased tissue formed due to the attack of pathogens.

Leucocidin: A substance which destroys lecuocytes.

Leucocyte: A type of white blood cells with beaded and elongated nucleus.

Leucocytosis: Increase in leucocytes in response to the injury or infection of a pathogen.

L-forms: A non-rigid stage of the bacteria without cell wall and defined morphology, can be propagated on the solid media and are also called "soft protoplasmic elements".

Lichen: A thallus made up of algal and fungal components having a symbiotic relationship. They form a separate group of organisms.

Lichen acids: Various organic compounds synthesized by lichens, which are useful in identification and classification of lichen in chemotaxonomy.

Life Cycle: The sequence of various stages (especially spore producing) following a particular phase or spore form which again reaches to the production of same spore form.

Lignification: Impregnation of cell wall with lignin, making it hard and woody.

Lignin: A complex organic substance or group of substances that impregnates the cell wall of xylem vessels and certain other plant cells. It is an amorphous, polymer formed from the condensation of some oxidation products of alcohols.

Liginase: An enzyme which breaks the lignin chain. Some of the pathogenic fungi are known to produce these enzymes for their penetration.

Lime sulphur: A mixture of calcium polysulphide and calcium thiosulphate, obtained by mixing varied proportions of rock lime and sulphur in water and boiling the mixture. It was widely used and is still used as a fungicide to control common fruit diseases and powdery mildews.

Line: (i) A homogenous strain produced by reproduction amongst themselves.

(ii) An isolate of the microorganism.

(iii) A further subdivision of a physiologic race of a pathogen.

Lipase: Group of enzymes which splits the fats.

Lipids: Substances consisting of glycerin and fatty acids and sometimes certain additional types of compounds.

Lipophilic: Substances which are easily soluble or become miscible with oil.

Lipovirus: Some mammalian viruses which are surrounded by membrane of lipids and proteins.

Loboulbeniales: Order of Loboulbeniomycetidae in Ascomycetes, which are specialized perithecial fungi, lacking a true mycelium and are obligate parasite of arthropods.

Localized infection: Infection restricted to a particular tissue or organ and does not spread throughout the entire body of host.

Local lesion: A localized spot produced on a leaf upon mechanical inoculation with a virus, which are used as a test for pathogenicity of many viruses.

Loculate: Usually referred to fruiting bodies etc. which have chambers or locules.

Locule: Cavity in the stroma containing asci and is not surrounded by the wall of archicarp or sexual structure origin.

Lomasome: A vesicular body of different forms found between cell wall and plasma membrane.

Long Cycled: Usually referred for rusts and smuts (uredinales) having the alternation of generation with all the five stages in the life cycle (*cf.* short cycled).

Loose smut of wheat: Internally seed borne disease of wheat caused by *Ustilago nuda tritici,* in which grain is converted into a mass of black powder which is a mass of fungal chlamydospores.

Lophotrichous: Bacteria having a bunch of flagella at one or both the ends.

Loranthus: An angiospermic parasite of trees (Mango, Ficus, Albizzia, Dalbergia etc.). Also known as "Banda" and "Giant Mistletoe".

Lower fungi: A loosely used term for members of Mastigomycotina and Zygomycotina or Phycomycetes of older classification.

Low volume Spray: A spray in which 200-600 l of solution is sprayed per ha of area for bushes and about 50-200 l per ha per ground crops.

Lumen: The cavity of the cell within the cell walls.

Lyases: Enzymes which split their substrate by eliminating a molecule of water from them. Many plant pathogen degrade component of cell walls (mainly pectin) through such enzymes.

Lycomarasmin: A pathotoxin produced by the wilt pathogen *Fusarium lycpoersici* which is specific to tomato plants.

Lyophilization: A long term preservation of microorganisms or tissues (especially the cultures) by quick freezing and dessication under high vaccum. It is also called freeze drying.

Lysin: An enzyme capable of disrupting the cells.

Lysis: Disruption of the cells (especially bacteria or erthrocytes) by the action of specific antibody plus compliment or virus attack or an enzyme.

Lysis: Destruction of the host cells during the multiplication and release of bacteriophage.

Lysogenic bacteria: Bacteria carrying the phage particle as a genetic material attached to their DNA (episome) during their temperate stage, which may cause the lysis of the bacterial cell under certain conditions and again start behaving like lytic types.

Lysogenic conversion: Change in the characteristics (especially the pathogenicity) of the bacterial cell due to the association of the phage DNA (pro-phage) with the bacterial chromosome.

Lysogeny: Phenomenon of association of the phage DNA with the host cell chromosome till it becomes virulent after separation causing lysis of the bacterial cell.

Lysozyme: An enzyme which hydrolyses the bacterial cell wall; usually produced during the release of the phase particles from the host cell after multiplication inside it.

M

Macerate: To soften or make loose. Referred to maceration of plant tissue as a result of fungal or bacterial attack in rotting diseases.

Macroconididum (Pl. Macroconidia): A condidum as distinguished from the microconidium on the basis of its size.

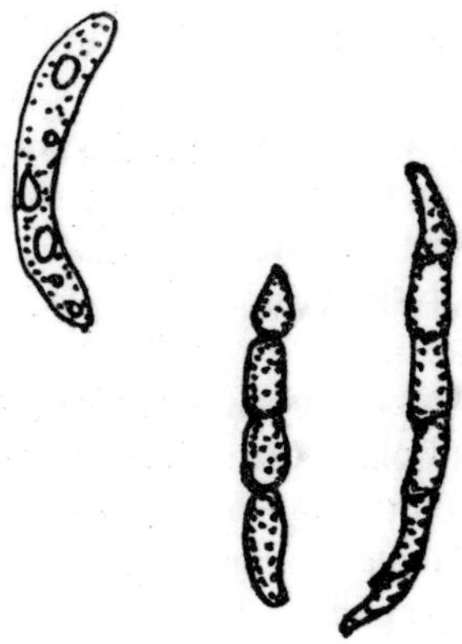

Macroconidia

Macrocyclic: Used for rusts which complete their life cycle in five different stages and produce binucleate spores other than teileospores. Also see long cycled.

Macrocyst: Protoplasmic, walled usually multinucleate portion of a myxomycetes sclerotium or walled cell masses of Acrasiales.

Macrofungi (Macromycetes): Fungi producing large fruiting bodies.

Macronematous: The stage when conidiogenous cell is specialized and can be differentiated from the rest of the mycelium, then it is called Macronematous conidiogenous cell.

Macroscopic: Visible without the aid of a magnifying lense or a microscope.

Magnification: The number of times the size of any object is increased by a microscope.

Malignant: A cell or tissue which divides and enlarges unlimitedly and its growth can not be controlled by the normal functioning of the organism on which it is growing.

Maltase: An enzyme which is capable of hydrolysing the maltose releasing glucose from it.

Mammiform: Morphologically looking like breast.

Mancozeb: A mixture of Zinc and Maneb which is widely used fungicide for the control of foliar diseases. It is marketed as "Dithane M-45", "Zebtane", "Zebra" etc.

Maneb: A widely used fungicide of dithiocarbamate group (Mangnous ethylene bis dithiocarbamate) used as a broad spectrum seed and foliage protectant. It is marketed with Zinc as Dithane M-45.

Marbled: Phenomenon of showing irregular strokes of dye in the staining.

Masked fungicide: Chemicals which is usually nontoxic to pathogen *in vitro* but on application to plants releases the molecules prefereably at infection site which are fungitoxic.

Masked symptoms: Symptoms usually caused by the virus infection but remain absent under certain environmental conditions and may appear when host plant is exposed to certain specific conditions of light and temperature.

Masked virus: Virus present in such a plant which itself does not show any symptom but transmits the virus into another susceptible host.

Mass median diameter (MMD): The figure dividing a total volume of spray into two equal parts, half the mass of spray having the droplets of smaller diameter than the MMD, half of droplets being larger diameter.

Mastigomycotina: A division of kingdom Myceteae (fungi) whose members typically produce flagellate cells.

Mastigonemes: Feather like projections on the main rachis of "Tinsel" type flagellum of fungal spore/ gametes. They are also called "flimmers".

Matrix: An undifferentiated mass in which an organ or the whole organism is embeded.

Maturation: Assembling of components (Nucleic acid and protein coat) to form a complete, mature phage particle.

Mazaedium (Pl. Mazaedia) Gr.: Stalked ascocarp of genus *Onygnea Trichocoma* and *Dendrosphaera* of Plectomycetidae, which on maturity produce a powdery mass of spores and sterile threads (capillitium) due to breaking of the peridium.

MBC: Abbreviated chemical name of a systemic fungicide, methyl-2 benzimidazole carbamate or "carbendazim".

Mebnil: Common name of systemic fungicide of carboxamide group which is chemically O-methyl benzanilide. It is effective against yellow rust of wheat and barley and brown rust of barley as spray and is also reported to be effective against *Rhizoctonia* and *Sclerotium rolfsii*.

Mecarbinzid: A systemic fungicide of benzimidazole group. Chemically it is Methyl-1-(methyl thioethyl carbamoyl) benzimidazole-2yl carbamate. Probably, on hydrolysis it produces MBC like benomyl.

Mechanical inocuation: Inoculation of plant with a virus through transfer of sap from virus infected plant to healthy plant.

Medicarpin: An antifungal phytoelexin accumulated in leaf diffusates of alfalfa plants inoculated with pathogen and non-pathogens.

Medium: A substance (mixture of a number of compounds) which provides nutrition for the growth and multiplication of microorganisms. They are used for artificially growing and maintaining the cultures of the microorganisms. They have a balanced chemical composition and may be used in liquid or solid states.

Medullery excipulum: The inner portion of the apothecium.

Meiocyte: Cell in which meiosis takes place.

Meiosis: A type of cell division occurring at different stages of life cycle of various organisms, especially in the sexual reproduction, in which a quick succession of two nuclear division occurs, one of which is reductional. Finally, four hapoid (n) nuclei with reduced (half) the number of chromosomes are produced.

lateron differentiate into specialised tissues. They are usually found at the tip of the young stem or root, or other apical tissues.

Meristogenous: Origin of fruiting body from the division of simple cell of adjacent cells of the same hypha.

Merogamy: Copulation between special sex cells.

Meront: One of the daughter

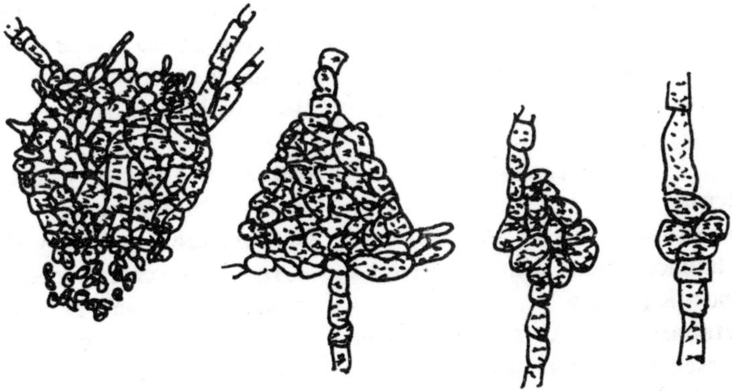

Meristogenous development

Meltatox: See "Dodemorph".

Memnospore: Spore which remain at its place of origin.

Mercury Fungicides: Fungicides having mercury compounds as an active ingredient. They may be inorganic (Mercuric chloride, Mercurous chloride or organic (Ethyl mercury chloride, Phenyl mercury acetate etc.). They are basically used for treating the seeds. It acts as protectant or eradicant.

Meristem: The undifferentiated tissue whose cells divide actively and

myxamoebae cut-off in succession by a parent myxamoeba.

Merosporangium (Pl. Merosporangia): A cylindrical sporangium or outgrowth from the swollen end of sporangiophore (in Mucorales) in which chain like series of sporangiospores are produced. eg. *Syncephalastrum*.

Mesophile: Bacteria or microbes which grow best at modrate temperature (25-40°C).

Mesophyll: Parenchymatous cells of the leaves between the two epi-

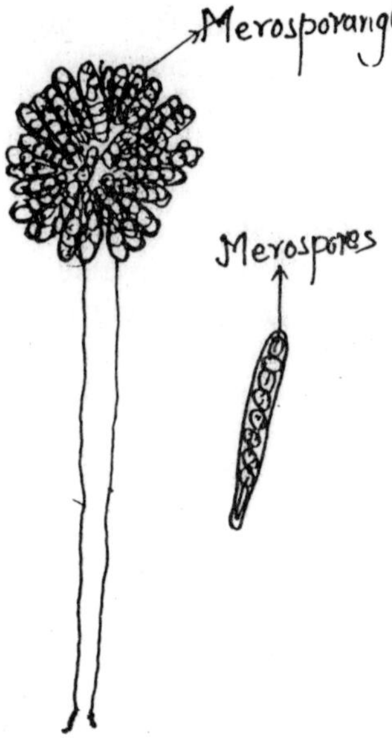

Merosporangia

dermis.

Messanger RNA: A chain of ribonuceliotides which codes for a specific protein.

Metabasidium: An alongate, transversely divided heterobasidium.

Metaboloism: The process by which cells or organisms utilize nutritive materials to build living material and structural component or breakdown of cellular material into simple substances to perform specific functions. All these processes involve intrique biochemical reactions.

Metabolite: Any chemical participating in the metabolism of an organism, usually referred to the chemical compounds produced as a result of the metabolic activity of an organism.

Metachromatic granules: A deeply staining body found in the protoplasm of some bacteria and yeast cells.

Metalaxyl: A systemic fungicide (common name) of acylanine group, quite effective against Oomycetous fungi, especially downy mildew and Phytophthora. Chemically it is methyl-DL-N (2, 6-dimethyl phenyl)-N-(2' methoxyacetyl)-alanin. It has both upward and downward movement in many plants. It is available as "Ridomil 25 WP", Aclan" and "Apron 35 SD" (seed dresser). In India, it is marketed in combination with mancozeb as "Ridomil MZ".

Metazoa: Animals whose bodies consists of many cells.

Methoxy ethyl mercury chloride: Organo mercurial seed treating fungicide available as 3 or 6% soluble form as Aretan, Agallol, Ceresan wet, Emisan etc.

Methyl bromide: A soil fumigating chemical active against nematodes and soil fungi as well.

Metula (Pl. Metulae): Apical part (branch) of the conidiophore (complex) which bears the philids (condidiogenous cells) e.g.; *Penicillium Aspergillus.*

Microaerophile: Microorganisms which grows best in the presence of small amount of atmospheric oxygen.

Microbiology: Study of small organisms, not visible with naked eye, specially bacteria with respect to their distribution, activities, relationships, their effects on plant, animal and human life etc.

Microconidium (Pl. Microconidia): A small conidium often acting as spermatium (cf. Macroconidium)

Microconidia

Microcop: Micronised tribasic copper sulphate (50% Cu)

Microcyclic: Short cycled, used for rusts producing only one type of binucleate spore (teleutospors) and does not have all the stages in its life cycle. (cf. Macrocyclic)

Microcyst: A small encysted protoplast usually encysted myxamoeba of myxomycetes.

Microfibrils: A bundle of cellulose molecule chains arranged in crystalline manner and is amorphous region of the cell wall.

Microfungi (Micromycetes): Fungi which produce small (usually microscopic) fruiting bodies.

Micromanipulator: A special equipment attached to a wide field microscope for picking up a single cell in the micropippeta from a hanging drop under the microscope.

Micron (μ) : A unit of the measurement of size of small microscopic objects. One μ is equal to 0.001 mm or 10^{-6} m.

Micronematous conidiogenous cell: Conidiogenous cell which is not different from rest of the vegetative mycelium.

Microscopic: Very minute in size and visibe only with the aid of a microscope.

Middle lamellae: A connecting layer between two adjacent cells, generally consisting of pectinaceous material, except in woody tissues where pectin is replaced by lignin. Enzymatic degradation of middle lamellae is an important step in the pathogenesis of many diseases involving rotting of tissues.

Migratory: Organisms which migrats from one place to another or from plant to plant.

Milcap: Combination of Ethirimol and Captafol.

Milcurb: See "Ethirimol" and "Dimethirimol".

Mildew: A fungal disease symtom of plants in which the mycelum and spores of the fungus are seen as a whitish growth on the host surface.

Milimeter (mm): A unit of length equal to 1/1000 (0.001) m or 1/10 of a centimeter; equal to 0.03937 of an inch.

Milo root rot: It is caused by a toxin produced by the fungus *(Periconia circinata)* growing in the root zone.

Milstem: See "Ethirimol".

Miltox: Trade name of a mixture of 20% zineb and 37% copper.

Minimum inhibitory concentration (m.i.c.) : Relative measure of fungitoxicity as the minimum concentration of fungicide at which no growth is obtained.

Mist blower: A sprayer producing a very fine air carried spray appearing just like the mist.

Mist spray: Dispersion of the droplats of 50-100 μ size.

Mist spraying : A method of spraying chemical of high concentration into a high velocity air stream under pressure, air acting as diluent and carrier. forming a mist of very fine droplets.

Mites: A small pest attacking and contaminating the fungal culture in the laboratory.

Miticide: Sen "Acaricides".

Mitic system: A system of taxonomic study of some basidiomycetous fungi based on the microscopic analysis of hypha types (generative, skeletal and binding) comprising the basidiocarp. There are mainly 3 types of basidiocarp, viz. "monomitic,", "dimitic" and "trimitic".

Mitochondria: A small, cytoplasmic cell organelle associated with the respiratory process of the cell.

Mitotic crossing over: Exchange of chromosome segment (genetic material) during the haploidization.

Molar solution: Solution or suspension containing one molecular weight of the solute in one litre of solution. It is abbreviated as 1 M solution.

Mold: A downy, profuse, wooly fungal growth on a damp, decaying substratum usually consisting of hyphomycete mycelium.

Molye disease: A disease of wheat and barley caused by nematode, Heterodera avenae, showing dwarfed, pale yellow plants with matted root system having white cyst attached to it.

Monera: See Prokaryotic organisms

Moniliaceous: Hyaline or brightly coloured fungi

Monoblastic: Blastic conidia produced only at one point on the canidiogenous cell.

Monocentric: A thallus radiating from a single point at which reproductive organs is formed or a rhizoidle system bearing only one reproductive structure (c.f. Polycentric)

Monoclinous: Form or state where antheridium is produced on the oogonial stalk.

Monoecious: Organisms having male and female reproductive organs on the same thallus. (cf. dioecious).

Monokaryon: A cell or hypha having genetically similar haploid (n) nuclei (cf. dikaryon, heterokaryon)

Monokaryotic: Condition of being monokaryan, having only one nucleus.

Monolayer: Uniformly thick, often single layered culture of cells grown in the laboratory out of individual clones.

Monomer: Simplest chamical compound or refer to simplest protein units of virus coat.

Monomitic: A condition when basidiocarp is made up of only one type of hyphae (cf. dimitic, trimitic).

Monomorphic: Organisms producing only one type of zeespheres.

Mononematous: Condition when conidiophores are produce solitary or in tufts of loose fascicles.

Monophagus: A pest restricted on a single species of the host plant.

Monophyletic: Referred to development from a single line of descent during the course of the evolution.

Monoplanetic: Having only single type of zoospores in the life period, e.g. Domycetes (cf. Diplenetism)

Monoplanetism: Formation of only one type of zooapores in its life period.

Monopodial: A type of branching in which branches come out only in one direction and one branch from the axis.

Monotrichous : Bacteria having a single flagellum.

Morels: Some of the edible of fungi of Morchallaceal eq. *Morchella.* They are also called Spong Mushrooms.

Morestan: See "Chinomethionat",

Morocide: See "Binapacryl".

Morpholines: A group of systemic fungicides having heterocyclic ring compounds. They have eradicant action especially against powdery mildews. Two important compounds are tridemorph and dodemorph.

Mosaic: A symptem of certain virus diseases of plants characterized by intermingled patches of normal and light green or yellowish colour.

Mottle: A type of specific arrangement of spots or colour blotches, being systems of virus diseases.

Mottle: An irregular pattern of indistinct light and dark area.

Mottling: A common symptom of virus diseases on plants showing a general chlorosis.

Mould: Common term used for the microfungi exhibiting well marked mycelial growth/spore mass, especially, economically important saprophyte.

Moult: The shedding or casting-off of the cuticle.

Mμ (milimicron): A unit of length equal to 0.001 micron or 10^{-9} meter; also called nanomster (nm).

Mucormycesis: A disease of animal or human beings caused by a number of fungi belonging to order mucorales.

Mucronate: Pointed, ending into a sharp point.

Mucronate Hyphopodia: Somewhat bottle shaped, erect hyphal structure of Meliolales found on host surface. Their neck is formed at advanced stage and they look like phialides (*cf.* capitate, hyphodia).

Multiline: A crop consisting of several genetically distinct groups or lines of a cultivar (composite cultivars) grown to achieve resistance against several races of the pathogen.

Multimeric: Protein units formed out of several monomers and themselves comprise of the capsomere.

Mummy: A dried, shriveled fruit.

Murogenous: Type of development in which expansion of conidiophere tip results into formation of condium (*cf.* blastogenous, phialogenous, porogenous).

Mushroom: A groups of basidiomyceteous fungi with large fruiting body having a stipe, pileus and basidium produced on upside down hanging gills. Usually, referred to fleshy, edible variety.

Mutant: An individual having new genetic characters as a result of mutation and consequently altered morphology/physiology.

Mutation: Accidental but stable change in the chromosome (genetic material) due to some external factors resulting into sudden appearance of a new character in an individual. The mutants of fungi, bacteria and yeasts are very useful in industrial production of fermented products.

Mycangium: A sac or cup shaped fungal repository (accumulation) of ectodermal origin located in or on an ambrosia beetle.

Mycelium (Pl. Mycelia): A Mass of interwoven hyphae comprising of a thallus.

Mycetangium (Pl. Mycetangia): A specialized pouch in some beetles which harbours their fungal symbionts.

Myceteae: A separate Kingdom consisting of fungi. It has recently been created, separating the fungi from Kingdom Plantae.

Mycetismus: Phenomenon of poisoning as a result of eating mushrooms.

Mycetozoa (sing. Mycetozoon) Gr.: Term used by de Bary for members of present day Myxomycetes (slime molds). De Bary believed that they originated independently from bacteria and treated them as a separate group. This name has been followed by Lister, Bessey, Kudo and Olive in their classification.

Mycology: Gr. mykes = mushroom + logos = discourse). Science dealing with the study of fungi.

Mycobient: The fungus component of a lichen thallus

Mycophagy: (i) Phenomenon of eating of mushrooms.

(ii) Lysis of fungal cells by a phage (virus)

Mycoplasma: A new type of organism somewhat in between virus and bacteria having a unit membrane, lacking a definite cell wall and are highly pleomorphic, usually cured by tetracyclines also called PPLO (Pleuropneumonia like Organism). They are responsible for yellow type diseases.

Mycorrhize: A symbiotic relationship between fungus and roots of the higher plants which may be of following type.

Ectotrophic-Fungus on the surface of roots.

Endotrophic-Fungus inside the

root tissues.

Pseudotrophic-Fungus becoming parasitic.

Tolypophagus-Fungus is killed and digested by the host.

Mycosis: A disease of animals caused by the fungi

Mycostatin: See "Nystatin".

Mycotoxicoses: Poisoning of animals/man due to feeding the food products having mycotoxins of fungal origin.

Mycotoxins: Toxic metabolite of fungal origin which have lethal effects on human and animal health.

Myxamoeba (Pl. Myxamoebae): An amoeboid cell particularly one of the myxomycetes.

Myxomycetes: A class of subdivision Plasmodiogymnomy cotina and division Gymnomycotina, whose members are flagellate swarm cells (myxamoeba) which fuse to form the zygote which develops into plasmodia, exhibiting reversible streaming. It results into various types of sporophores.

N

Nabam: A fungicide of dithiocarbamate group (Disodium ethylene bis dithiocarbamate) effective against foliar diseases and seedling infection of *Pythium, Rhizoctonia* etc.; marketted as Dithane D-14 and Dithane A-40.

NAD (Nicotinamide adenine dinucleotide): The oxidized form of an enzyme involved in various steps of respiration. $NADH_2$ is the reduced form of this coenzyme. Formerly it was known as DPN (Diphospho pyridine nucleotide).

NADP (Nicotinamide adenine dinucleotide phosphate): A coenzyme involved in the oxidation-reduction reaction of respiration. Its reduced form is $NADPH_2$. Previously it was known as TPN (Triphospho pyridine nucleotid).

Nanism: Dwarfism or phenomenon of being relatively small in size.

Nasee apicale (apical basket): A specialized structure found in the bitunicate asci of Loculoascomycetidae, which is an important character of the subclass.

Natality: The birth rate of any organism.

Necrophyte: Organisms which are capable of living on dead material (*cf.* perthophyte, saparophyte).

Necrosis: Death of cells or tissues mainly due to some pathological reasons.

Necrotic: Area or tissues showing the symptom of necrosis.

Necrotrophic parasite: A fungus parasitizing on another fungus and feeding on the dead cells of the host after killing it (*cf.* biotrophic parasite).

Necrotrophs: Parasties deriving energy from dead cells of the host (cf. autotroph, heterotroph).

Nectarthode: An opening at the base of flower from which nectar exudes out.

Needle cast: Defoliation of leaves (in conifers) caused by Phacidales.

Negative staining: A technique used for bacterial staining to avoid distortion from physical and chemical factors. In this bacterial suspension is mixed with India Ink or nigrosine before smear preparation. Dried smear is directly observed under the microscope showing transparent bacteria against a dark background.

Negri bodies: Minite crystalline structures (inclusion bodies) found in certain brain cells of animal infested with rabies virus, named after its discoverer-P. Negri.

Nematicide: Toxicant (chemical) effective in controlling nematodes by killing or inhibiting them.

Nematode: Generally microscopic, worm like animals that live saprophytically in water or soil or as parasite of plants and animals causing various diseases.

Neoplastic growth: Tumerous growth arising out of abnormal localised multiplication of certain types of cells, malignant; if the growing cells infilterate the surrounding tissue and are carried by blood or lymph to other location in the body, growing there more vigrously, otherwise benign, i.e. localized at a particular place only and less harmful.

Neotype: A specimen deposited in the herbaria in place of the original material of description (holotype) which would have been lost/damaged.

Neutralization: A serological test in which a virus suspension is neutralized by specific antibodies and loses its infectivity.

Nitrification: Conversion of ammonia to nitrates form of nitrogen especially by microorganisms.

Nitrogen fixation: The formation of nitrogen compounds from the atmospheric nitrogen (especially by the symbiotic bacteria in leguminous crops).

Noble rot: See Edel foulle.

Nomenclature: Method of naming a new fungi or organism which is governed by uniform International rules.

Non-infectious disease: A disease caused by factors other than a living pathogen like environmental factors, nutrition etc.

Non-parasitic diseases: See Non-infectious diseases.

Non-persistant virus: Viruses which can not remain infective for a longer period inside their vector (an insect).

Non-persistant: Chemicals which remain active for only short periods after application.

Nonsexual reproduction: Production of the new progeny without any sexual cycle or formation of sexual spores.

Nova Plantarum Genera: A classic book having description of plants and fungi, written by Italian scientist, Pier Antonio Michell in 1729.

Nozzle: A part of the sprayer which converts liquid into fine droplets forming various type of shape of the spray viz., air blast, a vivil, cone or swirl, hollow cone, fan, deflector, etc.

Nuclear Cap: A body containing a large quantity of RNA at the side of the nucleus in a zoospore or gamets (usually in Blastocladiales).

Nucleases: A type of hydrolytic enzymes degrading nucleic acids (DNA or RNA) into their components.

Nucleic acid: An acidic substance containing pentose sugar, phosphorus and purine or pyrimidine bases. It determine the genetic properties of an organism.

Nucleoid: The central nucleic acid portion of a virus particle also called 'Core'.

Nucleolus: A dense Protoplasmic body within the nucleus.

Nucleoprotein: Usually referred to viruses madeup of nucleic acid and a protein coat.

Nucleoside: Molecule of purine/pyrimidine base attached to a pentose sugar in nucleic acid.

Nucleotide: The phosphoric acid

ester of a nucleoside which form DNA and RNA.

Nucleus: The dense protoplasmic body essentially found in all cells and governs all the activities of cell or organ through its genetic material, in the form of chromosomes.

Numerical aperture (NA): Characteristic of objective lens of a microscope which is the product of refrective index of the medium between object and objective and the sine of half of the angle formed by the rays of light entering the front lens from a point on the object.

Nurelle: See "Pyroxychlor".

Nutriocyte: The inflated portion of the ascogonium of Pericystis finally developing into spore cyst.

Nyctotemperature: Night temperature.

Nysatin: An antibiotic also known as "Mycostatin" of "Fungicidin" produced by *Streptomyces neuresi* and used for clinical use and also for anthracnose of beans and downy mildews of Cucumber etc.

O

Obligate Parasite: A parasitic organism incapable of living as a saprophyte and lives only as a parasite (*cf.* facultative parasite).

Obligate saprophyte: An organism living strictly on dead organic matter and is incapable of infecting a living organism.

Occult Virus: Hidden virus, causing infection without expressing any symptom on the host, remaining latent.

Ocellus (Pl. Ocelli): An eye spot functioning as a lens and concentrating the light rays on a photosensitive spot.

Oidiophore: A hyphae which fragments into oidia from the tip towards base.

Oidium (Pl. Oidia): A thin walled cell formed by the breaking of a hypha. They function as a non-sexual spore and are usually produced in long chain e.g. members of Erysiphales.

Oidization: The fusion of an oidium with a somatic hypha, resulting in the dikaryotization.

Oil immersion objective: An objective lense of microscope which functions very close to the object with a medium of transparent oil (having refrective index value of more than one) between the object and lens thus increasing the numerical aperture of the lense and resolving power of the microscope.

Okazaki fragments: Segments of DNA that combine with one another to form a DNA molecule during chromosomal duplication.

Oidium

Oligodynamic action: Ability of certain metals like Silver, Copper etc. to exhibit a lethal or inhibitory effect upon bacteria at a very low doses.

OMC: Organo-mercurial compounds being widely used as fungicide for treating the seeds to protect against diseases. e.g. ethyl mercury chloride, phenyl mercury acetate, methoxyethyl mercury chloride.

Omnivorous: Organisms which can grow on any type of substrate and may parasitize a large number of plant species; also called polyphagus.

Oncogenic: Agents which are capable of inducing cancerous growth.

Oogamous: A type of fertilization in which two heterogametes come in contact and content of male gametes are transferred to female gametes through a fertilization tube.

Oogamy: Union (plasmogamy) of a large non-motile female gamete (egg) with a motile male gamete or with the cytoplasm of antheridium.

Oogonium (Pl. Oogonia): A female gametanium containing one or more female gametes (eggs).

Oomycetes: A class of fungi under division Mastigomycota. Its members have biflagellate, usually reniform zoospores with two (one whiplash and one tinsel) flagella placed laterally. They have oogamous sexual reproduction resulting in the formation of oospores.

Ooplasm: Protoplasm in the centre of oogonium which becomes oosphere.

Ooplast: A large membrane bound, cellular inclusion present in the oospores of Saprolegniaceae.

Oosphere: A large sphere of protoplasm functioning as egg or female gamete in the oogonium.

Oospore: A fertilized egg (or sometimes developed parthenogenetically) which develops a thick wall around it and functions as resting spore.

Operculate: An ascus or sporangium bearing an operculum or lid at its pore. (*cf.* inoperculate).

Operculum: A flap or lid like structure covering the opening of an ascus or sporangium.

Opsonin: A substance in the blood serum that renders microorganisms susceptible to the injestion by the phagocytes.

Orchinol: An antifungal substance probably phytoalexin, produced in *Orchis militaris* tissues infected with *Rhizoctonia repens*.

Organic amendments: Addition of certain organic manures like oilcakes of neem, karanj, mahua, castor etc. in the fields to control the soil borne pathogens, especially nematodes.

Organic mulching: Covering the top soil of crop fields with organic material of non-host origin, which helps in decreasing the intensity of soil borne diseases, especially nematodes. Pine needles, waste paper, wheat straw, paddy husk are common material used for mulching.

Organophosphorus compounds: A group of fungicides (mostly systemic) which includes triamiphos, pyrazophos, kitazin etc.

Organo-tin compounds: Organic compounds acting as fungicides in which a hydrocarbon (alkyl, aralkyl or aryl) group is attached to the metal atom of tin by means of carbon atom. Three main compounds of this group are duter, brestan and brestanol. Use of these compounds is very limited due to their phytotoxicity on many hosts.

Orobanche: An angiospermic root parasite of many crop plants, especially tobacco, mustard, brinjal, tomato, radish, potato etc. Roots of parasite penetrate the host roots and draw nutrition. Also known as "broom rape".

Osmophilic: Species which can grow at a highly concentrated solution.

Osmosis: Diffusion of solvent through a semi-permeable membrane due to the osmotic pressure.

Osmotrophic: Forms taking their nutrition through absorption, as found in fungi.

Ostiole: A tubular passage terminating in a pore which extends through the neck of a perithecium, or pycnidia or other such fructification through which spores are released.

Ovary: The female reproductive part of higher plant within flower, which contains egg.

Oxathiins: A group of chemical compounds which act as systemic fungicides. Vitavax and Plantvax are commonly used oxathiin compounds. Also See "Carboxamides".

Oxidation: Chemical reaction in which oxygen combines with another substance or hydrogen atoms or electrons are removed from a substance.

Oxidative phosphorylation: The utilization of energy released by the oxidative reactions of the respiration to form energy rich ATP bonds.

Oxycarboxin: Systemic fungicide of oxathiin group, widely used for the control of rusts, smuts etc. Chemically it is 2, 3-dithydro-5-carboxanilido-6-methyl-1, 4-oxathiin-4, 4-dioxide, abbreviated as DCMOD. It is available under trade name of "Plantvax 75 wp".

Oxytetracycline: Antibiotic effective against mycoplasma, bacteria and few fungi. Chemically, it is 4-(dimethylamin), 1, 4, 4a, 5, 5a, 6, 11, 12-oxytahydro- 3, 5, 6, 10, 12-12a, hexahydroxy-6-methyl, 1, 11 dioxo-2-naphthacenecarboxamide. It is available as "terramycin".

Oxythioquinox: See "Chinomethionat".

P

Paddy straw mushroom: Common name of edible macrofungus *Volvariella vovacea* of Basidiomycetes.

Paint spoiling fungi: Many fungi like *Aureobasidium pullulans Phoma herbarium* etc. attack the paint film and degrade it.

Palisade fungi: A common name used for basidiomycetous fungi.

PAN (Peroxy-acyl-nitrates): A chemical gas which is an air pollutant, produced as a by-product in the exhaust of internal combustion engines and is injurious to plants.

Pandemic: Referred to the diseases or pests found all over the world or world wide epidemic.

Pan-tropical: a species occurring widely throughout the tropics and subtropics.

Papilla (Pl. Pappillae): A nipple-like extension or bulging usually found on a zoosporangium, e.g. *Pythium Phytophthora.*

Paragynous: When antheridium is situated on the side of oogonium.

Paraphysis (Pl. Paraphyses): Elongated sterile cells found in between the asci in the hymenium. They grow upward and have free apicies.

Parasexual cycle: A sequence of heterokaryotization, diploidization and haploidization (or stages of plasmogamy, karyogamy and meiosis) entailing into new recombinants. Unlike sexual cycle this may occur at any point or throughout the life cycle, but not at specific points.

Parasite: An organism deriving food from another living organism usually (not necessarily) causing harm to its host. Also called "Biotrophs".

Parthenogenesis: Reproductive development of female gamete without fertilization into a parthenospore.

Parthenosome: Cap shaped, perforated, double membrane structure present on both the sides of a dolipre septum found in hyphae of basidiomycetes. Also called "pore cap".

Partial veil: A veil in some Hymenomycetes extending from the stipe to the margins of the pileus.

Part per million (ppm): A measurement of concentration or ratio of toxic chemical to liquid or solid. One ppm is 1 mg in 1 litre of water.

Passive immunity: Type of resistance against the pathogen in which the activity of the host do not participate but the protection is provided through the antibodies already injected in the system (blood or serum) of the host (*cf.* Active immunity).

Pasteur effect: A metabolic respiratory phenomenon in living organisms especially microbes in which the fermentation is suppressed or inhibited and respiration is increased in the presence of free oxygen or under aerobic conditions.

Pasteurization: Process of killing the microorganisms in a product commonly milk by heating at controlled temperature at which nature of the substance is not affected.

Pathodeme: Population of a host in which all individuals have a particular character of resistance in common.

Pathogen: A parasite or other factor causing disease in host.

Pathogenicity: Ability to produce a disease or capable of being a successful pathogen.

Pathogenesis: Sequence of events in the disease development from the initial contact between a pathogen and its host to the completion of the syndrome.

Pathotype: Further division of species usually on the basis of pathogenicity, in relation to host range. 'Pathover' is a similar term used in bacteriology.

Pathovar: See Pathotype.

PCNB: See "Penta chloro nitro benzene".

Pectic substances: See 'Pectin".

Pectin: A methylated, linear polymer of galacturonic acid linked through alfa 1-4 glycosidic linkages found in the middle lamellae and primary cell wall of plant cells. During the pathogenesis of many diseases this component is attacked by the pathogen and is degraded for penetration and further nutrition.

Pectinase: A group of enzymes which attack the pectin chain and breaks it during the pathogenesis of various plant diseases.

Pectin-methyl esterase: A demethylating enzyme attacking the native pectin of cell wall, removing the methyl group without altering the chain length which makes it susceptible to other chain splitting enzymes.

Pectin trans-eliminase (Pectin lyase): An enzyme attacking the glycosidic linkages of native pectin (polymer of methylated galacturonic acids) and breaking the chain by eliminating one molecule of water (trans-eliminatively).

Pectolytic enzymes: A group of enzymes produced by fungi and bacteria (especially pathogenic forms) which attack the pectin chain of the middle lammellae and cell wall and cause degradation of the tissue. These enzymes are also produced commercially for fruit processing and clearing of fruit juice.

Pedogamy: Pseudomixis of mature and immature cells. e.g. copulation between a yeast mother cell and its daughter cell (bud).

Pelleting: Making a coating of inert material around seed at the time of treating with fungicides for ensuring the uniform size and shape.

Pellicle: A film formed over the surface of the culture meida as a result of microbial growth. This film is the aggregation of bacteria or yeast on the surface of liquid media.

Penetrant: Oil or any material

added to the spray solution to enable it to penetrate the waxy cuticle of the plant surface.

Penetration: (i) The process of enterance of the pathogen inside the host tissue involving a sequence of physical and bio-chemical interactions.
(ii) The process of injection of phage particle's nucleic acid into the host bacterial cell.

Penicillin: The first discovered antibiotic having a wide use in medicinal world, it is produced by *Penicillium notatum (P. chrysogenum)* and also by certain species of *Cephalosporium* and *Aspergillus.*

Penicillus: A broom-like arrangement of phialides, metulae and branches at the apex of stipe as found in the conidiophores of *Penicillium* sp.

Pantaene G-8: An antifungal antibiotic isolated from the mycelium of *Streptomyces anandii*; it is effective against *Colletotrichum, Helminthosporium* and *Puccinia* sp.

Penta chloro nitro benzene (PCNB): A fungicidal chemical widely used as seed and soil treating agent. Also known as Quintozene, Brassicol and Terrachlor. Now, it is not in use in India.

Pentamer: Capsomere of the virus protein coat formed of five monomer units.

Pentose: Sugars with five carbon atoms, e.g. ribose.

Pentose pathway: Oxidation of glucose by elimination of one cabron atom as CO_2 and formation of five carbon sugar (pentose).

Peplomere: Unit of envelope that covers some virus particiles and which is usually a lipid membrane.

Peptidase: An enzyme attacking the peptide chain and releasing the amino acids.

Peptide: A long chain of several amino acid.

Peptone: Partially hydrolysed protein as nutrient in the culture media.

Peptonization: Conversion of proteins into peptones, e.g. solubilization of casein in curd by proteolytic enzymes.

Perennial mycelium: Mycelium overwintering or oversummering as such in a host plant.

Perfect stage: The portion of life cycle producing sexual spores like, ascospore, basidiospore (*cf.* imperfect stage). See "Teleomorph".

Periconin: A host specific toxin produced by *Periconia circinata* which increases respiration and disrupts the membrane functions.

Peridermium: The form genus of the aecial stage of rusts in which aecial lip is large and conspicuous, generally white and appears as thin structure (*cf.* Caeoma, Aecidium and Rosteline).

Peridiole: In Nidulariales, a seggregated part of gleba surrounded by a wall and having basidiospores, functions as a unit of dispersal and is called peridiole.

Peridium (Pl. Peridia): The outer-

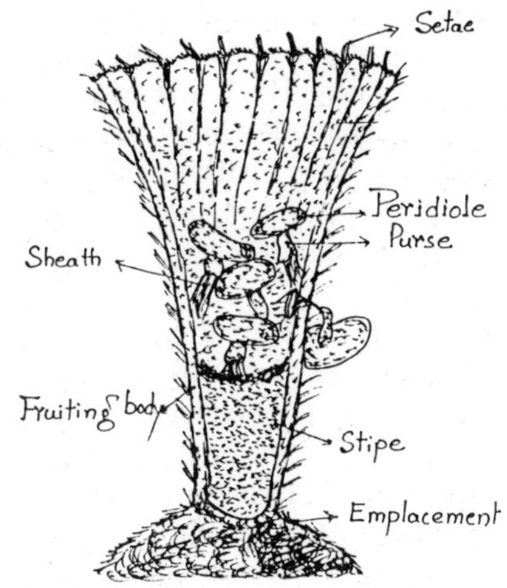

Peridiole

most membranous covering of a fructification.

Periphysis (Pl. Periphyses): Short hair-like filaments found in the osticle of some pyrenomycetes.

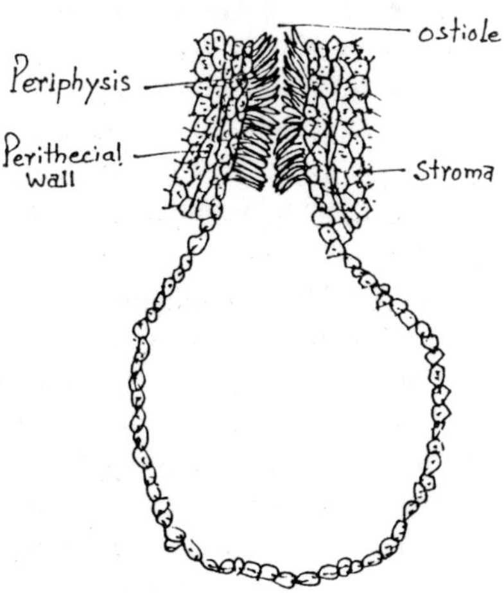

Periphysis

Periphysoids: Lateral periphyses originating all along the inner wall of an ascocarp and curve upward towards apex.

Periplams: In oomycetes the outer layer of protoplasm surrounding the oosphere within the oogonium.

Perispore: A membranous covering around the spore not permanently attached (*cf.* epispore, endospore).

Perithecium (Pl. Perithecia): A closed ascocarp having a regular opening. Its wall is developed from archicarp or other sexual structures.

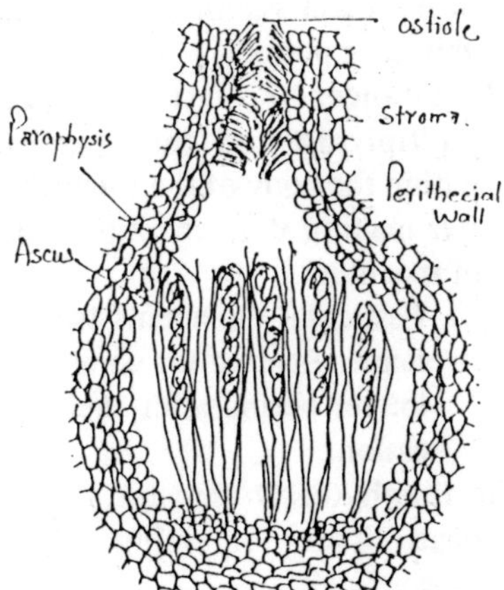

Perithecia

Peritrichous: Having flagella around the entire surface-referred to certain category of bacteria having such morphology.

Permeability: Ability of the biological membrane of being highly selective in permiting the diffusion of chemical substances through them. During the course of disease development the permeability of the membrane is affected leading to manifestation of various disease symptoms.

Peronox: Trade name for fungicide, cuprous oxide, available as 50 WP or 4-6% dust.

Peros: Through the mouth.

Perrennate: Survival of pathogen from one crop season to the other usually under unfavourable conditions.

Persistant virus: Virus which can remain in the infective state within its insect vector for a considerably long period.

Persistance: (i) Virus capable of multiplying and thus persisting within the body of the infected carrier making it an indefinite source of inoculum.

(ii) Pesticidal chemicals which remain effective for considerably longer period after their application.

Perthophyte: Organisms living on dead tissues of living host (*cf.* saprophyte).

Perthotrophs: Organisms which kill the host tissues in advance of penetration and then live saprophytically e.g. *Sclerotium rolfsii*.

Petri dish: A circular, flat dish of glass with vertical walls and a lid of same shape, used for growing the microorganisms in the culture. It was named after a German scientist R.J. Petri.

pH: A unit of acidity or alkalinity in a solution which is the log of 1/H+ where H+ is the hydrogen ion concentration in the solution.

Phage: A virus attacking other microbes and destroying or eating them e.g. bacteriophage, cyanophage.

Phagocyte: A special type of cells in the blood which has the ability to injest microorganisms or any other foreign material entering in a amoeboid fashion.

Phallotoxin: A group of toxins found in poisonous mushrooms, especially *Amanita phalloides.* Chemically they are cyclopeptides and principal toxin is "phalloidin". (*cf.* amatoxin).

Phaltan: Trade name of Folpet 75 WP. Also see "Folpet".

Phaneroplasmodium: A plasmodium made up of a well defferentiated advancing fan and conspicuous thick stands with well differentiated ecto-and endoplasmic regions in which the protoplasm is coarsely granular. e.g. Physarales. (*cf.* "Aphanoplasmodium", "Protoplasmodium").

Phase contrast microscopy: A special technique of studying the microorganisms under the microscope in which special phase contrast objectives and condenser assembly is used to provide controlled illumination on the specimen. Light rays are refracted from their original path according to the refractive index of the material through which they pass and thus such objectives can be distinctly viewed.

Phaseollin: A phytoalexin (antifungal substance) produced in detached open bean pods following inoculation with non-pathogens like *Monilinia fructicola*.

Phasmid: One of the pair of lateral caudal swellings of an nematode believed to be chemoreceptive.

Phallogen: The cambium tissue which give rise to cork and in some plants internally to the phelloderm. Also termed as Cork cambium.

Phenol Coefficient: the ratio between the highest dilutions of a test chemical and phenol which is able to kill a test organism (especially pathogenic microorganism).

Phenols: A group of chemicals compounds with one or more aromatic rings with hydroxyl groups which are supposed to be responsible for resistance in plants.

Phenotype: Visible characters (morphological) of an individuals which are usually controlled by a set of genes called genotype.

Phenotypic mixing: A process during which phage of two different types (usually close mutants) infect the same bacterial cell and induce the host to produce their components. During the assembly of phage particles some components of one type get assembled with other types thus producing a different type of phage particles.

Phenyl mercury acetate: Organo mercurial seed treating fungicide, available as 1% dry powder as "Ceresan dry".

Pheromone: Substance secreted by an individual which induces specific reaction in other individuals (especially of opposite sex) of the same species.

Phialide: A flask shaped terminal cell of conidiophore bearing conidia on or in it. e.g. *Aspergillus*.

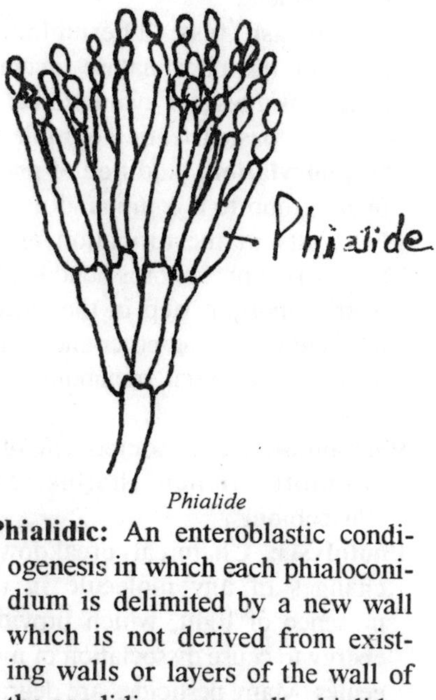

Phialide

Phialidic: An enteroblastic conidiogenesis in which each phialoconidium is delimited by a new wall which is not derived from existing walls or layers of the wall of the condidiogenous cell, phialide.

Phialogenous: A type of conidial development in which conidium is formed from a phialide (*cf.* blastogenous, murogenous, porogenous). Also called "Phialidic".

Phialospore: A spore produced from the phialide.

Phloem: Food conducting tissue of higher plants which are made up of sieve tubes, companion cells, phloem parenchyma and fibres.

Phorophyte: Host tree of an epiphyte.

Phosphatase: An enzyme liberating inorganic phosphorus from the organic compounds. During the pathogenesis of many diseases the alteration in the activity of this enzyme plays a vital role in disease development.

Phosphatase test: Thermolabile point of the phosphatase enzyme is used as a test for evaluation efficiency of pasteurization of milk.

Phosphorylation: Addition of phosphate group to a compound usually results in the formation of an energy rich phosphorus bond which is an important step in the oxidation, reduction and other energy reaction of the vital metabolic processes.

Photobiont: Algal component of a symbiotic lichen thallus. See "Phycobiont".

Photolysis: Chemical breakdown/ changes of any molecule in the presence of light, which provides energy to cause dissociation of molecules. Many pesticides are degraded in strong sun light.

Photoperiod: Optimum duration of the light and dark cycle required for the normal growth and reproduction of a plant in nature.

Photoreactivation: Celles of the microorganisms inactivated due to the exposure to ultraviolet light radiation may sometime recover on immediate exposure to visible light and this phenomenon is called Photoreactivation.

Photosysthesis: A vital metabolic process of the higher plant in which CO_2 and water combines in the presence of sun light and chlorophyll to form carbohydrate. This vital activity is also sometimes affected in the diseased plants due to various reasons and in various ways.

Phototrophs: Microorganisms (usually bacteria) which can utilize light energy for their metabolism.

Phototropism: Growth or bending of any organism in response to the stimulus of light.

Phragmospore: Spores with two or more cross septa eg. *Helminthosporium*) (*cf.* amerospore, didymo, didyscoleco, helico and staurospore).

Phragmospore

Phycobiont: The algal component of the lichen.

Phycomycetes: A group of lower fungi having cross walls. Now this class is not in use.

Phygon: Trade name for

"Dichlone".

Phyllody: A disease in which floral parts are replaced by leaf like structures, usually caused by a virus.

Phylloplane: Pertaining to the surface of leaf. Used for the microflora of phylloplane.

Phyllosphere: Area or zone immediately surrounding the leaf.

Phylogeny: Theory of evolution of groups of related organisms.

Phyragmobasidiomycetidae: A subclass of fungi under Basidiomycetes class. Its members have a transversely or longitudinally septate basidia (pharagmobasidia) produced on various types of sporophores or directly on mycelium.

Physiological specialization: Development of certain characters in fungi which, are not morphological but physiological and can be detected by observing their effect on the host.

Physiologic races: A group of parasites of same specids behaving differently (with specialization) for different cultivars of host species, or the group of biotypes which are similar in morphological appearance but different from each other in certain, cultural, physiological and pathological characters.

Physiotype: A population of pathogen in which all the individuals have a particular characters of physiology (but not pathogenicity) in common.

Phytiatry: Treating the plant disease especially with the chemicals.

Phytoalexin: A complex compound (usually phenolic in nature) which inhibits fungi and is produced in hest tissue in response to stimulus of the pathogen.

Phytoncides: Substances of the plant origin active against other organisms. They include phytoalexins, preformed and post-infectionally formed substances imparting resistance in plant pathogens.

Phytopathology: Study of plant diseases, covering all aspects of nature, develolpment, biology and control of plant diseases.

Phytoplankton: A term used for the group of plants or plant like organisms floating on the surface of water.

Phytosanitary certificate: A certificate to the effect of that the plants or plant products being exported, are disease free and remained so throughout the period of crop growth. It is necessary for exporting plant material under quarantine law.

Phytosanitation: Measures requiring the removal or destruction of diseased or infected plant material likely to form a source of reinfection or reinfestation.

Phytotoxic: A chemical causing damage or injury to the plant part.

Phytotoxin: Chemical compound produced by the microorganisms which is harmful to higher plants.

Picolinic acid: A toxin produced by *Pyricularia oryzae*, which affects the catalase activity and chelates iron in respiratory enzymes.

Pileus (Pl. Pilei): The expanded cap like portion of some basidiomycetes or ascocarps which supports the hymenium.

Pili: Neumerous, small hair like structure found around bacterial cell which are not involved in motility, and their function is not known but, probably help in stickking on the surface. Their production is supposed to be an inherited character. Some of the pili are involved in transfer of genetic material from one bacterial cell to another and are known as sex pili.

Pink mold rot: Rotting disease caused by *Cephalosporium roseum*.

Pink yeast: Some species of yeast give carotenoid pigmented colonies in specific media due to their ubiqutious distribution and broad nutritional activities.

Pincocytosis: A phenomenon of host resistance in which the virus particles are completely absorbed by the host cell antibody.

Piperazines: Hetrocyclic nitrogenous compounds acting as fungicide against powdery mildews of cereals, apple, and cucurbits; cereal rusts and apple scab. Common example of this group is "triforine".

Piricularin: A non-specific toxin produced by *Piricularia oryzae* responsible for increased respiration and inhibition of many oxidative enzymes.

Pisatin: A phytoalexin (antifungal substances) formed in pods of garden pea inoculated with fungi.

Pistil: Female and central part of the flower consisting of ovary, style and stigma. These parts sometime get infected by fungal pathogens.

Pit: A small membranous area of generally thick walled cells found in the conducting tissues of the higher plants.

Pith: Parenchymatous tissue in the central region of a stem which sometimes get infected with pathogens and become hollow as a result of disease development.

Plankton: A large group (mass) of microorganisms floating passively on the surface of the water without using their own energy for this purpose.

Planogamete: A motile gamete usually with flagella. They are produced by lower fungi (chytridio mycetes) during their sexual reproduction.

Planogametic copulation: During the course of sexual reproduction (especially in lower fungi) the fusion of naked male and female gametes takes place and one or both of them are motile. Such type of fusion is called Planogametic copulation.

Plantvax: A systemic fungicide used mainly for internally seed borne fungal infections like smuts. Also see "oxathiins" and "Oxycarboxin".

Plaque: When a bacterial culture on the solid media is attacked by a pathogenic virus (bacteriophage) clear transparent (usually circular) zones are formed in the culture as a result of lysis of the bacterial cells by the virus, and these clear zones are called 'plaques'.

Plasmalemma: The cytoplasmic membrane found just below the cell wall, which is madeup of mainly lipids and proteins and is very vital for various metabolic activities of the cell including the permeability of the cells. In many diseases the normal functioning of these membranes are altered as a result of host-parasite interactions leading to various toxic effects on the host.

Plasmalemmasome: Intracytoplasmic vesicles formed by invagination of plasmalemma, which are filled with tubular diverticula.

Plasmids: A small, closed, loop like molecule of DNA, other than chromosome, present in microbes. They carry specific genes for certain resistance and are also used as tools in genetic engineering.

Plasmodesma (Pl. Plasmodesmata): A fine protoplasmic thread connecting two protoplast and passing from one to other cell through walls. Sometimes it serves as a source for the spread of the pathogen from cell to cell.

Plasmodiocarp: A curved or branched, sessile vein like fruiting body of some myxomycetes. e.g. *Hemitrichia* (*cf.* aethalium, pseudoaethalium, sporangium).

Plasmodiophoromycetes: A class of fungi under division Mastigomycota. Its members produce motile cells with two anteriorly whiplash type flagella and are parasitic with multinucleate thalli (Plasmodia) without cell wall.

Plasmodium (Pl. Plasmodia): A naked, multinucleate thallus moving and floating in amoeboid fashion, which is characteristic of members of myxomycetes and Plasmodiophoromycetes.

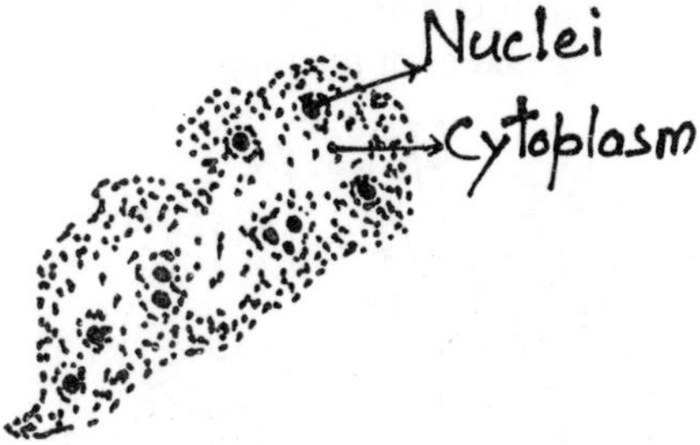

Plasmodium

Plasmogamy: During the course of sexual cycle especially in higher fungi the two sexual bodies (which may be gametes or other somatic cells) fuse with each other, but at this stage only cytoplasm of these cells fuse and not the nuclei. This process is followed by karyogamy or fusion of the two nuclei.

Plasmolysis: When a cell is placed in a solution of relatively of higher concentration, the water from the protoplast of the cell comes out as a result of exosmosis and cytoplams of the cell starts separating from the cell wall and gets shrinked on one side. This phenomenon is called plasmolysis.

Plastids: A specialized cell organelle which usually contain pigments.

Plectenchyma: In some of the fungal bodies the, fungal hyphae grow very close to each other, gets interwoven and form a compact mass of hyphae which appears just like the tissue of the higher plants and such tissues of fungal origin are called plectenchyma. They may be prosenchyma or pseudoparenchyma. It is found mainly in the fruiting bodies of the fungi.

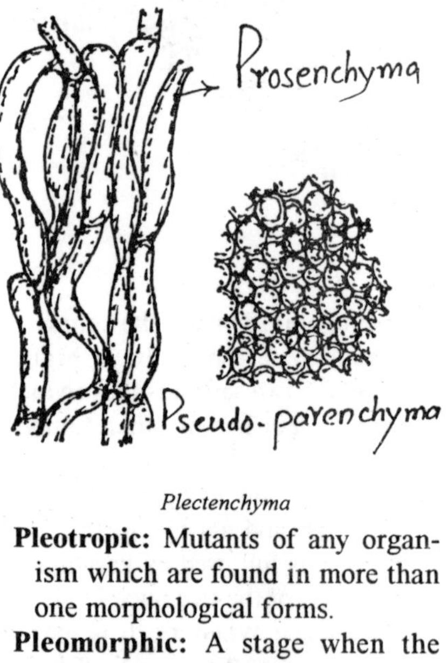

Plectenchyma

Pleotropic: Mutants of any organism which are found in more than one morphological forms.

Pleomorphic: A stage when the same species or strain of any microorganism is found occurring in the nature in different morphological forms. It is of common occurrence in some of the bacteria with flexible cell wall.

Plerome: The plant tissue inside the cortex.

Plerotic: Referred to a condition in which mature oospore lies free in the oogonial wall filling it.

Plug: A tightened cotton wool placed to close the mouth of the culture tube, which checks the entry of contaminant spores but allows the air for respiration of cultures.

Plugging: In case of wilt diseases, as a result of host-parasite interactions, some polymerized melanins and high molecular weight compounds are produced which get accumulated in the xylem vessels of the host plants and cause the blocking of the vessels leading to wilting. This phenomenon is known as plugging of vessels.

Podetium (Pl. podetia): An erect, columnar or branched reproductive structure of lichens bearing apothecia. e.g. *Cladonia cristalella* (cf. pseudopodetia).

Poisoned food technique: Method of evaluating the efficacy of fungicides by poisoning the nutrient medium with a definite concentration of the fungicide and then allowing a test fungus to grow on such medium. Both solid (agar) and liquid (broth) media may be used in this technique.

Polycentric: A type of microbial growth especially in fungi where the thallus emanates in a radial fashion from many points at which reproductive organs are formed, or a rhizoidal system or thallus bearing many reproductive structures (cf. monocentric).

Poygalacturonase: A hydrolytic enzyme which breaks the glycosidic linkages of pectic acid (polymer of galacturonic acid) in a hydrolytic manner, releasing galacturonic acids. This enzyme is involved in the pathogenesis of various diseases for degrading the pectinaceous substances of the host cell wall. This enzyme is specific for demethylated pectic acid and usually remains active under slightly acidic pH.

Polygalacturonate trans-eliminase (Pectate-lyase): An enzyme attacking the polymer of galacturonic acid and spliting the chain in a trans-eliminative manner by removing one molecule of water. This enzyme normally acts at alkaline pH and is implicated with the development of many plant diseases, especially with the penetration and degradation of the tissues.

Polyhedron: A spheroid particle or organ with many planes/faces-usually used for virus particle of such shape.

Poly-methyl galacturonase (PMG): A chain splitting hydrolytic enzyme specifically attacking the polymers of methylated galacturonic acids (native pectin) and is responsible for degradation of pectin component of cell wall and middle lamellae during the pathogenesis of many fungal and bacterial diseases.

Polyphagus: Organisms capable of parasitizing a wide range of host species (omnivorus).

Poyphyletic: Appearance of any particular species through several lines during the course of evolution.

Polyplanetic: A species of microorganisms having a number of swarming periods of only one type of zoospores during the course of its life cycle.

Polyplanetism: Condition of having many swarming periods and resting phases of zoopores found in polyplanetic forms.

Polypore: A generalized common name for the members of the polyporales of basidiomycetous fungi in which besidia are arranged inside a long pore of the fructification. These fungi are also named as 'bracket fungi'.

Polysaccharide: A large organic molecule consisting of many units of the simple sugars.

Polysome (Polyribosome): A cluster of ribosomes found in the cells which are responsible for protein synthesis and are mainly associated with the messenger RNA.

Polyzime: Commercial name of an enzyme preparation used for saccharification of rice starch and manufacture of soy-sauce which is obtained from the strains of *Aspergillus oryzae.* The main function of this enzyme preparation is to convert the native starch and carbohydrate to sugars.

Pore: A small opening or mouth of the tube, usually used for the opening of the fruiting structures of the fungi which serves the purpose of releasing the spores.

Porogenous: A type of conidial development in which conidium originates as a protrusion of protoplasm through pore in the conidiophore wall (*cf.* blastogenous, murogenous, phialogenous).

Porospore: A spore which is produced from the pore of the conidiophore.

Portal: Point of entry, usually referred for entrance of the pathogen, virus etc.

Potential host: Any organism which is highly specific and can be attacked by a specific parasite only.

ppm: See "part per million".

Pre-access interval: The time interval required between the last application of the pesticide to an area and safe access to the area for domestic livestock and persons.

Precipitin: An antibody that causes precipitation of soluble antigens.

Predacious fungi: Fungi parasitizing on amoeba, nematodes and other small animals.

Predator: An organism which captures other organism and uses it as food after killing it. Usually referred to the insects which prey upon other pests harmful to the crop plants and serve as a mean of biological control for that pest.

Predispose: Phenomenon of making the host more vulnerable for the attack of any pest or diseases. Various environmental or other factors may act as predisposing factor for a particular host-pathogen combination.

Pre-harvest interval: The minimum time interval required between the last application of the pesticide to the crop and safe harvesting of the edible crops for consumption, so as to avoid any toxic effect of pesticide on human being or animals.

Primary culture: The first tissue culture developed directly from the cells taken from the plant/animal source and not from other cultures.

Primary infection: The first infection of the crop plant in the season which takes place by the attack of the surviving propagule of the pathogens.

Primary inoculum: The overwintering or oversummering propagules of the pathogen in the form of spores or hypha etc. which become active, as soon as they get the new crop in the field and attack it causing the primary infection.

Primary symptoms: In case of such diseases where more than one type of symptoms are produced, the one appearing first during the crop growth is termed as primary symptom.

Primary zoospores: One of the two types of zoospores found in oomycetes. They are pear shaped bearing two flagella at apex and are invariably produced first from the sporangium. (*cf.* secondary zoospores).

Prions: Infectious particles of protein probably associated in human diseases of brain.

Primordium (Pl. Primordia): During the development or ontogeny of any structure or organ like conidia, conidiophore, sporangium, gametangium etc. the first cell or group of cells which initiate the development is known as primordium.

Probasidium: A dikaryotic cell in the basidiomycetous fungi, which lateron gets differentiated into a heterobesidium after karyogamy.

Prochloraz: Broad spectrum fungicide of imidazole group, especially active against Ascomycetes and Deuteromycetes; shows some systemic action and has translaminar translocation. Chemically it is 1-N-propyl-N-(2(2, 4, 6-trichlorophenoxy) ethyl) carbamoyl imidazole. Available under trade name "Sportak 45 EC" for spray and seed treatment. Its combination with carbendazim and manganese are also available.

Procymidone: Systemic fungicide of dicarboximide group, specifically active against members of Sclerotiniaceae like *Sclerotinia, Monilinia, Botrytis* etc. Chemically it is N-(3, 5-dichlorophynyl)-1, 2-dimethylcyclo-propane dicarboximide. It is translocated both upward and downward directions. Trade name is "Sumilex". Not marketed in India.

Progametangium: A cell or hyphal portion which lateron develops into a gametangium and suspensor cell.

Progamete: In case of Zygomyceteous fungi, the swollen hyphal tips with dens protoplast which lateron gets differentiated into a gemetangium at the tip, and a suspensor.

Progamones: Group of sex hormones produced in Zygomycotina which gets converted to trisporic acid, activating zygophore formation in opposite strain.

Prokaryote: Primitive organisms without any definite nucleus and nuclear membrane; also see Monera eg. bacteria, Cyanophycean algae.

Proliferation: A rapid and repeated production of new cells tissues or organs at the same place, usually referred to production of sporangia from the same hypha which has already produced spores and sporangium, and the remnants of the old sporangium like its wall still present. e.g. *Saprolegnia* sp.

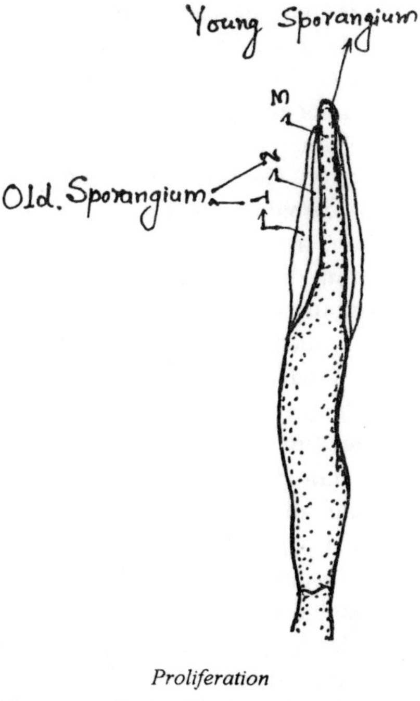

Proliferation

Promycelium (Pl. Promycelia): In rusts and smuts, the teleutospore germinates to give rise a long

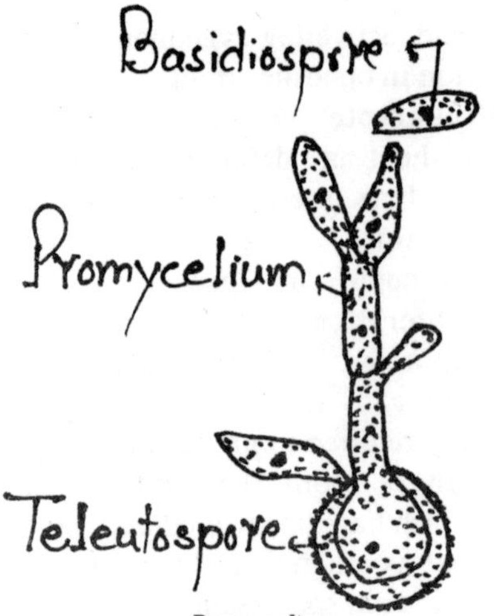

Promycelium

germ tube which acts as epibasidium and lateron becomes septate and bears basidiospores. This germ tube is termed as Promycelium.

Propamocarb: Systemic fungicide of aliphatic group, chemically being (propyl- (3-dimethyl amino)-propyl) carbamate hydrochloride. Effective against seedling diseases of cotton, black shank of tobacco, damping off and *Phytophthora* rot of soybean etc. Trade name is "Previcur N". Not marketed in India.

Properdin: The naturally occurring substances in the serum of the mammalian blood which are associated with the non-specific general resistance to any infection. They resemble antibodies, but have much broder spectrum of activity and less serological specificity.

Prophage: A non-infectious virion which multiplies with the bacteria but does not cause its lysis. They remain associated with the bacterial chromosome. They may cause lysis of bacteria under certain conditions, and start behaving like lytic type. Bacteria carrying such prophage are called lysogenic type.

Prophylaxix: Procedures and methods adopted to safeguard the plants/animals against any possible attack of the diseases, using various methods like clean cultivation, sanitation, avoidance, exlusion and chemical protection etc.

Prosenchyma: A false tissue made up of loosely interwoven fungal hyphae lying parallel to each other, usually found in the fruiting structures of the higher fungi, (*cf.* Pseudoparenchyma).

Prosorus (Pl. Prosori): A structure in some of the lower fungi which lateron gives rise to sorus after division e.g. *Synchytrium endobioticum.*

Protectant: Any substance/chemical (fungicide) which protects any organism against the attack by a pathogen. eg. Fungicides applied over the seed acts as protectant and saves from the attack of many seed and soil-borne pathogens, e.g. zineb, sulphur.

Protective fungicide: Fungicides used to safeguard the host surface against the attack of the pathogen and are applied prior to the actual attack of the pathogen.

Prothiocarb: Systemic fungicide of aliphatic group acting specifically against *Pythium, Phytophthora, Peronospora, Pseudoperonospora* etc. Chemically it is S-ethyl-N-(3-dimethyl amino propyl) thio-carbamate hydrochloride. Its trade name is "Previcur" and is stable in acid only. Not marketed in India.

Protista: A group of unicellular organisms from plants as well as animals.

Protocatechuic acid: A phenolic compound, antifungal in nature; prevents the spore germination and is responsible for resistance in many plants against various diseases.

Protopectin: A naturally occurring pectic substance, insoluble in water, releasing pectic acid upon hydrolysis. It is found in the cell wall and middle lammela of the plants and is attacked by various pectic enzymes produced by the fungal and bacterial pathogens during the course of their pathogenesis.

Protoperithecium: A mass of interwoven haploid hyphae at the early stage, which lateron develops into a perethecium after plasmogamy.

Protoplasmodium: A microscopic plasmodium without any differentiated fan shaped region or strands showing irregular and slow streaming, usually giving rise to a single minute fruiting body. (*cf.* "aphanoplasmodium", "phaneroplasmodium".

Protoplast: All the content of the cell inside the walls including the nucleus.

Prototunicate: A type of asci having a thin, delicate wall releasing their spores either by breaking apart or deliquescing. (*cf.* unitunicate, bitunicate).

Prototrophic: Microorganisms which are not fastidious and can be easily cultivated on the normal, standard culture media.

Pseudoaethalium Gr.: A type of fruiting body of myxomycetes in which several sporangia get crowded together appearing like a single sporophore or aethalium, but individual sporangia are clearly distinguishable. (*cf.* aethalium, plasmodiocarp, sporangium).

Pseudoangiocarpous: A type of basidiocarp development where the hymenium lateron becomes enclosed by the incurving of the pileus margin or sometimes by an outgrowth of the stipe. Hymenium thus remains enclosed till maturity of spores, when stipe expands and hemenium is again exposed (*cf.* gymnocarpous, hemiangiocarpous).

Pseudocapillitium (Pl. Pseudocapillitia): A fructification of many Myxomycetous fungi in which irregular threads, plates or other structures are found among the spores.

Pseudocolumella: A calcareous, spherical or rod shaped structure located in the centre of the spore bearing sac of myxomycetes, free of peridium. It consists of fused lime nodes and sometimes may be connected to capillitium. e.g. *Physarum* sp.

Pseudomycelium (Pseudomycelia):

A chain of similar cells attached in end to end fashion with each other and thus resembling a hypha, e.g. yeast.

Pseudoparaphysis: A sterile thread that grows downward in the cavity of some ascocarp usually becoming attached to the bottom.

Pseudoparenchyma: A false tissue madeup of compactly, interwoven fungal hyphae, having more or less isodiametric cells and resembling the parenchyma of the higher plants. In such tissues, the individual fungal hypha loose its individuality. Usually they are found in fruiting structures of fungi.

Pseudoperethecium: An ascostroma with a single locule having asci, and resembling like a perethecia.

Pseudoplasmodium: Aggregation of the amoeboid cells, consisting of the initial stage of the fruiting in the Acrasiales.

Pseudopodetia Gr.: A vertical structure developing from a vegetative thallus which produces an ascocarp premordium at its tip; found in lichens. (*cf.* podetium).

Pseudoseptum: A sieve like structure with numerous pores in some members of the lower fungi.

Pseudospore: A nonmotile, naked spore found in members of Acrasiales of Myxomycetes.

Pseudostroma: A mass of cells having cells of both fungal and host origin.

Pseudothecium: An ascostromatic ascocarp having asci in many unwalled locules e.g. Loculoascomycetes. Also see "Ascostroma".

Pseudotrophic mycorrhiza: See mycorrhiza.

Psilocin: One of the Psychoactive compound found in hallucinogenic mushroom. *Psilocybe*, which has hallucinogenic properties and is very much similar to 'Psilocybin' (*cf.* Psilocybin).

Psilocybin: Psychactive compound found in hallucinogenic mushroom, *Psilocybe* which is 4 substituted tryptamine derivative having hallucinogenic properties similar to LSD (d-lysergic acid diethylam-ide). (*cf.* psilocin).

Psychrophile: Microorganisms which have the ability to grow at 0°C or below.

Ptomaine: An alkaloid like substance produced during the microbial decomposition of the dead organic matter.

Puff balls: Common name for some basidiomycetous fungi of order Lycoperdales of Gasteromycetes. Most of them are edible. At maturity, they have a powdery gleba having light coloured spores and a well developed capillitium. Members of order Tulostomatales of Gasteromycetes are referred as "stalked puff balls".

Punctiform: Very small, but may be seen without magnification.

Pure Line selection: A method of selecting the genetically resistant plant material from the existing population in which seed from the selected plants are collected and tried separately and again subject-

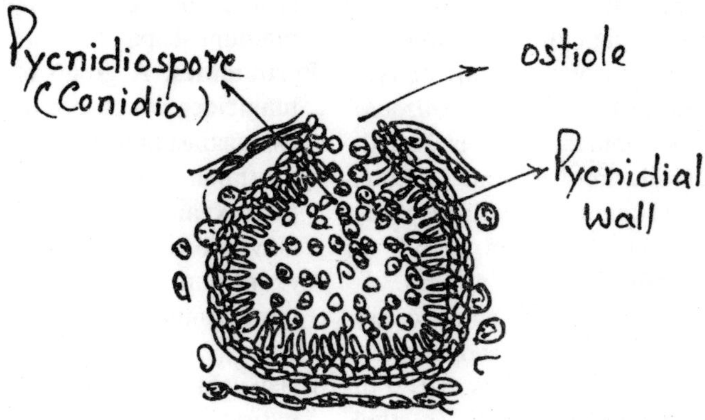

Pyenidium

ed to the same process of selection.

Purple seed stain: A disease of soybean appearing mainly on seeds, caused by *Cercospora kikuchii.* Affected seeds show purple coloured patches on seed coat, sometimes extending on the entire surface. Other plant parts are also affected with the disease.

Pus: A fluid produced as a result of infection and its inflamation which contains serum, bacteria, dead cells and leucocytes.

Pustules: Raised blister like spot on the plant surface formed due to bursting of the epidermal tissue through which the spores of the fungi comes out. e.g. uredio or teleutosori of the rusts.

Putrefaction: The process of the microbial decomposition of the protein giving rise to an unpleasant odour.

Pycnidiospore: A conidium produced inside a pycnidium.

Pycnidium (Pl. Pycnidia): A closed flask shaped, ostiolate fructification containing a cavity bearing conidiophores and conidia. It is an asexual reproducing structure of fungi. e.g. *Macrophomina, Phoma* etc.

Pycniospore: A spore produced inside a pycnium. They are also termed as Spermatiospores. e.g. *Puccinia* sp.

Pycnium (Pl. Pycnia): In Uredinales the stage zero consisting of male fertilizing elements (pycniospores) and female, flexous hyphae, also called Spermogon-ium.

Pycnosclerotium: A hard walled structure without any spores which very much resembles a pycnidium.

Pyemia: A type of septicemia where pyogenic organisms in the blood form secondary foci in the tissue.

Pyogenic: Peferred to those organisms which stimulate the formation of pus upon infection.

Pyracarbolid: Common name of a systemic fungicide of carboxamide (oxathiin) group having activity against rust, smut and *Rhizoctonia.* Its chemical name is 2-methyl-5, 6-dihydro-4H-pyran-3-carboxylic acid anilide. Available under trade name of "Sicarol 50 WP" or 75 SD. Not marketed in India.

Pyrazophos: Common name for systemic fungicide of organophosphate group. Chemically it is 2-(0-0-diethyl-thionophosphoryl)-5-methyl-6-carbethoxy-pyrazolo (1, 5a) pyrimidine. Available under trade name or "Afugan 30 EC". Effective against powdery mildew of vegetables, fruits, cereals and ornamentals as spray. Not marketed in India.

Pyriculol: A toxin produced from *Pyricularia oryzae.*

Pyriform: Usually referred to the pear shape, for the spores, sporangium shape.

Pyrimidines: A group of systemic fungicides which are quite effective against powdery mildews. e.g. triarimol; dimethrimol. They are hydroxyaminopyrimidines with alkyl substituted in the nucleus.

Pyrophilous: Forms growing on burnt ground, steam sterilized soil etc.

Pyroxychlor: Systemic fungicide of pyrimidine group, effective against soil diseases, especially *Pythium, Phytophthora, Bremia* and other Phycomycetous fungi as seed and soil treatment. It was the first chemical to show both downward as well as upward translocation from leaf to root. Chemically it is 2-chloro-5-methoxy-4-trichloro methyl pyridin. Trade name is 'Nurelle". Not marketed in India.

Q

Quarantine: A legal restriction imposed on the import and export of the plants and plant material to prevent the spread of diseases and pests. In case of regulatory quarantine, the imported material is kept in isolation and is rigorously inspected before making use and introducing it in the country. In India, the quarantine act is called the Destructive Insect and Pest (DIP) Act of India.

Quelkorper Gr.: A gelatinous mass of cells present inside the ascocarp of Coronophorales, which swells and make an opening of ascocarp looking like an ostiole but it is not ostiole.

Quinone fungicides: A group of fungicides having quinone compound as active ingredient. They are widely used as seed dressing fungicides. e.g. chloranil and dichlone.

Quintozene: See "Penta chloro nitro benzene".

R

Rabies: A highly infectious, acute viral disease of human being and animals affecting the brain, spreading through the bite.

Race: A group of individuals of the same species infecting same cultivar of the host but differing genetically, geographically and physiologically.

Ramus: A branch; a cell bearing verticil of metulae and phialides.

Ratoon stunting: A disease of sugarcane caused by bacteria. Symptoms include, stunted growth with short internodes and brick-orange discolouration of internal tissue of nodal region.

Reagin: A substance found in the serum and spinal fluid resembling antibodies and reacting with the antigens.

Recapticle: Basal structure (usually cup shaped) on which fruiting body is placed; in case of "Phallales" the structure on which gleba is displayed.

Receptive hypha: A flexuous hypha (on spermatium of rust fungi) or trichogyne, which act as female organ and get fertilized by microconidia/spermatia (male cell).

Receptor sites: Specific sites on the host (bacteria) surface on which the infecting virus (phage) particle get absorbed.

Reconstitution: Process of assembly of artificially disintegrated virus particles.

Recombination: Formation of offsprings of gene combination not present in either of the parents but are formed as a result of combination of both the parental genomes. In case of viruses, it involves exchange of fragments of replicating DNA molecules of the two different virions, simultaneously infecting the same host cell.

Recurrence: The reappearance of symptoms after certain period of absence.

Red bread mould: Common names of *Monilia sitophila (Neurospora* sp.) which causes infection in bakeries, as a result of it in a very short time the bread is covered with a pink loose textured growth of the fungus. It is widely used for genetical studies.

Red rot: An important disease of sugarcane caused by *Colletotrichum falcatum (Glomerella tucumanensis*). Main symptom include reddening of internal tissue of cane giving alcoholic smell and drying of canes.

Reductase test: A technique of determining the milk quality, in which the oxidation-reduction reaction of milk bacteria form the basis of test changing the colour of the indicators due to its reaction.

Release: (i) Liberation of virus particles from the lysed bacterial cell, which is disintegrated as a result of phage attack.

(ii) Liberation of spores from any fruiting structure.

Reniform: Having the shape of kidney (used for the shape of zoospores in Mastigomycotina).

Rennet Curd: Coaggulation of milk by the enzyme Rennin producing sweat curd.

Rennin: An enzyme transforming the soluble casein into insoluble paracasein.

Replication: Referred to the multiplication of the viruses.

Replicative form: Newly formed DNA strand complimentry to existing single stranded DNA synthesized during early stages of the replication cycle of the phage. The replicative form is synthesized using the phage DNA as template. It is primarily responsible for the formation of the new phage particles.

Reproduction: Production of new individuals by the parents which are similar to parents in all characters.

Residue: Amount of the pesticides remaining in or on the plant tissue (or in the soil) after a given time especially at the harvest time.

Resistance: The ability of the host to counteract the attack of the pathogen and remaining completely or partially free of the disease. This character is mostly genetically controlled and is inherited from the parents.

Resolving power: The ability of a microscope to distinguish between the two adjacent points which depends upon the wavelength of light and numerical aperture of the objective.

Resting Spore: A thick walled spore which remains in resting (dormant) period during the unfavourable period and germinates afterwards under favourable conditions of growth.

Resting sporangium

Restricted type: The bacteriophage mutant that are capable of infecting only a few strains of the bacterial host due to the presence of preventive nucleases in the bacterial strain. Restricted type can acquire the ability to infect on previous glucosylation.

Resupinate: A position of the thallus in which it lies flat on the substratum with hymenium on the free surface.

Reticulate: Appearance of the fungalmycelium, when it looks like a net, surrounded by net like ridges.

Reticuloendothecial system: A specific system in organs and tissues like spleen, liver and bone marrow which are involved in resistance and immunity reactions.

Retrovirus: An RNA containing virus which uses reverse transcriptase to synthesise DNA from RNA.

Reverse transcriptase: An enzyme mediating the synthesis of DNA molecule from the code supplied by RNA molecule.

stratum.

Rhizomorph: A thick strand of fungal hyphae which penetrates the intact root periderm and resembles the root of a higher plants e.g. *Armillaria* sp.

Rhizomycelium: An extensive rhizoidal system superficially looking like a mycelium.

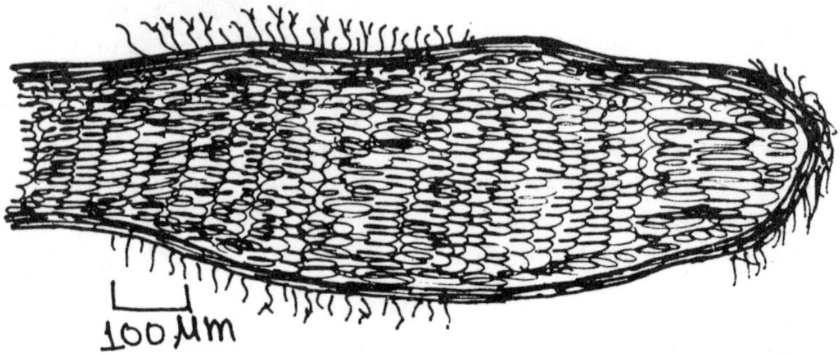

Rhizomorph

Rheotaxis: Accumulation of zoospores in response to a current of water, emerging from a capillary at a speed similar to the swimming speed of zoospores.

Rhizobitoxin: A non-specific phytotoxin produced by the *Rhizobium japonicum* in tobacco, producing chlorosis.

Rhizobium: A bacterium living symbiotically in roots of leguminous plants producing nodules and fixing the atmospheric nitrogen.

Rhizoid: Root like anchoring hypha, in chytridiales, a filamentous part of a thallus extending into the sub-

Rhizoplane: Usually referred to the microorganism present just on the surface of the root of plant.

Rhizoplast: A strand connecting the nucleus to the blepharoplast (base of the flagellum) in the motile cells.

Rhizosphere: The zone of the active microbial activity around the roots of a plant.

Riboflavin: A vitamin of the B group produced commercially from the fungus *Nematospora gossypii* and *Eremothecium ashbyi*.

Ribonucleic Acid (RNA): Nucleic

acid found in the cytoplasm and nucleus. It contains phosphoric acid, D-ribose and any of the four bases.

Ribosome: A cell organelle in the cytoplasm which is the site of the protein synthesis.

Rickettsia: An intracellular, obligate parasitic microorganism which is still smaller than bacteria (0.3-0.5 x 0.2-0.3 μ). It is found usually in the phloem tissues of the plants and are transmitted by the insects.

Ridomil MZ: Mixture of metalaxyl and mancozeb. Also see "Metalaxyl".

Rilan: See "Topsin-M".

Ringspot: A symptom of the virus disease in which a ring of chlorotic tissue is formed with green centre.

Ringworm: A fungal infection of the skin of human beings.

Rishitin: An antifungal phytoalexin formed in resistant potato tubers inoculated with avirulent race (O) of *Phytophthora infestans*.

Rizolex: See "Tolclofos-methyl".

Roguing: A technique of plant disease control, in which pathogen is excluded by removing the diseased and undesired plants from the field and destroying them to check the further spread of the pathogen.

Rohr: A long tubular cavity appearing within the encysted zoospores of *Plasmodiophora*, attached on the surface of host root hair. Within this tubular cavity there is a sharp dark staining rod, called Stachel. These structure help in the penetration of fungal protoplast inside host root cells.

Root knot: A common disease of vegetable crops caused hy nematodes *Meloidogvne* sp. showing stunted, sick growth of plants and formation of galls and knots on the roots.

Root nodule bacteria: Bacteria (mostly Rhizobia) living in the nodules of leguminous plants roots with symbiotic relationship, fixing the atmospheric nitrogen, thereby providing nutrition to the plant.

Root rot: A very common disease of many plants caused by soil-bore pathogens like, *Rhizoctonia, Macropho-mina Fusarium, Sclerotium* etc. in which root tissue become rotten.

Rostelia: A form genus of aecial stage of rusts in which aecial lip protrudes much beyond the surface of the host and soon becomes shredded often bending back and forming an ornamental fringe around the mass of spore chain (*cf.* Caeoma, Aecidium and Peridermium).

Rostrate: Organs, usually for spores having a beak.

Rostrum: A beak like process.

Rotting: The process of dissolution of the cell wall and middle lammellae of the plant tissue mainly by the attack of microorganisms through pectin and cellulose degrading enzymes, which make them soft and disintegrated. It also separates the fiber tissues form the fiber plants like flax or hemp.

Run-off: Shedding of the sprayed chemical solution from the plant surface during the and after the application due to the coalesion of the droplets.

Rust: Referred to a disease, as also the basidiomycetous fungi causing this destructive disease, in which the teleospores and urediospores of the fungus comes out after bursting the host epidermis in the form of small pustule.

S

Saccardoan system: A system of classifying the fungi of Coelomycetes and Hyphomycetes of Fungi imperfecti, based on the shape, colour and septation of the conidia. It was initially proposed by Saccardo in 1899.

Saccardo's Spore form: Morphological types of fungal spores described by Saccardo. See-amerospore, didymospore, phragmospore, dictyospore, scolecospore, helicospore and Staurospore.

Saccate: Having appearance like a sac or bag.

Saccharolytic: Micro-organisms which are capable of splitting or degrading the sugar compounds and living on them.

Saddle fungi: Common name for members of "Helvellacea" of Discomycetes.

Safenar: A chemical which reduce the phytotoxicity of another chemical e.g. lime acts as safener for copper sulphate.

Sake: Japanese food/beverages prepared as the fermentation products of *Aspergillus flavus-oryzae* group.

Saltant: A discontinuous variation of unknown origin or a mutant; the term is used mainly for the mutants in fungi.

Saltation: The appearance of different type of individuals (mutants) among the population of single isolate having the genetic purity.

Sandal spike: A disease of sandal tree caused by mycoplasma exhibiting reduction in leaf size, shortening of the internodes, looking like a spike.

Sandoz copper: Copper fungicide having 50% cuprous oxide.

Sanitation: A cultural practice to keep the fields clean and free from debris which helps in controlling many infectious diseases.

Sanitize: Bringing down the micro-organism population of any article (especially those used for eating purpose) to the safe levels prescribed by the public health departments.

Saporobes: See Saprophyte.

Saprol: See "triforine".

Saprophyte: Micro-organisms which draw their nourishment from the dead organic matter.

Satellite Virus: A very small virus of only 17 nm which can not multiply independently and remains attached to some other virus and multiplies alongwith it.

Saxicolous: Forms growing on rocks.

Scab: A plant disease symptom having hyperplastic scab like lesions.

Schick test: A skin test in which susceptibility of an individual for diphtheria is evaluated.

Schizont: A stage in the asexual life cycle of malarial parasite.

Schultz-Charlton Reaction: A skin test for determining the scarlet fever.

Sclerotium (Pl. Sclerotia): A hard fungal body made up of interwoven compact mass of hyphae,

Sclerotium

which acts as a resting body and on germination give rise to a new mycelium, fructification or stroma.

Scolecospore: An elongated needle, thread or worm like spore (*cf.* amero, didymo, dictyo, phragmo, helico spore).

Scolecospore

Secondary inoculum: After infection fresh crop of spores or infective propagules produced on diseased plants, which causes fresh infection and helps in further spread of the disease.

Secondary symptoms: Symptoms appearing at the later stages after primary symptoms—in case of the viral diseases, whose more than one type of symptoms are found.

Secondary zoospores: One of the two types of zoospores produced in Oomycetes. They are kidney shaped bearing two oppositely directed flagella placed laterally in a deep groove on concave side of zoospore. Usually they are formed by the germination of encysted primary zoospores. (*cf.* primary zoospores).

Sectoring: Appearance of sectors in culture plate with altered characters of growth due to the mutation.

Seed borne disease: Disease caused by the pathogen present on the surface or inside the seed of the host, and usually perpetuates with seed.

Seed borne pathogen: Infectious agent associated with the seed, capable of causing the disease.

Seed transmission: Transportation of a pathogen through the agency of seed which results into the spread of disease in newer areas.

Selective media: Addition of specific chemicals in the normal media which prevents the growth of undesired micro-organisms and helps in isolation of specific organisms.

Self compatible: A thallus which does not require any other partner for the sexual reproduction and sex organs produced by it are able to combine with each other, also called 'self-fertile'.

Self-incompatible: A thallus can not reproduce sexually in itself and require another thallus, may be of other strain for completing the sexual cycle. Also called 'self-sterile'.

Senescent: Growing old, or reaching to the maturity.

Sensitivity: The phenomenon of showing very sharp reaction of symptom expression on infection with a pathogen.

Septal pore cap: A dome like configuration of endoplasmic reticulum occurring directly over the pore in a septum, found in basidiomycetous fungi.

Septate: A thallus of fungus with cross walls or septa.

Septicemia: Systemic disease caused due to the infection and multiplication of pathogenic micro-organisms in the blood stream.

Septum (Pl. Septa): A cross wall in a hyphae of the fungi or in the spore, which may be primary or adventitious.

Serology: A branch of science which uses blood serum or other body fluids in antigen-antibody reactions, for determining the presence of a specific micro-organisms or other such foreign material.

Serotype: Population of microbes (usually bacteria or virus) in which all the individuals have common serological properties.

Seta (Pl. Setae): A bristle like hair found over the surface of certain fructification of the fungi e.g. Setae on the acervulus of *Colletotrichum* sp.

Sewage: Liquid or solid waste products of domestic and industrial nature, coming in the sewers, which is then subjected to microbial degradation.

Sexduction: A type of sexual reproduction (recombination) in bacteria, where genes get attached to the F factor and are transferred from donor to receipient cell alongwith this factor.

Sexual reproduction: Production of progeny or new individual involving the process of karyogamy and meiosis at the regular stages in the life cycle of the organism. Plasmogamy may or may not take place. The new individuals produced by this process show the variability as a result of recombination of genetic material.

Shan: See "dithianon".

Sheath blight: Rice disease caused by *Corticium sasakii,* in which straw coloured spots are formed on leaf sheath above water level, on which sclerotia are formed.

Sheathing of hyphae: A type of defence mechanism in plants in which the hyphae penetrating the cell wall are enveloped in a sheath which may be the extension of the cell walls or deposition from the cytoplasm, thus further spreading of the pathogen is prevented.

Shift cultivation: Growing crops on land reclaimed from bush for few years and then moving to new area and reverting the old area to bushes. It helps in containing the severity of many soil borne diseases and nutritional deficiencies as well.

Shitake: Common name for the large, edible, basidiomycetous fungus, *Lentinus edodens.*

Short cycled: Referred for rusts which have less than five spore stages in their life cycle (*cf.* long cycled).

Shot holes: A disease symptom in leaf spots, in which the dried itssue of affected spots fell due to external forces like wind and small holes, usually with dark margins are formed on leaf lamina.

Shoyau: Fermented food product of Japan prepared by fermenting soybean and whole wheat with *Aspergillus oryzae* or *A. soyae* and afterwards with *Saccharomyces rouxii* after adding brine. The preparation is also known as Soy-sauce.

Sicarol: See "Pyracarbolid".

Silica Gel culture: A method of maintaining the culture for longer period in which 5% spore suspension in the skimmed milk is added to anhydrous silica gel in screw cap bottles and stored at low temperature.

Simple interest diseases: A disease which increases mathematically like simple interest and has only one generation in the crop life. It spreads from one common source of inoculum e.g. soil borne diseases like wilts.

Sirenin: A reproductive hormons secreted by the female gametes of *Allomyces* sp. which attracts the male gametes.

Skeletal hyphae: A type of hyphae forming the basidiocarp which are formed from the generative hyphae. They are sterile, without clamp connection, thick walled, aseptate, usually unbranched, functioning to strengthen the basidio-carp. (*cf.* generative, binding type).

Slide germination method: A technique to test the efficacy of fungicide against spore germination. In this slides are sprayed with fungicide or a drop of fungicide is allowed to dry on slide and then a drop of spore suspension is placed on it and germination percentage is recorded after proper incubation.

Slime layer: A gelatinus covering around the bacterial cells, which has polysaccharides and polypeptides secreted by the bacterium.

Slime mold: The common term to designate the members of myxomycetes.

Slopes: Culture tubes having the sterilized media solidified by keeping the tube at an angle from the surface providing larger surface for the growth of the microorganism. They are used for growing the microorganisms for study and storage.

Sludge: The semisolid portion of sewage sedimented at the base and acted upon by the bacteria.

Slurry: Chemicals, usually seed treating fungicides having a paste like consistency, which are used to treat the seed in slurry treating plants.

SMDC: Sodium methyl dithiocarbamate, used as soil fumigant to control fungal and nematode diseases. It is marketed as "Vapum".

Smear: A thin layer of bacterial culture suspension spread over a glass slide for microscopical observation.

Smut: A disease and symptom of cereals and also some other crops caused by Ustilaginales, in which the floral parts get converted into black powdery mass of fungal spores.

Soft rot: A symptom of many diseases in which the superficial decay of tissue is caused by the fungal activity through their cell wall degrading enzymes.

Soil amendments: A cultural practice of adding organic materials like dried grass, alfaalfe, wheat straw or oil cake etc. to the soil which helps in reducing the incidence of many soil borne diseases.

Soil fungicides: Fungicides applied to the soil as drench or fumigant to control soil borne diseases. They may be applied as pre-plant before sowing (e.g. chloropicrin, vapam etc.) or in standing crop (e.g. captan, PCNB, thiram etc.).

Soma (Pl. Somatic): The vegetative portion of an organism distinguished from the reproductive organ/phase.

Somatogamy: Fusion of vegetative (somatic) cells, during the plasmogamy.

Sooty mold: Fungi belonging to Capnodiaceae growing on honeydew excreted by aphids.

Soralia: Various types of pustules found on lichen thallus which contain asexually reproducing bodies, Soredia.

Sorbic acid: A popular preservative used for wrapping and packing material. Its water solube potassium salt is also used so preservative for tobacco products.

Soredium (Pl. Soredium): Small, rounded, bud like outgrowth (structure) on the upper surface of lichen thallus having few algal cells surrounded by fungal hyphae. They may be detached from the thallus and are able to germinate and produce new thallus.

Sorelium: A localised pustule like area of lichen thallus where asexually reproducing structures (Soredia) are located in groups. e.g. *Parmelia, Physcia.*

Soreshin: An important disease of cotton caused by *Rhizoctonia bataticola.* Main symptoms are defoliation and sudden drying of plants in patches and rotting of roots. Disease is also called "root rot" or "dry rot".

Sorocarp: The fruiting body of the Acrasiales.

Sorus (Pl. Sori): A cluster or mass of spores inside a structure e.g. Plasmodiophorales.

Spathulosporales: A small order of Loboulbeniomycetidae which is parasitic on marine algae and lacks true mycelium.

Spawn: The mycelium of the edible mushrooms which is used as the starter to inoculate the fresh bed for growing the mushrooms.

Species (Pl. Species): A unit of the classification in the taxonomy,

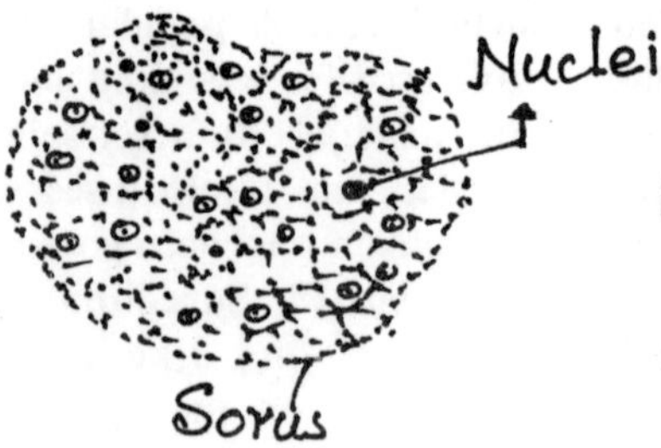

Sorus

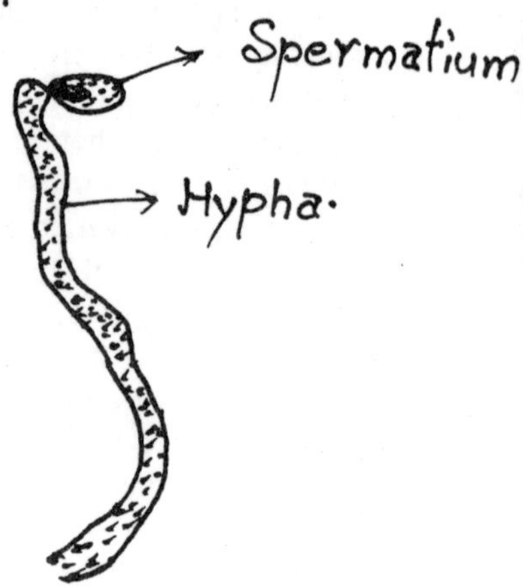

Spermatization

used for a group of individuals having close resemblence in genetic characters and placed in a single genus. It is used as second name in the binomial system of nomenclature.

Species plantarum: Name of a classic book of taxonomy written by renowned Swedish botanist Carl Linnaeus in 1753, who gave the binomial nomenclature and is also called 'father of botany'.

Spermatiophore: A specialized hyphae producing spermatia at its tip, found in *Puccinia*.

Spermatium (Pl. Spermatia): A non motile, uninucleate, male gamet which,usually becomes detached before plasmogamy.

Spermatization: Plasmogamy of a spermatia with a female receptive structure.

Spermogonium (Pl. Spermogonia): A pycnidium like structure producing spore like structures which may function as spermatia

Spermosphere: Area/zone around seed of higher plants. Referred for microflora around seed surface (spermosphere mycoflora).

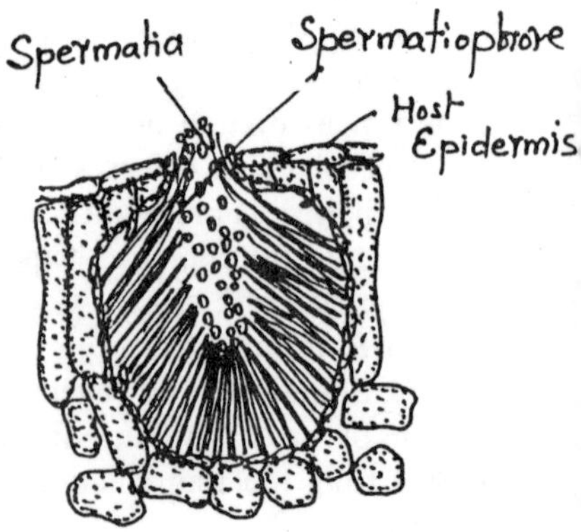

Spermogonia & Spermatia

Sphareocyst Gr.: Large, globose or oval cells, scattered among the elongated cells in the trama of basidiocarps of Hymenomycetes.

Spheroplast: Bacteriole without any proper cell wall or only bacterial protoplast produced in medium having penicillin and are very soft and non-rigid.

Spirochete: Spiral shaped bacterium, most of them are parasitic in nature.

Spirilum: A spiral or screw shaped bacterium.

Spitzenkorper: A small body found in the apical region of actively growing septate hyphae, which remains situated near to cytoplasmic membrane and is deeply stained.

Sponge mushrooms: Common name for edible fungi of Morchellaceae. Also called "Morels".

Spontaneous: Any event happening automatically without any external stimulus.

Spontaneous generation: A theory proposed by earlier scientists like Needham that microorganisms could arise themselves spontaneously in the nature.

Sporabola: The path or curve followed by a ballistospore during the course of its discharge from its sterigmata.

Sporadic: Disease appearing only on limited number of plants or area at very irregular intervals.

Sporangiolum (Pl. Sporangiole): A small sporangium; or a number of small sporangiole attached to one sporangium, usually in Mucorales.

Sporangiophore: A hyphae bearing sporangium.

Sporangiospore: A non-sexual spore produced within a sporangium.

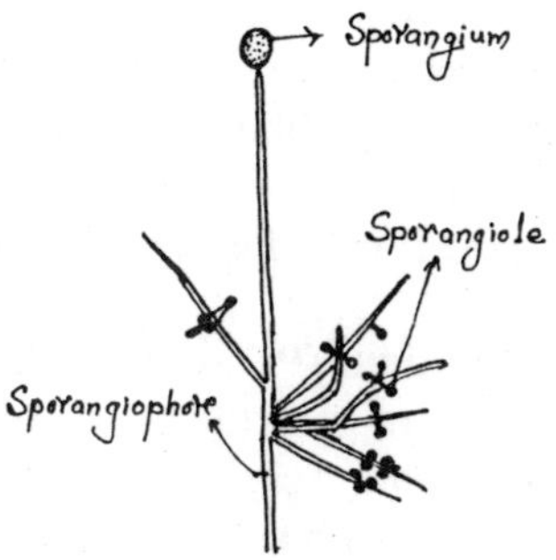

Sporangiole

Sporangium (Pl. Sporangia): A sac like structure at the apex of the sporangiophore, which have endogenous non-sexual spores.

Spore: A discrete sexual or non-sexual reproductive unit having a rigid wall and is capable of being disseminated. Unlike seeds of higher plants, the spore does not contain an embryo.

Spore ball: A group of teliospores of smuts cemented together. The cells of spore ball may be distinguished into fertile and sterile cells.

Spore germination stimulant: Properly filtered orange juice is added in equal amount with the spore suspension to stimulate the germination. Juice of fresh oranges is filtered through cheese cloth, filter paper and then Berkefield cylinder and filtrate is diluted ten fold with distilled water and 10 ml solution in each corked vial is stored below freezing temperature. It is mixed with spore suspension before it is put for germination test.

Spore print: When basidiocarp (esp. of Agaricales) is placed on a coloured sheet of paper, the basidiospores are liberated below the lamella or pores and this deposit of spore is called spore print of that fungi.

Sporocladium (Pl. Sporocladia): A special type of fertile branch of sporangiophore bearing the merosporangia, found in Mucoroles.

Sporidium (Pl. sporidia): Referred to basidiospores of rusts and smuts.

Sporocladium (Pl. Sporocladia): A special type of fertile branch of sporangiophore bearing the merosporangia, found in Mucoyales.

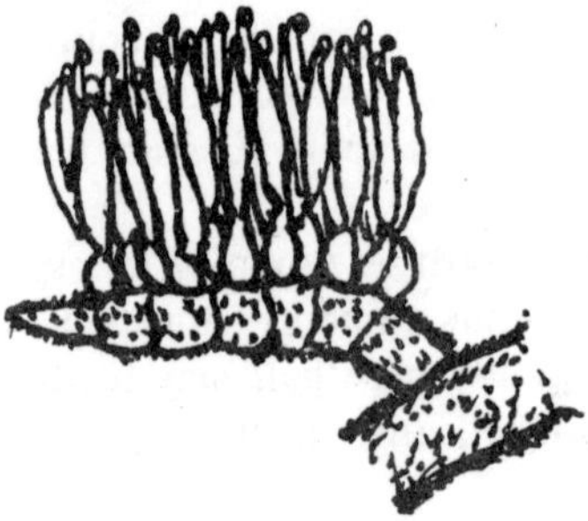

Sporocladium

Sporocyte: A spore bearing structure found in Labyrinthulales, usually surrounded by mucilaginous envelope and are found in slim filament (tube).

Sporodochium (Pl. Sporodochia): A cushion shaped packed cluster of conidiophores arising from a stroma or mass of hyphae. e.g. *Fusarium* sp.

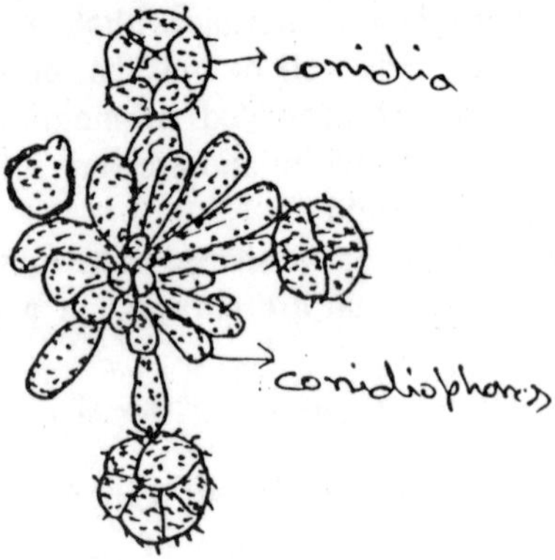

Sporodochium

Sporogenous cell: A cell which gives rise to spores.

Sporophore: Any structure bearing spores.

Sportak: See "Prochloraz".

Sporothallus (Pl. Sporothalli): Thallus producing spores (*cf.* gametothallus).

Spreader: The inert material which are added to the spray solution mixture, to improve the contact between fungicides and plant surface and increase the even distribution and coverage of the chemical e.g. soap, glycerol, oils, resins, gums etc.

Squamulose: Referred to a growth form (type) of lichen thallus which are scale like in appearance and consist of many small lobes or squammules.

Stability: Ability of the chemical/fungicide to resist the degradation over a period of time in storage or after application.

Stachel: A sharp rod shaped structure found within Rohr of encysted zoospores of *Plasmodiophora.* It punctures the wall of host root hair and thus help in the penetration of fungus inside host cell. Also see "Rohr".

Stage 0: A stage in the life cycle of "rusts" which produces spermogonia bearing spermatia and receptive hyphae. Since function of spermatia stage was not discovered till other stages were well established, it was given zero "0" status.

Stage I: Stage of rust life cycle producing aecia and aeciospores.

Stage II: A stage in life cycle of rusts producing uredinia bearing urediniospores.

Stage III: Stage in the life cycle of rusts producing telia bearing teliospores.

Stage IV: A stage in life cycle of rusts which produces basidia and basidiospores.

Staining: For the contrast in the microscopic studies, microorganisms are coloured with various dyes (stains) like, cotton-blue, piciric acid, orange-G, acid fuschin, trypan blue etc.

Stalked Puffballs: See "Puffballs".

Staling: Sometimes, the metabolic products of the micro-organism show inhibitory effect on their growth itself and thus reduce the growth rate in the culture and this effect of poor rate of growth is called staling.

Staphlococci: Spherical (coccoid) bacteria found in irregular, grape like culusters.

Statismospores: Basidiospores, which are not projected (discharged) violently (in Gasteromycetes). *cf.* ballistospores.

Staurospore: Star shaped spores or sometimes, fork shaped spores (*cf.* amero, didymo, dictyo, phragmo and helicospores).

Staurospore

Stelioger: A hemispherical cytoplasmic area formed in the amoeboid cells of Protosteliomycetes of Gym-nomycotina during the process of sporulation. eg. *Nematostelium ovatum.*

Stem Gall: A disease of coriander plant caused by *Protomyces macrosporus* in which veins of the leaves, stem and fruit show tumor like swellings.

Stenotherophile: Organisms growing at 60°C or above higher temperatures; true or obligate thermophile.

Sterigma (Pl. Sterigmata): A pointed peg like structure over the basidium on which basidiospores are produced or over the other thalli bearing sporangium or conidium.

Sterile: (i) Structure which do not produce any spores.

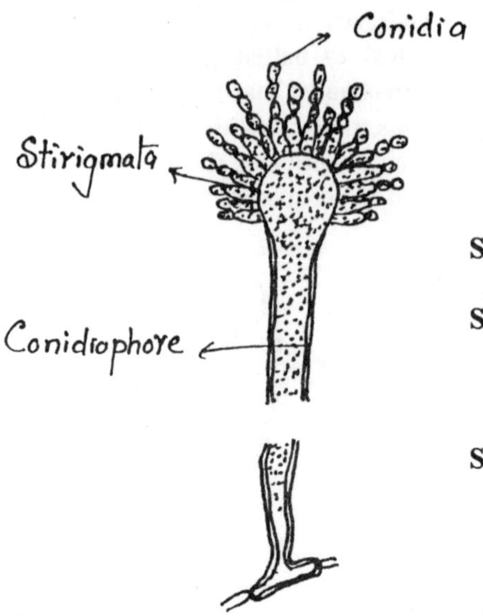

Sterigmata

(ii) Material free from any micro-organism (for use in cultural studies).

Sterile threads: Sterile, non-spore producing structures within the fruiting body especially ascocarp which are used as taxonomic criteria. Some of these are paraphyses, periphyses, periphysoids, pseudo-paraphyses etc.

Sticker: Inert material which is added to the fungicidal sprays for improving the adherence to the plant surface eg. flour, starch, gum, dextrin etc.

Stinkhorns: Common name for basidiomycetous fungi of order Phallales of Gasteromycetes. They give a fetid odor upon exposure of the gleba.

Stinking Smut: It is also called 'Hill Bunt'. A disease of wheat, occurring mainly in hilly regions, in which the grains get converted into smut balls and give foul smell of trimethyl amine-a volatile compound. It is caused by the fungus *Tilletia caries* and *T. Foetida.*

Stipe: A stalk like portion of fruiting body, ascocarp or basidiocarp.

Stolon: A horizontal hyphae anchored in the substratum at points of contact, giving rise to sporangiophores. e.g. Mucorales.

Strain: (i) A group of individuals from the same race, isolate or form.

(ii) Group of viruses having common type of antigens.

Streptococci: Coccoid bacteria which divide in a way that a chain of cells is produced.

Streptomycin: An antibiotic produced by *Streptomyces griseus*, which is highly effective against many bacteria.

Striga: Phaenorogamic plant parasite of many crop plants like sorghum, sugarcane etc., belong to family Scrophulariaceae. It is parasite on host roots and draws nutrition from them.

Stroma (Pl. Stromata): A compact mass of vegetative tissue (hyphal mass), sometimes interwoven with host tissue which often bears fructifications either within or upon its surface.

Stunting: A disease symptom appearing mainly as a result of virus and nematode attack in which growth of plants is checked and

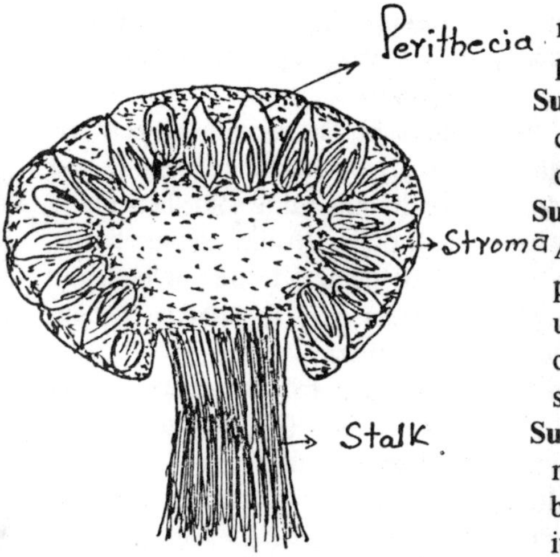

Stroma

they remain dwarfed with shortened internodes.

Stylet-borne: Referred to the viruses found in the stylet of vectors (insects or nematodes) and are transmitted through the stylet, during feeding.

Stylospores: An elongated pycnidiospore often with curved end looking like the walking stick, of unknown function as they donot germinate and occur alongwith normal small oval, round, functional pycnidiospores e.g. *Phomopsis* sp.

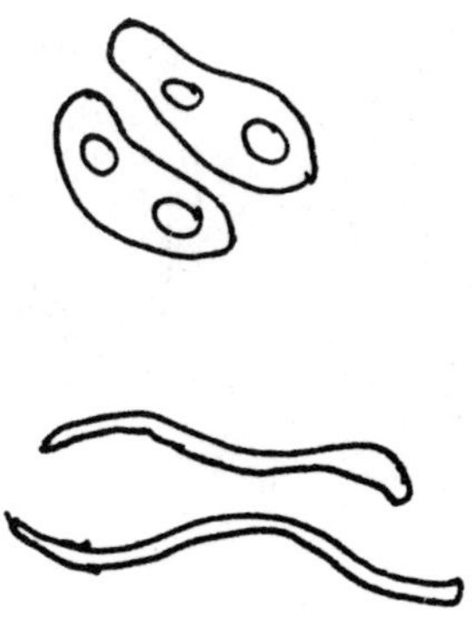

Stylospores

Subculture: The culture of the microorganism prepared from the originally isolated culture.

Subhymenium (Pl. Sybhymenia): A differentiated layer of tissue (hyphal mass) just below the hymenium which gives rise to elements of hymenium like asci, basidia or sterile structures.

Subiculum (Pl. Subicula): A loose mattress of the fungal hyphae bearing fructifications within or on it.

Suboperculate: An operculate ascus having an interrupted thick ring beneath the opening.

Substrate: (i) An object which provides anchorage and nutrition for the growth of microorganisms.

(ii) The chemicals on which an enzyme acts.

Subterminal: Next to the extreme end or apical.

Succession: Progressive ecological change in the fungal population of a substratum or habitate.

Sufu: Fermented food items of Orient prepared from soybean using *Actinomucor elegans* as fermenting organism. (*cf.* Tempeh).

Sugary disease: A common disease of sorghum (jowar) caused by *Sphacelia sorghi,* in which a stickly pink, sugary fluid is exuded from gynoecium of flowers which becomes brown or black. In some spikelets fungal sclerotia are formed. It is also called "ergot" disease.

Summer spore: The spores which germinate without any rest period and have a shorter life.

Sun scald: Superficial injury to fruits/plants by the strong sun light.

Suppuration: The process of pus formation in the tissues as a result of infection.

Surfactant: Wetting agent or spreader, mixed with the fungicide to help in spreading the chemical over plant surface.

Susceptibility: Inability of a host plant to resist the attack of a pathogen or other disease causing agent and become diseased.

Suspensor: A specialized hypha bearing the gametangium (and zygospore after the fertilization) at its tip e.g. Mucorales.

Swarm cell: Cells with flagella; usually motile cells of myxomycetes and Plasmodiophoromycetes.

Swath: The width of target area sprayed at one pass.

Symbiont: A partner in a symbiotic relationship eg. *Rhizobium* is a symbiont.

Symbiosis: The relationship between two organisms living together and usually providing some benefits to each other. It may also be harmful or neutral to one or both the partners. e.g. *Rhizobium,* lichen, mycorrhiza.

Symphogenous: Originating from many bodies (hypha) growing together, usually a fruiting body formed from a number of interwoven hyphae eg. *Zythia* (*cf.* simple and compound meristogenous).

Symplastic: Movement of chemicals within living plant parts, i.e. in the sieve tubes of phloem tissues in an active manner, requiring metabolic energy. It is usually in downward direction from leaves to root. Most of the systemic fungicides do not have symplastic movement. Only metalaxyl has

Symphogenous development

been reported to move symplastically.

Sympodial: A type of branching, in which the zig-zag type of growth produces conidia alternately on both sides with youngest at the apex.

Sympodium: A kind of false axis in which after formation of a terminal sporangium/spore the sporangiophore extends further beyond sporangium and produces second terminal sporangium/spore and it is repeated many times.

Symptom: Expression of physiological or morphological abnormalities during the course of disease development.

Symptomless carrier: Plants which carry some pathogen especially viruses but exhibit no symptoms of the disease themselves.

Synchronous: In this type of condial development, the conidiogenous cell is swollen and bears conidia synchronously all over the surface e.g. *Botrytis* sp.

Synascus: A compound ascus (spore sac) emerging from the thick walled resting cells in Protomycetales. It is supposed that in this multinucleate structure each nuclei divide meiotically forming 4 ascospores, in a wall less ascus and thus a number of such asci in this spore sac make it a synascus.

Syndrome: All the symptoms (symptom complex) providing complete picture of a particular disease.

Synergism: A phenomenon where the efficacy of two factors (pathogen, organisms, chemicals, environmental factors etc.) is enhanced if they act simultaneously in comparison to the sole effect of any of them.

Syngamy: Fusion of male and female cells to form the zygote.

Synkaryon: Nucleus (esp. of zygote) having double (2n) chromosomes, formed after karyogamy.

Synnema (Pl. Synnemata): A fassicle of conidiophores usually in upright direction.

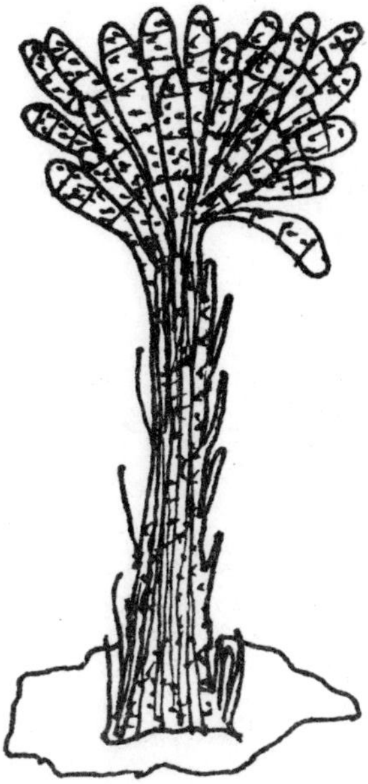

Synnema

Synonym: Other names given to the same organism.

Syphilis: A venereal disease caused by *Treponema pallidum*.

Systema Mycologicum: An old classic book on descriptions of fungi, written by a Swedish mycologist Elias Magnus Fries during 1821 to 1832. He for the first time started the nomenclature of fungi and is regarded as 'Liannaeus of mycology'.

Systematics: The branch of the science in which the organisms are grouped systematically according to their similarities of characters.

Systemic: Movement of pathogen or chemical (fungicide) throughout the plant system from one part to the other (usually from root to leaves).

T

Tabtoxin: A toxin produced by *Pseudomonoas tabaci* which is a type of wild fire toxin.

Taint: An unwanted odour in the food material due to a pesticide used on growing crop.

Takadiastase: Commercial enzyme preparation of fungal origin from *Aspergillus oryza* used for saccharification of rice starch.

Target Organ: Location of the initial infection of the viruses, particularly animal viruses. After getting established, virus particles may further spread to other regions of the body.

Taxonomy: The science of the classification of the plants or animals.

TBZ: Abbreviation for thiabendazole (fungicide).

Teleomorph: The ascosporic or basidiosporic, sexual or perfect stage of the fungus. See "Perfect stage". (*cf.* Anamorph).

Teleutospore: See Teliospore.

Teliomycetidae: A subclass of fungi under basidiomycetes class. Its members produce thick walled spores which germinate to give rise a finite germ tube called promycelium or metabasidium, that bears the basidiospores. Basidiocarps are absent in the group.

Teliospore (Teleutospore): A thick walled resting spore which is the site of karyogamy and produces the basidium after germination on the germtube or promycelium.

Telium (Pl. Telia): The final sorus (stage III) of the teliospores on the host surface in the life cycle of the uredinales.

Tempeh: A fermentation food product of soybean, popular in far-east countries, utilizing certain species of *Mucor, Rhizopus* and *Neurospora*, for fermenting (*cf.* 'Sufu').

Temperate phage: Bacteriophage which upon entery into a host bacterium does not cause its lysis, but remains associated with its chromosomes and multiplies alongwith host. It may start behaving as a virulent type causing the lysis of bacteria under certain conditions.

Tenacity: Property of the fungicide to act against removal of chemical by weathering.

Tentotoxin: A non-selective phytotoxin produced by *Alternaria tenui,* which causes reduction in the chlorophyll content and stimulates the stomatal closure.

Tenuazoic acid: A vivotoxin produced by *Alternaria longiceps* and *A. alternata* and inhibits the protein synthesis.

Teratology: Science of studying the abnormalities in shape and structure of any organism which may be of pathological nature.

Terminal Disinfection: Complete disinfection of a room or the building after being vacated by a patient suffering from a contagious disease.

Terminal infection: An infection by the pathogenic microorganisms which culminates in the death,

Terrachlor: See PCNB.

Terramycin: See "oxytetracyclin".

Terrazole: Trade name of "Ethazol". See "Ethazol".

Terrestrial: Forms with habit of growing on land or away from water (opposite of aquatic).

Terricolous: Forms (esp. lichen) growing in soils in hot area with scanty rains.

Tetanus: A serious disease of human being and animals caused by the bacterium *Clostridium tetani*.

Tetracycline: An antibiotic produced by *Streptomyces* species, which is very effective against a number of bacteria and mycoplasma diseases.

Tetrads: A morphological shape of the bacteria in which four cells are arranged as a square.

Tetrapolar sexuality: A type of sexual compatibility controlled by multiple allelic genes at two loci on separate chromosomes. In such type of cases each of the four basidiospores on a basidium are of different strains. (*cf.* bipolar sexuality).

Thallic: A type of spore production in which the whole conidiogenous cell forms the conidium, and conidial initial does not enlarges before septa are laid down. Conidia so formed are called thallic conidia or arthroconidia (*cf.* blastic).

Thallospore: Conidia produced by the conversion of whole cell or segment of hypha with no enlargement till septa are laid down.

Thallus (Pl. Thalli): Any simple vegetative plant body lacking root, stem and leaves and having a thalloid structure, e.g. Algae, Fungi etc.

Therapeutants: Chemicals which have a curative action on diseases.

Therapeutic: The ability of the chemical (fungicide) to cure the disease, which has already established on the host. Only few chemicals, mainly systemic fungicides and antibiotics have the curative action in case of plant diseases.

Therapeutic index: A scale for measuring the selectivity of the systemic fungicide to co-exist with living cells of host and effect on pathogen. It is calculated by dividing the minimum curative dose by the maximum tolerated (by host) dose.

Thermal death point: The critical temperature at which a particular species of the micro-organism are killed.

Thermal inactivation point: The lowest temperature which completely destroys the infectivity of a virus or the activity of an enzyme if, applied for 10 min.

Thermoduric: An organism or chemical which have the capacity to tolerate the exposure to high temperature.

Thermolabile: Referred to the micro-organisms or the enzyme preparations which are susceptible to the temperature and are destroyed by exposure to high temperature, which is even lesser than 100°C.

Thermophile: Micro-organisms which have their optimum growth at a temperature of 50°C or above.

Thermophilic Gr.: Fungal forms or other micro-organisms which have their optimum growth at relatively higher temperatures and can grow above 50°C, but can not grow below 20°C. e.g. *Aspergillus fumigatus, Chaetomium thermophile*.

Thermostable: Micro-organisms or enzyme preparations or chemicals which are resistant to heat and are not destroyed upto a temperature of 100°C.

Thermotolerant: Species which can grow on a wide range of temperature especially towards higher side. e.g. *Aspergillus fumigatus*.

Thiabendazole: Systemic fungicide of benzimidazole group, chemically being 2-(4'-thiazolyl)-benzimidazole; abbreviated as TBZ. It is thermostable and can be autoclaved at 121.5°C for 20 min. It has a broad spectrum of activity but is not effective againt Phycomycetous fungi and few dark coloured Deuteromycetes. Marketed as "Mertect" and "Tecto". Not marketed in India.

Thiolutin: A pyrothin type of antibiotic produced by *Streptomyces albus;* used for the control of late blight of potato and downy mildews.

Thiophanates: A group of heterocyclic compounds based on thiourea acting as systemic fungicide. They have a broad spectrum of activity.

Thiophanate-methyl: Systemic fungicide of thiophanate group having broad spectrum of activity. Chemically it is 1, 2-bis (3-methoxy carbonyl) 2-thioureido) benzene. Available under trade name "Topsin-M" or Cercobin-M or Rilan as 70 WP. On application, they are hydrolysed to MBC, which ultimately affects the fungi.

Thiram (Tetramethyl thuram disulphide): A fungicide of dithiocarbamate group widely used as seed dresser and spray. The trade name of this chemical is Thiride.

Thiride: Trade name of fungicide "Thiram".

Thymol: A volatile solid biocide used to protect small objects like books etc. from the attack of microorganisms.

Tikka disease: A leaf-spot disease of groundnut caused by *Cercospora personata* and *C. arachidicola*.

Tinsel Flagellum: A flagellum having small hairs over its surface (*cf.* whiplash flagellum).

Tirak: A non-parasitic disease of cotton, caused mainly due to adverse conditions of increased alkalinity of above 0.15 to 0.5% or deficiency of N in light sandy soils. Symptoms include, yellowing and redenneing of leaves, followed by defoliation at reproductive stage. Cracking of bolls and seeds.

Tissue: An aggregated mass of cells usually of similar morphology. In case of fungi also, the hyphae get aggregated forming a compact mass, looking like tissues of higher plants.

Tissue culture: A technique of simplifying organisms and growing them rapidly under laboratory conditions on defined media, so that they are readily available for experimental use or otherwise and may be handled more efficiently. This technique is used for faster multiplication/propagation of plants under aseptic conditions.

Titre: Standard or stock suspension of the virus particle in an appropriate medium, which is used for maintaining the virus cultures.

Tolclofos-methyl: A fungicide used for soil and seed treatment, especially against *Rhizoctonia solani, Pythium, Phytophtora, Sclerotium, Fusarium, Cercospora* etc. Chemically it is 0, 0-dimethyl-0-(2, 6-dichloro-4-methylphenyl)-phosphorothioate. Available under trade name "Rizolex" as 50 WP and 20, 10 or 5D.

Tolerance: Ability or characteristics of the host plant to endure the infection or infestation by the pathogen without showing the severe symptoms of distress or marked loss of yield.

Tolerance permitted: Maximum amount of toxicant allowed in the foodstuff for the human or animal consumption.

Tolerant: Individual organism which endure the infection without showing severe symptoms and much looses.

Tolypophagus mycorrhiza: See mycorrhiza.

Topsin-M: Trade name of "thiophanatemethyl". Also see "thiophanatemethyl".

Toxemia: Expression of various symptoms due to the presence of toxin in the blood.

Toxic: Property of being harmful or poisonous to plant, animal or man.

Toxicant: Any toxic substance, chemical or metabolic product of micro-organisms.

Toxin: Usually referred to non-enzymatic substances/chemicals produced by the micro-organism during their metabolism, which are lethal to plants, animals or man.

Toxoid: A toxin whose harmful properties have been destroyed by various treatments, without hampering its antigenic properties.

TPTA: See "Triphenyl tin acetate".

TPTC: See "Triphenyl tin chloride".

TPTH: See "Fentin hydroxide".

Tracer: A radioactive or fluorescent chemical molecule used to locate a pesticide in the plant system or other such chemicals in the metabolic pathway of the plant system.

Tracheomycosis: A fungal disease (wilt) in which pathogen remains mainly confined to xylem tissue of the plants.

Trama (Pl. Tramae): A layer of cylinder or sterile tissue which supports the hymenium in the members of Hymenomycetes.

Transduction: A process of sexual recombination in bacteria, in which DNA is transferred through a temperate virus without any destruction of the bacteria i.e. lysogenic state.

Transformation: A process in which the DNA portion from a related strain or virus is transferred and incorporated into the genetic material of the bacterium.

Translaminar: A fungicide which passes through the leaf tissues from one surface of lamina to the other and acts against the pathogenic fungi on that side of the leaves also.

Transmission: Dispersal or movement of the pathogen from the diseased plants to healthy one through various agencies like propagating material, wind, water, animals, insects etc.

Trap crop: Some crop plants are useful in reducing the pathogen population in soil by stimulating the dormant stage to become active and then checking their further life cycle. They are called "trap crops e.g. some crops produce chemicals which stimulate the nematode eggs to hatch, but after some times the larvae die, as they do not develop further.

Tree wood dressing fungicides: Fungicides which are used on the wounds or cut ends of trees to control diseases. e.g. Bordeaux paste, Chaubattia paste etc.

Tretic: Enteroblastic conidium formed and limited by the extension of the inner walll of the conidiogenous cell.

Triadimefon: Systemic fungicide of triazole groups, chemically being 1-(4-chlorophenoxy)-3, 3-dimethyl-1-1 H-(1, 2, 4-triazole-1-yl) butane-2-one. Effective against rusts, smuts, powdery mildews, bunts and *Helminthosporium, Septoria* etc. Trade name is Bayleton 50 WP or 100 EC. Also see "Bayleton".

Triamiphos: An organo-phosphorus systemic fungicide, effective against powdery mildews of ornamentals. Chemically it is 1-bis (dimethyl amino) phosphoryl-3-phenyl-5-amino-triazole-1, 2, 4. Due to high mammalian toxicity and phytotoxicity it has a low acceptability and use. It also has insecticidal and acaricidal properties. Trade name is "Wepsyn". Not marketed in India.

Triarimol: Common name of a systemic fungicide of pyridimidine group. Chemically it is alpha-(2-4 dichlorophenyl)-alpha phenyl-5-pyridine methanol. It was a broad spectrum fungicide, effective against apple scab and powdery mildew, but due to its toxic effect on animals, its further development was stopped.

Triazol: An organic compound having systemic fungicidal activity. Chemically it is 4-n-butyl-1, 2, 4-triazol and is available under trade name "Indar" 70 LC or 25 WP. It is selectively effective against wheat rust.

Trichogyne: Usually long and filamentous part of the archicarp (neck of the archegonium), which initially receives the male proto-

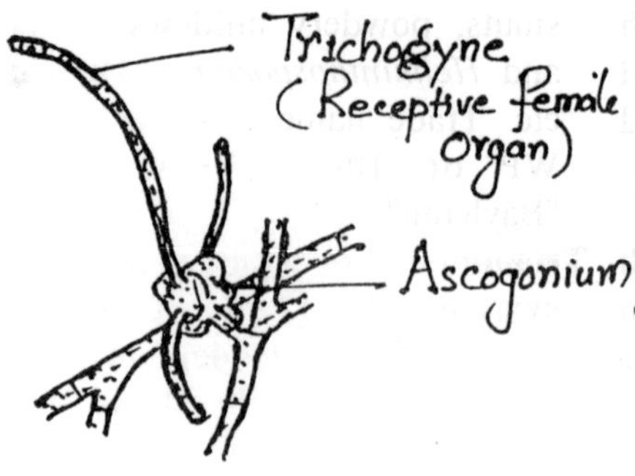

Trichogyne

plasm.

Trichome: A uniseriate, multicellular organ, just like hairs found on the surface of plants.

Trichomycetes: A class of fungi under Zygomycotina. Its members are obligate symbionts or commensals of arthropods having very limited mycelium; reproduce asexually by amoeboid cells, arthrospores or sporangiospores. Sexual reproduction is not confirmed.

Trichospore: A special structure found in order Harpellales of Trichomycetes which is exogenous, dehiscent, elongated sporangium, containing a single uninucleate sporangiospore and has one to many basally attached filamentous appendages close to the sporangial wall.

Tridemorph: Systemic fungicide of morpholine group. Chemically it is 2, 6-dimethyl-4-tridecylmorpholine. It is highly effective against *Mycospharella, Erysiphe, Botrytis, Phomopsis, Diplodia, Cladosporium, Septoria, Helminthosporium, Rhizoctonia* etc. Available under trade name "calixin" 75 EC or 5D. Also see "Calixin".

Trifolirhizin: An antifungal phytoalexin, glucoside, isolated from the roots of red clover.

Triforine: Common name for a piperazine derivative acting as systemic fungicide against powdery mildews, rusts and apple scab. Chemically it is N, N'-bis-(formamido-2, 2, 2-trichloroethyl) piperazine. Marketed as "Saprol" or "Funginex".

Trimitic: A condition when a basidiocarp consists of all the three (generative, binding and skeletal) hyphal types, (*cf.* monomitic, dimitic).

Tripan blue: A micro-scopic dye (stain) which specifically stains the hyphae of ustilaginaceous fungi (smut) inside the plant tissue; very useful in detection of smut infection present in the plant tissue.

Triphenyl tin acetate (TPTA): Organotin fungicide effective against *Cercospora, Septoria, Mycospharella, Pyricularia* etc. Available under trade name of "brestan" 40 or 60 WP. It is phytotoxic to many fruits, ornaments etc.

Triphenyl tin chloride (TPTC): Organo tin fungicide having activity similar to TPTA. Available under trade name "Brestanol" 45

WP. It has been reported to be systemic in paddy and sugarbeet seedlings.

Trisporic acid: A low molecular weight, diffusible, hormone found in members of Mucorales, which induces the formation of zygospores.

Trophocyst: An enlarged cell; or swollen portion of the sporangiophore of *Pilobolus* sp. a member of Zygomycetes.

Tropism: A bending or growth movement of plants or other organism like fungi etc. in response to some unidirectional stimulus of light, chemical gravity etc. e.g. Chemotropism, geotropism phototropism.

True slime molds: Commonly referred to the members of Myxomycetes. (*cf.* "cellular slime molds".

Truffles: Commonly used name for members of Tuberales under Discomycetes, some of which form highly prized food in Europe.

Tubercle: (i) A nodule on the surface of plant tissue.

(ii) The specific lesion of tuberculosis.

Tuberculin: Extract of *Tuberculosis bacilli* which induces inflamatory reaction in animal body having been sterilized by the presence of living or dead tubercle bacilli.

Tundu: A bacterial disease of wheat caused by *Corynebacterium tritici* showing twisting of the stem, distortion of the earhead, rotting of spikelet with profuse oozing of yellow liquid from affected tissue. In nature this disease occurs alongwith the ear cockle nematode of wheat.

Tungro: A viral disease of paddy caused by Rice Tungro Virus in which plants remain stunted and produce only few ears and leaves exhibit yellow-brown discolouration.

Turban: See "Ethazol".

Turbid: Referred for many solution/liquid which is not clear and transparent and has some fine suspended particles making it opaque or cloudy.

Turbidimetry: A technique of estimating the bacterial population in terms of degree of opacity (turbidity) of the suspension.

Twort-d' Herelle phenomenon: The process of clarification of the bacterial cultures (broth) by addition of bacteria free filtrate of sewage and, ability of the clarified broth filtrate to dissolve another lot of bacterial culture. The phenomenon occurring due to the activity of the bacteriophage, was first discovered by d'Herelle.

Type specimen: See Holotype.

Tylose: A bladder shaped structure arising from the wall of a vessel (conducting tissue of higher plant) in the wilt infected plants.

Tylosis: Phenomenon of tylose formation.

Tyndalization: A method of sterilizing the media and chemical which can not withstand the temperature of more than 100°C. They are heated at 100°C, consecutively for three successive days with incubation period in between enabling the resistant spores to first germinate during the incubation period and then destroy them by heating. Steam-Arnold apparatus is used for this process. The process is also called Fractional Sterilization.

U

Udbatta: A disease of paddy caused by *Ephelis oryzae* in which a brownish white, cylindrical, hard rod shaped structure is formed in place of normal ear. Diseased ear do not produce any grain.

Ultracentrifuge: A high speed centrifuge used for determining the particle size of the virus and protein, and also for separating the virus particles.

Ultrafiltration: Method of filtering out all but very small particles like viruses.

Ultra low volume spray: A spray in which concentrated solution of the pesticide is sprayed and only less than 50 litre of solution is sufficient for one ha of area in case of ground crops. In this type of spraying the chemical is released in large volume of blown air forming a sort of mist or fume of chemical in the atmosphere. It is also used for sterilizing the atmosphere by fumigation.

Ultrasonic waves: Sound waves of high intensity, beyond the audible range, which are used for destruction of micro-organisms, especially bacteria, and thus are helpful in sterilization process.

Ultraviolet radiation: A technique of sterilizing the laboratory, culture chambers etc. with a source of ultraviolet radiation with UV waves of 390-200 m μ wavelength, which kills the fungal spores and bacteria. Sublethal exposure of the UV rays causes mutation in fungi and bacteria.

Undulant fever: A disease caused by Brucella group of bacteria and is also termed as 'Brucellosis'.

Unilateral Hymenium: A type of hymenium found only on one side of the sporophore in Aphyllophorales. (*cf.* Amphigenous).

Unitunicate: Referred to the sacus having a single layer of wall or if double layered, both the layers are rigid and can not be separated from each other (*cf.* Bitunicate, prototunicate).

Universal veil: A thin membranous veil enveloping the entire basidiocarp during the early stages of the development of the fruiting structure. It is found in some Hymenomycetes.

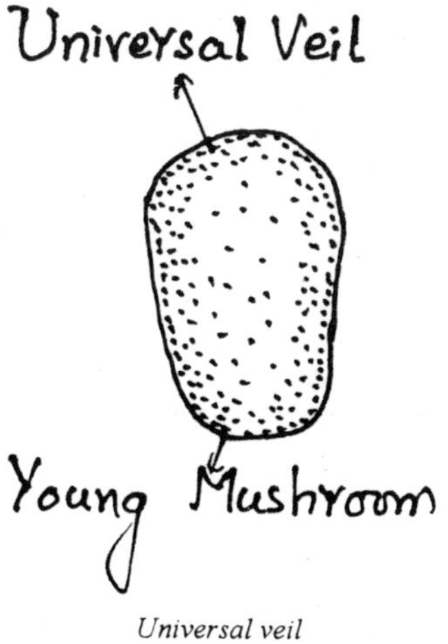

Universal veil

Univoltine: Organisms producing only one generation in one season.

Unrestricted type: Mutant of the bacteriophage which are capable of infecting a variety of bacterial strains. Restricted type may be induced to become Unrestricted type through various treatments. (*cf.* Restricted type).

Urease: An enzyme hydrolysing the urea; produced by many microorganisms.

Urediospore: A binucleate, spore borne in the uredium; found in members of Uredinales.

Uredium (Pl. Uredia): The sorus which bears the urediospore (stage II) in the Uredinales.

V

Vaccination: The process of introducing the vaccine in the host to protect it against the attack of a particular pathogen, whose vaccine has been used.

Vaccine: A suspension of weakened (attenuated) or killed viruses or other pathogens which on injection into the body of a susceptible host especially in case of the animals; imparts the immunity against the same type of pathogen or even against the toxins by stimulating the production of abtibodies but, it does not cause disease itself.

Vacuola: A clear space in the cytoplasm of the cell which is supposed to be participating in the gaseous exchange of the cell and some other vital metabolic processes of the cell.

Vapam: A fungicide of dithiocarbamate group used for soil treatment. It also acts against nematodes. Chemically, it is sodium methyl dithiocarbamate (SMDC).

Variant: Organisms showing some degree of variations from the parent culture.

Variety: A subdivision of the species, differing from each other in minor morphological, cultural and other characteristics.

Vector: Agents which transmit the propagules of the pathogens from infected host to healthy ones. e.g. Insects transmitting the viruses.

Vegetative: A cell or structure, which does not produce the reproductive structures usually in the assimilative stage.

Vegetative transmission: Transmission of the pathogen along with the vegetative propagating organs like tuber, rhizome, grafts etc.

Vein banding: A symptom of the virus disease, which show formation of dark green bands along the vein in the infected plants.

Vein clearing: A symptom of the virus diseases, in which vein of the young leaves have yellow margins.

Venutricidin: An antifungal antibiotic isolated from strains of *Streptomyces* sp.

Verruclose: Very finely verrucose.

Verrucose: One having small rounded projections, outgrowth, warts.

Vert de gries disease: An infection of commercial mushroom cultivation caused by the fungus *Myceliophthora lutea,* producing yellow to brown or dark green colour.

Vertical Resistance (Differential Resistance): Resistance of the host plant completely effective against some limited races of the pathogen, but not against all other races. It is usually controlled by a single major gene and is also called Monogenic resistance.

Verticil: A whrol of three or more branches of the conidiogenous cells.

Vesicle: (i) A bladder like sac or evanescent bubble within which, the zoospores mature.

(ii) The terminal bulbous head of conidiophore in case of *Aspergillus* sp.

Vestigial: Poorly developed, degenerate or nonfunctional organ or part of any organ.

Viable: Any organ, cell, tissue, seed or culture which is having the capacity to perform the metabolic activities, or in the living state.

Vibrio: A type of bacteria which have slightly curved body resembling with a comma.

Victorin: A toxin produced by *Helminthosporium victoriae* which is host specific and acts by altering the membrane permeability and respiratory process.

Vigil: See "Didobutrazol".

Vincent's angina: An ulcer of the tonsil and gums caused by *Spirillium* and Fusiform bacteria.

Virescent: Referred for the colour turning green.

Viriod: A ribonucleic acid of low molecular weight without protein coat which has the pathogenic capabilities and multiplies within the host cell, or may be called a single infective, naked nucleic acid.

Virion: A fully infective virus particle having a central core of nucleic acid and a protein coat; also called a Nucleoprotein unit.

Virosis: A virus disease.

Viruleferous: Referred to a insect vector containing the infective virus.

Virulence: A measure for the aggressiveness of a pathogen to cause disease on its host; degree of pathogenicity.

Virulent: Strongly pathogenic.

Virus: A submicroscopic, self replicating, non cellular obligate parasitic organism essentially composed of a protein coat surrounding a central nucleic acid molecule of DNA or RNA. They are able to pass through the bacteria proof filters. They are commonly termed as 'Nucleoprotein'.

Viscidity: Property of stickyness in the broth culture due to the bacterial sediment.

Vitamin C: A complex compound produced by some strains of *Aspergillus niger* which is essential for the normal growth and development of the living organism. Chemically it is Ascorbic acid.

Vitavax: A systemic fungicide of oxathiin group effective as seed dresser against smuts, bunts etc. having internal infection. Also see "Carboxin".

Vivotoxin: A toxic substance produced in the infected host by the pathogen and/or its host which functions in the generation of a disease and is responsible for all the symptoms. It is not itself the initiating agent of the disease.

Vogas-Proskause reaction (VP Test): A test to distinguish between species of Coliform group of bacteria in which presence of acetyl methyl carbino is detected.

Volunteer: Crop plants growing accidently from shed seed, without intentional cultivation.

Volva: A cup like structure at the base of a fructification found in basidiomycetes.

W

Wart: A disease of potato tuber in which abnormal, cauliflower like outgrowth are seen on the tuber and aerial parts also. The disease is caused by *Synchytrium endobioticum.*

Wassermann test: A complement fixation test for syphilis.

Water dispersible powder (WDP): A formulation of any fungicide/pesticide which are easily wetted and disperse well in water. Usually a wetting agent is present in these formulations. A frequent agitation is required in the tank to keep them dispersed in the solution. Such formulations generally have 5-50% fungicide, 1-10% non-ionic wetter like, polyethylene oxide condensate, thickening agent like, carboxy methyl cellulose and a hydrophilic diluent, such as bentonite. They are also called "Wettable powder" (WP).

Water storage: A method of storing the fungal cultures, in which the fungus growing on weak medium is cut into small pieces and transferred asceptically to sterile distilled water in screw cap bottles and stored at low temperature (15°C).

Wet rot: Rotting of the wood etc. requiring high humidity and moisture content above 50%, usually caused by basidiomycetous fungi like *Fibroporia, Coniophora, Phellinus* etc.

Wettable powder (WP): See "Water dispersible powder (WDP)".

Wetting agent: The material added to any spray solution for reducing surface tension and the contact angle of the liquid on the surface, and ensures the removal of air between solid and liquid, e.g. polyethylene, flour, ester of fatty acid.

Whiplash flagellum: A smooth long flagellum of motile cells without any appendanges or hairs on its surface (*cf.* tinsel flagellum).

White rust: A disease of the crucifers caused by *Albugo candida* showing white pustules of the conidiophores on underside of the leaves.

Whyerone acid: An antifungal substance found in broad been leaves infected with *Botrytis fabae*.

Widal test: A test used for the diagnosis of typhoid fever by agglutinating typhoid bacilli with patient serum on the slide.

Wild fire: A bacterial disease of soybean and tobacco, caused by *Pseudomonas tabaci*, exhibiting irregular, brown, necrotic spots on leaves.

Wild fire toxin: A non-specific pathotoxin (also called Tabtoxinine) produced by *Pseud-omonas* sp. which inhibits the glutamine synthetase activity.

Wilt: A common disease of many plants in which the pathogen enters in the vascular tissues of the host plant and blocks them leading to loss of the turgidity and drying of the plants.

Winter Spore: Fungal spores which remain dormant for certain period of unfávourable conditions and then germinate in the next season of favourable conditions after a rest period. They are also called 'Resting Spore'.

Witche's broom: A symptom of the yellow type of mycoplasma diseases in which a number of shoots are produced (abnormal proliferation), giving rise to broom like appearance.

Woronin body: A rounded organelle found near the septa in some Ascomyceteous fungi and also Fungi imperfectii.

X

Xanthotoxin: A phytoaexin isolated from the turnip root disc inocuated with *Ceratocystis fimbriata* which is a non-pathogenic fungus.

Xerophiles: Microorganisms which can grow at lower relative humidity (65-75%) like *Aspergillus glaucus* group *(A. penicillioides), A. candidus, Wallemia sebi (Sporendonema sibi), Monoascus bisporus.*

Xyloglucan: An important hemicullulose found in the primary wall of plant cells. It is made up of beta 1, 4 linked glucopyranose chain with terminal branches of 1, 6 linked d-xylopyranose.

Y

Yeast: A unicellular member of Endomycetales, sometimes used for a nonmotile unicellular stage for dimorphic animal parasites.

Yellow mosaic: A viral disease of legumes, viz. green gram, black gram, soybean etc. caused by bean yellow mosaic virus, exhibiting bright yellow and green patches on foliage. It is transmitted by white fly.

Yellow Sac: Membrane covering the yolk.

Yellows: A group of plant disease in which yellowing is the main symptom. Many of them are caused by mycoplasma.

Z

Zerlate: Trade name of Ziram 76 WP. Also see "Ziram".

Zeugite: Structure/organ where dikaryotic phase ends with the karyogamy/fertilization, like ascus or basidium.

Zineb: A dithiocarbamate fungicide (Zinc ethylene-bis-dithiocarbamate) widely used as seed and foliage protectant, commonly marketed as Dithane Z-78.

Ziram: A dithiocarbamate fungicide (Zinc dimethyl dithiocarbamate) used for many foliar diseases, commonly marketed as Zerlate and Cuman-L.

Zonate leaf spot: A symptom/disease in which the spot on foliage has several alternating zones or rings of light and dark colours. The disease is common in sorghum and is caused by *Gleocercospora sorghii*.

Zonation: A phenomenon of formation of alternate zones of sporulating and non sporulating mycelium caused mainly due to diurnal periodicity of light or different nutritional requirement of certain fungi for vegetative and reproductive phase.

Zone line: Narrow dark brown or black line in decayed wood (hard wood) generally caused by fungi.

Zoochory: Dispersal of pathogen through the agency of animals.

Zoological masses: Group of microorganisms in the form of film in a liquid.

Zoonosis: Animal diseases which may be transmitted to human also.

Zooplanktons: Group of nonphotosynthetic organisms (specially of animal kingdom) floating on water surface (*cf.* phytoplankton).

Zoosporangium (Pl. Zoosporangia): A sporangium producing zoospores within it.

Zoospore: A nonsexual motile spore bearing flagella.

Zygomycetes: A class of fungi under Zygomycotina subdivision. Its members are saprophyte or parasite, reproduce asexually by aplanospores and sexually by gametangial fusion forming zygospores.

Zygomycotina: A subdivision of Amastigomycota division. Its members are typically coenocytic, reproduce asexually by sporangiospores and sexually by gametangial fusion resulting in the formation of zygospore.

Zygophore: Specialised branches of hypha (in Mucorales) producing copulating structures.

Zygosporangium (Pl. Zygosporangia): A sporangium containing zygospores.

Zygospores: A thick walled resting spore resulting from the fusion of the gametangia, found in Zygomycetes.

Zygote: A diploid (2 n) cell resulting from the union of two haploid (n) gametes.

Zymophage: Viruses which infect and destroy the yeast.

SECTION III

NEMATOLOGY

A

Adanal: Connected/related with the bursa which does not cover the complete tail; near the cloaca/anus.

Adenophora: A class (taxonomic rank) of Nematoda, characterized by the presence of caudal adhesive glands, well developed mesenterial tissues and absence of phasmids.

Aerolation: Transverse line entering on lateral region of nematode body.

Alimentary canal (Gut): The tubular structure of the digestive system starting from the mouth or stroma (oval aperture) and terminating into anal opening or cloaca. In nematodes it is divided into 3 regions, (i) foregut or stomodaeum, (ii) midgut or mesenteron, and (iii) hindgut or proctodaeum.

Allotype: In the original description of the species, indicated as the sex other than that of Holotype; a type of paratype.

Ambifenestrate: Presence of two aperture on vulval cone which are separated by vulval bridge. It is found in some species of *Heterodera*.

Amphid aperture: Opening of the amphids present on labial or postlabial region. Amphid aperture is internally connected with a "pouch" of "fovea" which in turn is connected with the "fusus" of "sensilla pouch" through an amphid duct. The fuscus are connected to amphidial nerve through a nerve process. They are circular, hooked, stirrup, spiral or pore-like in different nematodes.

Amphidelphic: In female nematodes, a condition when the 2 sets of reproductive organs are located opposite to each other, one anteriorly and other posteriorly to vulva and vulva is located towards middle of the body (*cf.* Prodelphic).

Amphids: Lateral organs or appendages (2 in number) of cuticular origin present in the cephalic region of the nematode serving as sensory organs (mainly chemoreceptor). Shape and size of these amphids vary from one group of nematodes to other.

Amphigonus (dioecious): Condition of having male and female reproductive organs on separate individuals and those individuals are accordingly male and female.

Ampulla: Flask shape sac or swelling.

Anabiosis: State of remaining dormant with very slow metabolic activities under environmental stress.

Anguina: Commonly referred as "Gall nematode" attacking mainly wheat crop, forming hard dark coloured galls instead of grains and disease is referred as "ear cockle" disease.

Anisocytous: Nematode intestine having cells of different heights as seen in transverse section (*cf.* isocytous).

Annulations: Rings of thinner cuticle between regions of thicker portions looking like series of transversely running fine lines across the whole body of the nematodes, which are used for identifications. They impart flexibility to the reinforced body wall.

Annule: The space between two annulations of the nematodes, which may be capable of contraction and have outgrowths like, scales, spines etc. over them.

Anus: Terminal opening of the alimentary canal for discharge of faecial matter. It may be pore-like or slit-like.

Aphelenchoides: Foliar nematode, See "White tip disease".

Apolysis: First stage of the moulting in nematodes during which cuticle becomes separated from epidermis or hypodermis. (*cf.* cuticle formation, ecdysis).

Apophysis: A swelling or expansion in Z organs of nematode species of *Xiphinema* genus.

Arch: In nematode genus *Meloidogyne*, the portion of the perineal pattern situation on dorsal to lateral lines/region.

Arcuate: Referred to the shape looking like an arch.

Areolated: Divided into small sectors/regions. Commonly related with the cuticle of nematode.

Ascaris: A common nematode/parasite in intestine of animals, measuring 20 cm or more and 0.5 cm in width.

A Value: Value obtained as a ratio by dividing the length of the nematode from its maximum width. This value is used in calculating the de Mans formula.

Awl nematode: Common name for nematodes of *Dolichodorus* genus.

B

Basal bulb: See "Bulb".

Basal glandular region: Posterior most region of the nematode oesophagus which has many glands and is modified in different ways.

Basal lamella: Inner layer of nematode body wall which separates the rest of the cuticle from inner most layer of cuticle.

Belonolaimus: Generic name of a nematode commonly known as "Sting nematode" attacking cereals, legumes and cucurbits.

Bilaterally symmetrical: Condition when two portions of body divided in sagittal plane are exactly identical with each other; one looking like the image of other.

Bimodal: Performing 2 types of functions, e.g. internal labial papillae of nematodes act both as chemoreceptor and touch receptor (tactile).

Biology: Study of the different stages of the life cycle or life history of nematode in its natural sequence.

Biotypes: Sub-division or group of a species which are identical in genetical characters but differ in their morphology and physiology.

Blastomeres: Two equal cells or segments formed after the first cleavage of the egg during the embryo development of the nematodes.

Blastula: During the embryogenesis the early stage of embryo with fluid and lined with only single layer of cells.

Blind sac: Portion of the nematode intestine extending into the tail region forming sac-like structures. Presence/absence of blind sacs is an important character in identification of nematode species.

Bloat: Common name for an onion disease caused by *Ditylenchus dipsaci* nematodes.

Body regions: Portions/divisions of nematode body, created without sound morphological background. Usually they are (i) head, (ii) neck, (iii) tail, etc.

Body wall: Nematodes have a strong outer wall enclosing the whole body called body wall. It has 3 layers; outer cuticle, middle hypodermis and inner layer of muscles or musculature.

Buccal cavity: See "Stoma" and "Mouth".

Bud nematode: Common name for species of *Aphelenchoides* nematode, attacking the above ground plant parts, especially buds.

Bulb: One or more muscular swellings present in the middle or base of the oesophagus of nematodes which may be "true bulb" if they have valves, whereas in the absence of valves they are called "pseudobulb". Those present in the middle of oesophagus are called median oesophageal bulb, while those at the base are called "basal bulb".

Bulb nematode: See "Ditylenchus".

Burrowing nematode: Common name for *Radopholes* nematode attacking banana, lemon, coffee, sugarcane, etc.

Bursa: See "Caudal alae".

B Value: A ratio depicting the relative length of oesophagus of nematodes. This value is obtained by dividing the full body length by the length of apical/anterior end to the base of the oesophagus. This value is used in de Mans formula.

C

Caenorhabditis: Generic name of a nematode which feeds on bacteria. It is a model organism to study the development of the nematodes.

Cap cell: Apical region or germinative zone of the nematode ovary where oocytes are formed.

Carbofuran: A systemic granular nematicide used for soil treatment to control root knot nematode of vegetables. It is also used as a common insecticide. Commercially it is available under the name of "Furadon".

Cardia: A muscular structure of nematodes usually located between oesophagus and intestine.

Carnivorous: Feeding on animals.

Caudal alae (Bursa): Lateral wing like cuticular expansion around tail of male nematode which helps in holding the female nematode during copulation.

Caudalids: Sensory organs found in front of tail and represent paired anolumber commissures, connecting the lumber ganglion with preanal ganglion. Their exact functions are not known, but probably they are similar to cephalids or hemizonids.

Central Nervous system: See "Nerve ring".

Cephalic papillae: Four cuticular projections serving as sensory organs present only on submedian lips (labia), covering latero-dorsal and latero-ventral regions.

Cephalic sensory organs: Sensory organs present on the anterior region (especially head or lip region) of the nematode.

Cephalids: Band-like, highly refrective, 2 pairs of sensory organs forming a full ring around body, present in the anterior region on ventral or dorsal sides. First pair is located just after the head region and second pair follows it. Probably they act as pressure receptors.

Cervical: Region of the nematode body near or on the neck.

Cervical cell: See "Renette".

Cervical papillae: See "deirids".

Cheilorhabdions: Cuticularized walls of cheilostom of nematode mouth.

Cheilostome: Anterior portion of the nematode mouth formed by the invagination of outer cuticle and is not surrounded by oesophageal tissues (*cf.* Oesophagostome).

Chemoreceptors: Sensory organs which are sensitive to and receive the chemical stimulus, e.g. internal labial papillae which are tactile also (*cf.* tactile, touch receptor, mechanoreceptors).

Chloropicrin: A volatile, non-inflamable aerosol formulation used as soil fumigant mainly to control nematodes. Its chemical name is "Nitrochloroform" or "Trichloro-nitro methane", CCl_3NO_2.

Circomyarin: One of the 3 types of somatic muscle cells of nematodes which are rounded and contractile fibres are exactly in the centre of its circumference. Such cells are rare and are found in special muscles of vulva, spicules, etc. (*cf.* Platymyarian, Coelomyarian).

Citrus nematode: Common name for nematode genus *Tylenchulus* attacking mainly lemon, grapes, olive, etc.

Coarse root: A disease symptom in which growth of lateral roots is checked due to nematode attack when they are very young. In such cases, root system lacks lateral branches on main root, e.g. *Belonolaimus longicoudatus* attack on maize.

Coarse striations: Striae which are quite deep and pronounced on the nematode body and form relatively bigger intestinal region.

Coconut Nematode: Common name used for nematode *Rhadinaphelenchus cocophilus.*

Coelome: Body cavity of animals between the body wall and alimentary canal.

Coelomyarian: One of the 3 types of somatic muscle cells which have narrow base and contractile elements extending upto margins of cells towards pseudocoelome particularly surrounding the sarcoplasm. (*cf.* Platymyarian, Circomyarian).

Complex diseases: Diseases caused by the mixed infection of more than one pathogens.

Constricted ring: A ring formed by the hyphae of nematode trapping fungi which is just like non-constricted ring, but is 2-celled. Entering of nematode inside the ring stimulates the ring cells to suddenly swell to 3 times as a results of which nematode is tightly held inside the ring.

Conus: See "Spear cone".

Coparasitism: More than one parasites, parasitizing the same host.

Copulatory complex: Structures like spicules, gubernaculum present in the cloaca of male nematodes and the caudal alae or bursa present in the tail are main components of copulatory complex which take part during the copulation.

Cortical layer: External layer of nematode cuticle.

Criconematid: A type (modification) of oesophagus in nematodes, where procorpus and metacorpus get fused and isthmus is small, below which is a small basal bulb.

Crusa: Sclerotized lateral pieces of the gubernaculum which are connected with the cuneus through a corpus. The crusa and corpus form a groove on which the spicules slide.

Crustaformeria: Large secretory glands of nematode uterus located posterior to spermatheca. When they are arranged in 3 rows they are tricolumella, and if in 4 rows, quadricolumella.

Cuneus: Middle portion in between the two grooves of gubernaculum which is connected with the crusa through a transverse lobe or corpus.

Cuticle: Outermost layer of nematode body wall which is secreted from the epidermal cells situated beneath it. Cuticle is non-cellular, colourless, non living, elastic outermost layer which provides protection to internal tissues. It usually has 3 layers but some authors have described 4 layers. Nematode cuticle is made up of proteins like, albumins, glucoproteins, collagen and keratin like fibres. In certain nematodes (*Heterodera, Globodera*) the cuticle of cysts have quinones and polyphenols, which impart brown colour to cysts.

Cuticle formation: Second stage of moulting in which new cuticle is formed over epidermis. (*cf.* apolysis, ecdysis).

c Value: Value (ratio) depicting the relative tail length of nematode, calculated by dividing the body length by tail length. It is used in de-Mans formula.

Cyathiform: Shape looking like a cup.

Cyst: Oxydised cuticle of dead adult females of *Heterodera* nematode which provides protection to the eggs inside it.

Cyst nematode: Common name for genus *Heterodera* and *Globodera* of nematodes, attacking tobacco, soybean, sugarbeet, cereals, etc.

D

Dagger nematode: Common name for nematode genus *Xiphinema* attacking trees, grapes, soybean, etc. and is vector for many viruses of Nepogroup.

Dauer larva: A specialized third stage larva of nematode formed under unfavourable conditions which is resistant to dessication and harmful chemicals. Its epicuticle is relatively thicker and they have a striated layer in basal zone.

D-D: It is 50:50 mixture of 1, 3-dichloropropene and 1, 2-dichloropropane and is used as soil fumigant to control nematodes. It is available under trade names of "D-D", Nemafume, Telone, etc.

Deirids: A pair of small protuberances present in the middle of lateral region near excretory opening. They are sensory organs of cuticular origin without any opening and are also called "cervical papillae" as they are identical to cephalic papillae present on submedian lips. Probably, they act as mechanoreceptors.

de Mans formula: A system proposed in 1884 to describe the measurements of nematodes, used for taxonomic identification wherein n-specimen studied; L-body length in mm or microns; a-body length divided by maximum width; b-body length of oesophagus; c-body length/length from cloaca to the tail end; C-tail length/body width at cloaca; V-length from anterior end to vulva x 100/body length; T-distance from cloaca to anterior end of testis x 100/body length; stylet length in micron, stoma length, length of spicule and gubernaculum.

Denticle: A small teeth or outgrowth found in the mouth of few nematodes.

Deutoplasm: See "yolk".

Didelphic: Having two sets of reproductive organs in female nematodes (*cf.* monodelphic).

Digitate: A finger-like projection.

Diorchic: Nematodes having two testes (male gonads). (*cf.* monoorchic).

Ditylenchus: Nematode (generic name) attacking stem or bulb of onion and few other crops; commonly known as "bulb or stem nematode". Also see "Ufra or dark pora disease".

Diverticulum: Chamber or sac in alimentary canal of nematodes, where oesophageal glands are situated.

Dorlone: A commercial preparation containing E.D.B. and D-D being used as nematicide for soil treatment.

Dormant: Stage of an individual when it stops or slows down most of its metabolic activities due to unfavourable environmental conditions.

Dorsal side: Longitudinally, the side of the nematode body which lies just opposite to the ventral side.

E

Ear cockle disease: Nematode disease of wheat caused by *Anguina tritici* in which stem base is enlarged, leaves are crinkled and twisted and may have galls. Ear formation is advanced by 30-40 days and affected ears are more broad and short remaining dark green for longer periods and have galls or cockles in place of seeds. Such ears have less developed awns or glumes. Also see "Anguina".

Ecdysis: Third and last stage of moulting in nematodes, in which old cuticle is liberated. (*cf.* apolysis, cuticle formation).

Ectoparasite: Parasitizing on the surface of host and feeding on surface layers (*cf.* endoparasite).

E.D.B.: Chemically it is ethylene di bromide, $C_2H_4Br_2$ which is colourless, volatile liquid and is used as fumigant to treat the soil for nematode control. It is also used to fumigate stored grains.

Egg: First stage of the life cycle. In nematodes fertilized eggs have zygote which develops into larvae.

Egg sac: Slimy seretions of rectal glands in *Meliodogyne* where eggs are collected.

Ejaculatory duct: The distal posterior portion of the male gonoduct of plant parasitic nematodes, next to vas deferens, becoming gradually narrower and forms the cloaca in conjuction with rectum. It helps in liberation/ejaculation of sperms during copulation/fertilization.

Embryology: Study of the development of embryo from the egg.

Endocuticle: The inner most layer (of the 4 layers) of nematode cuticle described by Magentii, which is covered by mesocuticle. (*cf.* epicuticle, exocuticle, mesocuticle).

Endoparasite: Individuals (especially nematode) which enter inside the plant (host) body and draw their nutrition from there. Body of such parasite gets embeded in the host tissues.

Endozoic fungi: Fungi entering and parasitizing the nematodes, e.g. *Nematophthora gynophila* fungus parasitizing on *Heterodera* spp.

Epiboly: Growth of another body around the first for Gastrulation during the development of embryo in nematodes.

Epicuticle: Outermost of the 4 layers of nematode cuticle described by the Magentii, which covers exocuticle. (*cf.* exocuticle, mesocuticle, endocuticle).

Epidermis (Hypodermis): Second layer of nematode body wall just beneath the cuticle, made up of living cells and is syncytial. It becomes thick on all the 4 sides of body forming 4 (dorsal, ventral and 2 lateral) longitudinal hypodermal chords.

Epiptygma: A cuticular, membranous structure present on vulva of certain nematodes.

Epithelium: A curtain like layer of cells forming the internal lining of some organ or covering of outer layer.

Excretory pore: External opening of excretory system of nematodes situated towards base near oesophageal region.

Exocuticle: Of the 4 layers of nematode cuticle, the second one from the outer side which is covered by epicuticle and covers the mesocuticle. (*cf.* epicuticle, mesocu-ticle, endocuticle).

Exteroreceptors: Sensory organs present on the outer surface of nematode body, like phasmids, amphids, etc. (*cf.* interoreceptors).

F

False root knot nematode: Referred commonly to genus *Nacobbus* of nematodes.

Father of Nematology: The renowned nematologist Bastian (1837-1915) is called "father of Nematology" in view of his outstanding contribution in the field of nematology during its infancy. He wrote the "Monograph of Anguilluildae".

Fecundity: Capacity to reproduce.

Feeding apparatus: Mouth or stoma of nematode in connection with oesophagus forms the feeding apparatus.

Female sexual branch: Female reproducing organs of nematode consisting of ovary, oviduct, uterus, vagina and vulva.

Fenestra: A window-like opening of any organ or passing through the organ in nematodes. It is found in vulval cone of *Heterodera.*

Fensulphothion: A nematicide of organophosphate group, available as 5 or 10% granules under the trade name of "Dasnit". It is used @ 4-8 Kg. a.i./ha to control root knot nematode. Chemically it is 0-0, ethyl-0 1(methyl sulfinyl) phenyl) phosphoro-thionate).

Fibrillar layer: Inner most layer of nematode cuticle, which forms basal lamella in inner region connecting the epidermis.

Fiery Serpents: Common name of some animals (probably Guinea worms) described in ancient "Herbew laws of Sanitation" affecting peoples in deserts.

Fine striations: Striae which are very shallow and close to each other on the body of nematode forming very small interstrial region.

Flame cells: Commonly used term for solenocytes.

Foliar nematode: Referred to nematode genus *Aphelenchoides* attacking leaves of paddy, coconut, guldaudi, begonia, etc. Also see "white tip disease".

Foregut: See "Stomodaeum".

Fovea: See "Pouch".

Fusus (Sensilla pouch): See "Amphid aperture".

G

Gall nematode: See *"Anguina"*.

Ganglion: A definite group of nerve cells forming the central nervous system of nematodes.

Gastrula: Early stage embryo of nematode, which is formed from the hollow, double layered, cup shaped structure.

Genital papillae: Sensory organs located on tail of nematode, which help in reproduction.

Genital tubes: Middle portion of the nematode reproductive system lined with epithelium, connecting the gonads with genital aperture and through which germ cells are transported. They are vas-deference in male and oviduct in females.

Germinal zone: Apical region of the telogonic type gonads where germ cells are produced which is followed by growth zone where germ cells develop and divide.

Giant cell: See "Syncytia". Also called "Lysogenoma".

Golden nematode of potato: Nematode disease of potato caused by *Globodera rostochiensis.* Affected plants become weak and dwarfed in patches and start wilting during day time. Root system does not develop and secondary roots develop from collar region itself. Yellow cysts of nematode may be seed on roots.

Gonads (Reproductive glands): Main organ of the male or female reproductive systems where germ cells or gametes are produced.

Gonoduct: Tubular structure lined with epithelium connecting the gonads either directly outside the body through a genital opening (in female) or via the cloaca (in male).

Gonopore: External openings of genital glands, i.e. vulva in female and cloaca in male nematodes.

Gravid: Full of eggs (in nematodes).

Gubernaculum: A sclerotized, plate like or some time trough-like, structure of copulatory complex of male nematode formed by dorsal wall of the spicular pouch and located dorsally to the spicules. It has two grooves, which guide the spicules during copulation and protect them from breaking.

Guiding ring: Cuticularized structure, surrounding the stylet found in certain nematodes.

Guinea worm: Probably the first described nematode, *Dracunculus medinensis* living in arms and legs of human being and causing severe pain and swelling.

Gut: See "Alimentary canal".

H

Hatching: Liberation of juvenile nematodes from the egg after breaking the egg shell.

Hatching factor: Factor (chemical etc.) whose stimulus is essentially required for the hatching of certain nematodes, e.g. root exudates of Solanaceous plants stimulate the hatching in *Globodera rostochiensis*.

Head: Anterior portion of the nematode body comprising of mouth, lips and stoma. Some nematologist extend the Head portion even upto nerve ring.

Helminthology: Study of nematodes, parasitic on animals and humans.

Hemizonids: Belt or girdle-shaped, highly refractive, biconvex structure forming a semicircle connecting the nerve ring with ventral nerve cord representing the subventral commissure. It is located in between cuticle and epidermis near excretory opening terminating on lateral fields. They are sensory organs and are probably related with neurosecretory function.

Hemizonion: Sensory organ similar to hemizonids but relatively smaller and located posteriorly to hemizonid. Its function is also similar to hemizonid.

Hermaphrodite: Individual having both male and female sex organs in its body and functioning both as male and female.

Heterocytous: Made up of cells of different shape and size (*cf.* homocytous).

Heterodera: see "Cyst nematode".

Hexaradiate: Structure/organ having six radially arranged structures like the mouth of nematode which has 6 lobe like structures called labia.

Hindgut (Rectum or Protodeum): Small flat tubular and posterior (last) portion of the nematode alimentary canal starting from the end of midgut and terminating into cloaca, formed due to invagination of endoderm during embryogenesis. Its lumen is lined with the cuticle secreted by epidermis and is separated during moulting. It opens outside through anus or cloaca.

Hirschmanniella oryzae: See "Rice root nematode".

Hologonic: A method of germ cell proliferation where germ cells are produced along the full length of the gonads. (*cf.* telogonic).

Homocytous: Madeup of cells similar in shape and size. (*cf.* heterocytous).

Hoplolaimid: A type of nematode oesophagus in which metacorpus is well developed and valvular apparatus has big, well developed crescent shaped plates and basal glandular region is lobe like.

Hypodermal chords: See "Longitudinal epidermal chords".

Hypodermis: See "Epidermis".

I

Incisures: Synonym for lateral ridges.

Inner body tube: The alimentary canal in nematode is also called inner body tube.

Instars: Stages during the development of nematodes which are formed as a result of moulting of the earlier stage (instar).

Interchordial zones: Four longitudinal regions of nematode body in between the 4 longitudinal epidermal chords. Somatic muscle cells are arranged in these zones.

Interoreceptors: Sensory organs present inside the body of nematode, especially on the cuticular layer of digestive and reproductive system. (*cf.* exteroreceptors).

Intersex: Nematode individuals exhibiting both male and female characters. Usually nematode intersex are females having male characters like rudiments of spicules, gubernaculum, etc. In some species of *Meloidogyne* male intersex exhibit rudiments of female organs like vulva and vagina, etc.

Intestinal fasciculi (lateral canals): Side canals of the nematode intestine which may spread in tail cavity, if blind sac is present.

Interstrial region: Area between the two striae on the surface of nematode body.

Intestine: See "Mesenteron".

Intestinio-rectal valve: Valve made up of intestinal cells, separating the rectum from intestine. This valve is closed through a unicellular annular sphincter muscle. This valve is also called "pylorus".

Involutures: Synonym for "Lateral ridges".

Isocytous: For intestine exhibiting cells of uniform height in transverse section (*cf.* anisocytous).

Isthmus: A narrow region below the metacorpus of the oesophagus, which goes upto basal bulb of the oesophagus.

J

Juveniles: Very young stages of the individuals, especially of nematode, during their development, before they attain adulthood. May also be termed as "larvae". Nematodes usually have 4 larval stages, L_1, L_2, L_3 and L_4.

L

Labia (Sing. Labium): Six lobe-like structures arranged radially and forming the mouth of nematode. They are also termed as "lips". Of these, 2 labia are arranged laterally, 2 subdorsally and rest 2 subventrally in position. They have labial paillae on their surface.

Labial disc: Raised region around oral aperture (mouth) of nematodes, especially in *Criconemoides.*

Labial papillae: Small cuticular projections on the labia or lips of nematode mouth, which act as sensory organs. There is an inner circle of 6 papillae present on the inner surface of lips towards mouth, whereas 6 more papillae are found as outer circle on the outer side of lips.

Lamellae: Slightly raised structures above the body surface running along the length of nematode body, normally situated outside the lateral regions. They are also called "ridges" and are important characters of identification in some orders.

Lance nematode: Common name of genus *Hoplolaimus*, attacking maize, sugarcane, cotton, etc.

Larva (pl. Larvae): See "Juveniles".

Larvicide: Chemicals killing the larvae of nematodes. Most of the soil fumigating chemicals act as larvicide and are not effective against eggs and cysts.

Lateral fields: See "Lateral lines".

Lateral lines: Longitudinal striations of the nematode cuticle, present on the lateral sides of nematode body. They are usually raised above the body surface and such raised lines are also referred as ridges or lamellae. They are usually 1 to 5 in number and help in locomotion. They are also termed as lateral fields, "incisures or "involutures".

Lectotype: A specimen of original category selected after the actual description of the type.

Lesion nematode: *Pratylenchus* (generic name) nematode attacking many crops.

Lips: See "Labia".

Longitudinal epidermal chords: In nematodes, the epidermal region of the body wall becomes thick on all 4 regions of the body forming 4 chords on dorsal, ventral and lateral sides. The lateral chords are more developed than dorsal and ventral ones.

Longitudinal nerves: Nerves emanating from the nerve ring of nematodes extending in anterior and posterior regions of the body. Six nerves move anteriorly for innervation of cephalic sensory organs. Nerve from nerve ring also go posteriorly on dorsal, lateral and basal sides.

Longitudinal striations: Striations or lines on the cuticle of nematodes which extend along the length of body (*cf.* Transverse striations).

Lumen: Tubular structure or inner

cavity of any duct, especially of intestine of nematode.

Lysigenoma: See "Giant cell".

L-value: Complete length of nematode. This value is used in deMans formula.

M

Manubrium: Nearer portion of the spicules of nematodes.

Matrix: (i) Gelatinous substance found in *Meloidogyne* and few other nematodes in which eggs are deposited forming an egg-mass.
(ii) Middle layer of the nematode body cuticle. Also called "Homogeneous layer.

Mechanoreceptors: See "Tactile".

Median bulb: See "Metacorpus".

Median oesophageal bulb: See "Bulb".

Meloidogyne: See "Root knot nematode".

Meromyarian: Type of arrangement (of the two) of somatic muscle cells in which only 2 to 5 muscle cells are present in each interchordal zone. (*cf.* Polymyarian).

Mesenteron (Midgut or Intestine): The middle portion of the alimentary canal between stomodeum and proctodeum. It is a hollow, tubular structure madeup of a layer of endodermal epithelial cells and is usually divided into: (i) anterior ventricular, (ii) mid-intestine, and (iii) pre-rectum.

Mesocuticle: The third layer of the nematode cuticle from the outer side. It covers endocuticle and is covered by exocuticle. (*cf.* epicuticle, exocuticle, endocuticle).

Metacorpus (Median bulb): Middle portion of the oesophagus connected to procorpus on anterior side and basal bulb posteriorly through isthmus.

Metarhabdions: Cuticularized walls of metastome.

Metenchium: See "Spear cone".

Methomyl: Its chemical name is S-methyl N-1 (methyl carbamoyl) oxyl thioacetamidate, and is used as systemic nematicide for soil treatment, seed, root, sucker treatments. It is available as ‘Lannate’ or 'Dunet'.

Methyl bromide: A soil fumigant for controlling the nematodes in soil.

Microphagous: Feeding on microbes (especially for nematodes).

Micropyle: Minute hole in the egg through which spermatozoa (male gamete) enter inside the egg for fertilization.

Microvilli: Minute projections from the plasma membrane forming the lining of lumen in nematode intenstine. They increase the absorption area of intestine.

Midgut: See Mesenteron.

Migratory nematodes: Nematodes which remain capable of moving from one place to other during their complete life cycle. They may be endo-or ecto-parasitic, e.g. *Pratylenchus, Anguina, Trichodorus* (*cf.* Sedentary nematode).

Molya: A disease of wheat and barley caused by nematode, *Heterodera avenae* in which host plants become dwarf, hard, thin, yellow and stunted in patches with narrow leaf blades. Such plants rarely produce ears.

Monodelphic: Having only one set

of reproductive organs in female nematodes (*cf.* didelphic).

Mono-opisthodelphic: Condition when only one set of the female reproductive organ is located posteriorly to the vulva (*cf.* mono-prodelphic).

Monoorchic: Nematodes having only one testes (male gonad) (*cf.* diorchic).

Mono-prodelphic: Only one set of female reproductive organs located anteriorly to the vulva (*cf.* Mono-opisthodelphic).

Moulting: Separation of the older cuticle layer of nematode body and formation of new cuticle after a definite period of age. Normally from larva to adult stage nematode moults 4 times. This phenomenon is also called "Ecdysis".

Mouth: The oral opening of the digestive system, consisting of six lobe-like structures called "labia" or lips. Also see "Stoma".

Mucosa: A mucos membrane.

Mucronate: Ending into a sharp point; with pointed ends.

Mycophagous: Nematodes feeding on fungi.

N

Neck: In nematodes the region between head and base of the oesophagus.

Needle nematode: Common name for nematode, *Longidorus* which may act as vector for viruses.

Nematoda: Phylum (taxonomic rank) of animals, in which all the nematodes have been placed for taxonomic purpose.

Nematode Gr.: Worm-like, usually microscopic, triploblastic, bilaterally symmetrical, multicellular, non-segmented, pseudocoelomate, invertebrate animals, living saprophytically in soil or water or may be parasitic on animals and plants. Also called "Eelworm", "Roundworm"—etc.

Nematode trappers: See Nematophagous fungi.

Nematode-Virus complex: Nematode act as vectors of viruses attacking plants and this association of virus and nematodes during transmission of viruses is called Nematodevirus complex. Mainly 4 nematodes, *Xiphinema, Longidorus, Trichodorus* and *Paratrichodorus* are known to transmit the plant viruses.

Nematophagous fungi: Fungi predating on nematodes in soil and thus help in reducing their population, e.g, *Arthobotrytis, Trichothecium, Dactylaria, Dactylella,* etc. They are also called "Nematode trappers".

Nemax: Soil treating commercial nematicide, containing mixture of D-D and chloropicirin.

Nemin: Substance/chemicals which stimulate the nematode trapping fungi to produce the trap to capture the nematode.

Neotylenchid: A type of modification of tylenchid oesophagus in which median bulb (metacorpus) is without valvular plates and oval, pseudo-bulb may or may not be present.

Neotype: The original specimen actually collected from the original site of description which has been lost or destroyed.

NEPO Virus: Nematode transmitted viruses with polyhedral particles. This group of viruses are transmitted by spp. of *Xiphinema, Longidorus* and *Paralongidorus.* There are mainly 15 viruses under this group.

Nerve ring: A circumoesophageal, circum-intestinal or circum-isthmus ring having only nerve fibre but no nerve cell, forming the central nerveous system or brain of nematodes. Longitudinal nerves originate from nerve ring.

NETU Viruses: Nematode transmitted viruses with tubular shaped particles. They are also called "Tobra viruses". There are two viruses under this group, tobacoo rattle virus and pea early browning virus, which are transmitted by spp. of *Paratrichodorus* and *Trichodorus.*

Non-constricted rings: Some

nematode trapping fungi like *Dactylaria candida* produce a 3 celled ring of the hyphae which catches the nematode entering inside it and hyphae enter inside the nematode body. (*cf.* Constricted ring).

Nurse cells: Specially modified cortical cells of host formed as a result of invasion of nematodes which have dense protoplasm, bigger nucleus and nucleolus and lack vacuoles. They surround the invading nematode and provide nourishment to it, e.g. Infection of female nematode of *Tylenchulus semipanitrans* on citrus roots.

O

Odontium: A teeth located in the mouth.

Odontostyle (odontostylet): Buccal stylet or a big teeth-like structure originating from the oesophageal wall.

Oesophageal glands: Unicellular, uninucleate glands present in the basal region (one in dorsal sector and 2 in subventral sectors) of oesophagus which open in the lumen of oesophagus through a duct. Before opening duct form an ampulla fitted with one way valve to control the flow of secretions of glands. These glands are also termed as "salivary glands" and their secretions as saliva.

Oesophago-intestinal valve (Cardia): Valve in between the oesophagus and intestine of nematodes, which is a part of stomodaeum (foregut) and controls the passage of food from oesophagus to intestine and checks regurgitation. This valve may be of different shapes, like conoid, flat, round, collar-like, disc-like, etc. in different species.

Oesophagostome: Posterior portion of the nematode mouth, formed by the apical portion of the oesophagus and is covered with oesophageal tissue (*cf.* cheilostome).

Oesophagus (Pharynx): Second portiôn of the nematode alimentary canal next to the mouth and connected with the intestine on other end serving for food transportation, acting as pump for sending the food from low pressure mouth to high pressure intestine. Internally it is lined with cuticle and externally is separated from pseudocoelom by a basal lamella.

Olygocytous: Made up of only few cells (*cf.* polycytous).

Onchiostyle: A type of stylet found in certain nematodes where apical portion of stylet is formed from the specialized cells in the apical region of the oesophagus (*cf.* Stomatostyle).

Oocyte: A pre-matured female gamete.

Oogonium: Place for the early development of female gamete.

Operculum: Lid or door-like structure.

Opisthodelphic: When female reproductive organs of nematode are located posteriorly to vulva. If only one set of organs is present and is at posterior position to vulva, it is called "Mono-opisthodelpic" and is such cases vulva is located either at the middle position or on anterior side.

Ovary: A hollow, tubular structure lined with flat, membrane or epithelial cells which is the female gonad of nematodes and contains few to many oocytes. It has 3 zones: (i) germinative zone or cap cell, (ii) multiplication zone, and (iii) growth zone.

Oviduct: A narrow, tube made up of stout, epithelial cells which connects the ovary and uterus.

Mature eggs first come in the oviduct and then move further to uterus through it.

Oviparous: Nematode which lay eggs outside the body and further embryonic development takes place there.

Ovoviviparous: Nematodes in which eggs have shells but are not laid outside the body and juvenile nematodes develop intra-uterinary.

P

Paralongidorus maximus: A nematode known to be of maximum length (more than 1 cm) amongst the plant parasitic nematodes.

Paratype: A specimen related to original series but not the type described originally by the author, but created by the author.

Parthenogenesis: Development of egg into nematode juvenile without fertilization due to absence of male nematodes. It is found in many plant parasitic nematodes.

Penial tube: A tubular structure formed by the lips of cloaca in some male nematodes which is also called copulation tube.

Perineum: Region surrounding the cloaca and opening of genital organs.

Pharynx: See "Oesophagus".

Phasmids: Lateral sense organs of nematodes found in pair on posterior end or on both sides of tail. They are very small, cuticular appendages but are important taxonomically and their presence forms the basis of dividing them into major classes (Secerneta and Adenophorea). They open outsite through a minute pore, which is connected through a pouch to caudal nerve. They mainly act as chemoreceptors but glands associated with them secrete some chemicals which attract other members of their species.

Phenamiphos: An organophosphate nematicide available as 10% granule formulation as "Nemacur". Its chemical name is ethyl-4-(methyl thio)-m-toyl-isopropyl-phosphoramidate. It is effective against root knot of vegetables.

Phorate: Soil treating, granular fumigant used to control nematodes and insects. Its chemical name is 0, 0-diethyl S-(ethyl thio methyl phosphorodithioate).

Phytonematode: Nematode parasitic on plants.

Phytonematology: Branch of science dealing with the biology, host-parasite relationship and control, etc. of nematodes affecting higher plants.

Phytophagous (Herbivorous): Feeding on plants.

Pin nematode: Common name for nematode genus *Pratylenchus*.

Pin worm: Common name for nematode, *Enterobius vermicularis*, first described by Hippocrates in 430 BC.

Planarian: Common name used for flat nematodes belonging to order Turbellaria.

Platymarian: A group (of the 3 described) of somatic muscle cells which have broad, flat base and in which contractile elements are limited only upto the base, nearest to epidermis. The fibres do not extend upto the cells lying towards the psudocoelom. (*cf.* Coelomyarian, Circomyarian).

Polycytous: Made up of many cells (especially for nematode intestine). (*cf.* oligocytous).

Polymyarian: Of the 2 types of arrangement, the one in which more than 5 (normally many) somatic muscle cells are found in each interchordal zone. (*cf.* Meromyarian).

Population threshold: Population or density of the nematode in a unit area at which tolerance of the host plants ends and nematodes start doing economic losses to the crop.

Post-uterine sac: Extension of the uterus in nematode which spreads posteriorly to the vagina/vulva and may function for the collection of sperms.

Post-vulval branch or sac: See "Post-uterine sac".

Potato root nematode: Commonly used name for nematode *Globodera rostochiensis.*

Pouch (Fovea): See "Amphid aperture"

Pratylenchus : "Lesion nematode".

Predaceous fungi: Fungi predating on other organisms, especially insects and nematodes. Also see "Nematophagous fungi".

Predacious nematode: Nematode feeding on other parasitic nematodes.

Prerectum: Portion of the nematode digestive system in between intestine and rectum, found in some spp. of *Xiphinema* and *Dorylaimus.*

Prestoma: See "Vestibules".

Probolac: Distinct and developed cuticular extensions on the lip region of certain nematodes.

Procorpus: Anterior portion of original or modified oesophagus of nematodes, which is connected to median bulb or metacorpus fitted with crescent valvular plates.

Proctodaeum: See "Hindgut".

Prodelphic: Arrangement of one or both female reproductive organs in a position anterior in relation to vulva and vulva is located at terminal position. When both the sets are anteriorly positioned it is called "didelphic-prodelphic" e.g. *Heterodera, Meloidogyne*; and when only one set is found on such position, it is called "Mono-prodelphic" (*cf.* amphidelphic, opisthodelphic).

Protostom: Middle portion of the nematode mouth, located between buccal cavity and telostom.

Pseudobulb: See "Bulb".

Pseudocoelom: Body cavity of nematode, filled with a fluid; keeps nematode turgescent. Due to lack of mesodermal layer, it is called pseudocoelom instead of Coelom.

Pseudocoelomocytes: Cells found inside the Pseudocoelom.

Q

Quadricolumella: Secretory cells of ovary near spermatheca arranged in 4 rows (*cf.* tricolumella.).

Quiescent: A stage of relative biological inactivity (dormancy) produced due to some physical factors like low temperature, etc.

R

Rachis: A protoplasmic core around which oocytes gets aggregated. It keeps oocytes together, e.g. *Anguina.*

Radopholus similis: See "Burrowing nematode".

Rectal glands: Glands present in rectum wall and opening in rectum. Normally 3 in female and 6 in male. In *Meloidogyne*, 6 very big glands secrete a gelatinous matrix to hold the eggs.

Rectum: See "Hindgut".

Regulatory control: A method of checking the spread of plant diseases by regulating the movement of plant material from one country to other or one region to other to check the spread of the pathogens with them.

Renette: Excretory cells present usually laterally in post-oesophageal region, connected to excretory pore through a cuticularised excretory duct. Also called "Sinus" or "Cervical cell".

Reniform nematode: Rotylenchus nematode, attacking cotton, papya, tea, tomato, etc.

Reproductive glands: See "Gonads".

Rhabdion: Cuticularized portion of the mouth cavity.

Rice root nematode: *Hirschmanniella oryzae* attacking paddy roots, reducing the growth, tillering and yield.

Ridges: See "Lamellae".

Ring nematode: Referred to genus *Criconema* attacking woody plants.

Root knot nematode: Common name for *Meloidogyne*, attacking most of the crops, especially vegetables.

Rosette: A symptom of plant disease in which apical growth is checked and a dense crown of leaves is produced on it.

S

Salivary glands: See "Oesophageal glands".

Secernenta: One of the 2 major classes (taxonomic rank) of Nematoda which have pore or slit like phasmids, well developed excretory system; poorely developed mesenterial tissues, but lacks caudal adhesive glands.

Sedentary nematode (Sessile): Such nematode draw their nutrition from one place (either inside or from the surface of host) and remain confined to that place. They are not capable of movement and may be endo-or ecto-parasite, e.g. *Rotylenchulus, Heterodera,* etc. (*cf.* Migratory nematode).

Sclerotized: Hardened structure, especially of cuticle.

Segments: Portions or regions of the body. Nematodes are not internally segmented, but few free living nematodes have apparent segmentation confined only upto cuticle.

Semiendoparasite: Parasite nematode which is only partially established inside the host tissues.

Semifenestrae: A pair of opening on the vulval cone of bursa in *Heterodera*, which are separated by vulval bridge.

Seminal vesicle: Portion of the male gonad which functions for the temporary storage of sperms.

Sensilla pouch: See "Fusus".

Sensory organs: External receptors like, amphids, phasmids, deirids, hemizonids, hemizonions, cephalids and caudalids, cephalic sensors etc., of cuticular or epidermal origin present on the outer surface of nematode body. They receive the stimulus and transmit them to nerveous system.

Sessile nematode: See "Sedentary nematode".

Seta: A hard, hair like projection functioning as sensory organ in nematodes.

Sexual dimorphism: Occurrence of male and female individuals having different morphology.

Shaft: See "Telenchium".

Sheath nematode: Referred to genus *Hemicycliophora.*

Sinus: See "Renette".

Somatic muscle layer (Somatic musculature): The inner most layer of nematode body wall, present just below the epidermis. It is not continuous throughout the body, but is arranged longitudinally only in interchordial zones. Muscle cells are spindle shaped or rhomboidal and have been divided into 3 groups according to their shape and presence of contractile elements.

Spear: See "Stylet".

Spear cone (Spear tip or conus): Anterior conical part of the stylet or spear having pointed sharp anterior end and a fine spear lumen in the middle opening outside through a spear aperture. It is also called "Metenchium".

Spear guiding apparatus: A tubular extension of the vestibule or buccal cavity connected with mouth at one end and anterior conical part of spear on the other.

Sperm: Male gametes, which are mobile, exhibiting amoeboid motion and may be spherical, discoid, conical or elongated. Sperms present in vas-deferens are morphologically different than those present in female spermathecea.

Spermathecea: Distal portion of uterus, rounded or elongated in shape where sperms are stored in living condition. Its size and shape vary from species to species of nematode.

Spermatocytes: Male germ cells which give rise to sperms.

Sphincter: A ring like muscle which functions for the opening and closing of natural openings of nematodes.

Sphincter-Z: In some species of *Xiphinema*, a narrowing or contraction between uterus and spermatheca.

Spicular pouches: Outgrowth of the dorsal wall of rectum where spicules are formed and live.

Spicules: Tubular, sensory structures of copulatory complex of male nematodes, covered with sclerotized cuticle, formed and present inside the spicular pouches of the rectum. They are supposed to be the envaginations of cuticle as cuticle is undifferentiated with the spicular pouches. Spicules open the vulva of female for transfer of sperms during the copulation. Spicules have 3 regions: (i) head or capitulum, (ii) shaft or corpus or calomus, and (iii) lamina or blade.

Spinneret: Caudal cement gland of some nematodes.

Spiral nematode: Referred to nematode *Helicotylenchus* and *Rotylenchus*, attacking a number of crops.

Stem and bulb nematode: See "Ufra or dak pora disease".

Stem nematode: See "*Ditylenchus*".

Sting nematode: See "Belonolaimus".

Stoma: Mouth or oral aperture of the alimentary canal of nematode, consisting of 6 lobe like labia. Also see "Mouth". It is also called buccal cavity. Nematode mouth has 2 parts; anterior cheilostome and posterior oesophagostom.

Stomatostyle: Stylet of nematode formed as a result of contraction, collapse and fusion of walls of buccal cavity or stoma and is found in members of Tylenchida (*cf.* onchiostyle).

Stomatostylet: See "Stylet".

Stomodaeum (Foregut): Anterior portion of the nematode alimentary canal, consisting of stoma (mouth), oesophagus and oesophago-intestinal valve and goes upto mesenteron (midgut).

Striae: Transverse groove or line on the cuticle of nematode, which form ring-like annulations on body.

Stubby roots: A symptom appearing on root as a result of nematode attack in which growth of rootlets is checked when they attain moderate growth, as a result of which root system has many short branches arranged in bunches. eg. *Trichodorus christiei* on maize roots.

Stunt nematode: See *"Tylenchorhynchus"*.

Stylet (Stomatostylet or spear): A hypodermic needle-like apparatus of nematode mouth formed as a result of modification of stomal walls. It is characteristic of plant parasitic nematode and usually has 5 parts, viz. cephalic framework, spear guiding apparatus, conus or metenchium, shaft or telenchium and spear muscle.

Stylet nematode: Common name for *Tylenchorhynchus* nematode.

Subcrystalline layer: Outer surface layer present on some bursa of nematode species of *Heterodera*.

Syncytia (Giant cell): Multinucleate cell formed in the cortical region of roots as a result of invasion of nematode. It may be due to repeated nuclear division (e.g. *Meloidogyne*) or due to dissolution of cell walls (e.g. *Heterodera*).

Syncytial: Having many nuclei.

Syngonic: Condition when the same gonad produces both male and female gametes, i.e. spermatozoa and eggs.

T

Tactile (Touch receptor): Sensory organs, which are sensitive to and receive the mechanical or touch stimulus. They are also called "Mechanoreceptors".

Tail: Portion of body between posterior end and anus of nematodes.

Tapering: Shape (of nematode) with pointed ends.

Tardigrades: A group of organisms feeding on diverse type of food and may be helpful in control of certain nematodes, e.g. Control of *Meloidogyne incognita* by *Hypsibius myrops.*

Telamon: A part of the gubernaculum formed by the cloacal wall.

Telenchium (Shaft): Posterior cylindrical portion of the nematode stylet (spear) having one dorsal and 2 subventral knobs.

Telogonic: One of the 2 processes of germ cell proliferation in nematodes in which germ cells proliferate only at the anterior end of the gonads. Most of the plant parasitic nematodes are telogonic type. (*cf.* hologonic).

Telostom: Posterior portion of nematode mouth.

Tesselated: Looking like chess-board or having many segments.

Testes: Male gonads (reproductive organs) of nematodes. Nematodes may have one or 2 testes. (*cf.* monorchic, diorchic).

Thiozazin: Chemically it is 0, 0-diethyl-0-2 pyrazinyl-phosphorothioate, and is used as nematicide against root knot of vegetable. Available under the names of "Nemaphos" or "Zinophos".

Transverse commissures: Semicircular structures of nerveous system, transversely connecting the nerve of different regions of the body.

Tricolumella: Secretory cells of nematode ovary, behind spermatheca (crustaformeria) arranged in 3 rows. (*cf.* quadricolumella).

Triploblastic: Having 3 generative layers; exoderm, mesoderm and endoderm.

Tundu (Yellow ear rot): A bacterial disease of wheat, caused by *Corynebacterium tritici*. The disease appears when the bacteria are found in association with nematode, *Anguina tritici*. Diseased plants show twisting of lower and middle leaves and slimy or gummy exudation on ears. Also see "Ear cockle".

Tulip root: Referred for a symptom/disease on *Avena* plants caused by nematode *Ditylenchus dipsaci*.

Twisting: A disease symptom in which foliage becomes distorted, twisted and hard mainly as a result of virus and nematode attack. Commonly observed on wheat due to *Anguina tritici* attack.

Tylenchid oesophagus: A modified oesophagus having an anterior, cylindrical procorpus, which is connected to a narrow isthmus extended upto basal bulb, through a median bulb (metacorpus) having crescent valvular plates. As the basal bulb has oesophageal glands, it is also called "basal glandular region". Due to modification of this glandular region, oesophagus may be of: (i) Neotylenchid, (ii) Cricone-matid, or (iii) Hoplolamid types.

Tylenchorhynchus: Commonly called "Stunt nematode", attacking mainly tobacco, maize, cotton, etc.

Tylenchulus: A nematode causing "slow decline" or "die-back" disease of citrus. It is also called "citrus nematode. Also see "Citrus nematode".

U

Ufra or Dak pora disease: A nematode disease of rice caused by a stem and bulb nematode, *Ditylenchus angustus*. Diseased plants show deformation of ears in which panicle remain inside the leaf sheath with out grain formation or may come out with grain formation only in apical region.

Under bridge: Structure/organ extending from one to another end of vulval cone of bursa of nematode, *Heterodera*, which are beneath the vulval bridge and parallel to it.

Uterus: Largest and complicated part of the female reproductive organ, of nematode. Activity of fertilization, egg shell formation and egg laying takes place in uterus. It is situated between oviduct and vagina.

V

Vagina: A small, hollow, tubular structure of female reproductive organs of nematodes, lined with cuticle and made up of well developed muscle layer, opening outside through vulva.

Vapam: Commercial name of "Metham sodium" or SMDC, which is chemically sodium-N-methyl dithiocarbamate. It is used as nematicide and fungicide as soil drench.

Vas-deferens: In Tylenchid nematodes, the anterior portion of the male gonoduct next to the testis, which is glandular in nature and stores the sperms formed and liberated from the testis. It is undifferentiated from the testis. At the other end it is connected with the ejaculatory duct.

Vas-efferens: Tubules going from testis towards vas-deferens in nematodes.

Vector nematodes: Nematodes which transmit the plant parasitic viruses from diseased to healthy plants. Also see "Nematode-virus complex".

Velum: A thin, flange-like extension over the spicules.

Ventral side: In nematodes, the longitudinal side of the body on which anus and vulva are located.

Vestibules (Prestoma): In members of Tylenchida the internal labial papillae have pore-like openings or pits present around the stoma or inside an invaginated area which are termed as vestibules.

Viscera: Internal organs of the body.

V-value: A value used in de-Mans formula of nematodes. It determines the position of the vulva and is represented as percentage of total body length of the nematode.

Viviparous: Nematodes in which eggs develops inside the body of female and young larvae come out of the body. In such cases, eggs are without shell.

Vulva: External opening of the female reproductive system, which may be transverse slit-like or rounded pore-like.

Vulval bridge: A narrow joint across the fonestra of the vulval cone of bursa found in some *Heterodera* sp., which forms two semifenestrae.

Vulval cone: A distant sharp angle on the dorsal side of bursa of *Heterodera* nematode.

W

Wheat nematode: Common name for *Anguina tritici* causing ear cockle disease. Also see "ear cockle".

White tip disease: Disease of rice caused by foliar nematode *Aphelenchoides besseyi*. Leaves of affected plants become white or light green at apical region becoming necrotic. Nematode attacks many Graminaceous plants, viz. maize, pearls millet, etc.

Wool: A group/mass of dessicated *Ditylenchus dipspi* nematodes inside plant tissue.

Y

Yolk (Deutoplasm): Nutritive portion of the oocytes which help in the growth of oocytes.

Z

Z-organ or Organe-Z: A structure in between oviduct and uterus of nematode species of *Xiphinema*.